Electric Sounds

Film and Culture Series: John Belton, General Editor

Electric Sounds

Technological Change and the Rise of Corporate Mass Media

Steve J. Wurtzler

Columbia University Press   *New York*

Columbia University Press
Publishers Since 1893
New York   Chichester, West Sussex

Copyright © 2007 Columbia University Press
All rights reserved

Library of Congress Cataloging-in-Publication Data
ᴍᴸ
Wurtzler, Steve J.
Electric sounds : technological change and the rise of corporate mass media /
Steve J. Wurtzler.
p. cm. — (Film and culture)
Includes bibliographical references and index.
ISBN-10: 0–231–13676–5 (cloth : alk. paper)
ISBN-13: 978–0–231–13676–1 (cloth: alk paper)
1. Mass media—Technological innovations—United States—History—20th
century   2. Mass media—Ownership—United States—History—20th century
3. Sound—Recording and reproducing—History—20th century   I. Title   II. Series
P96.T422U639   2006
303.48'330973—dc22   2006020922
⊗

Columbia University Press books are printed on permanent and durable acid-free paper.

Printed in the United States of America

c 10 9 8 7 6 5 4 3 2 1

For Lois Wurtzler

Contents

Acknowledgments

At Columbia University Press, I had the pleasure to work with an understanding and patient editor, Juree Sondker, and an excellent copyeditor, Roy Thomas. The series editor, John Belton, provided unflagging support and insightful guidance. This book culminates a consideration of film technology begun decades ago in one of John's courses at Columbia University. John first got me engaged with technology as an object of study, and then he encouraged me and supported my career countless times during the intervening years. Comments from the anonymous readers of my manuscript also helped me to make this a leaner, more focused book. During the manuscript's preparation, the Graduate College of Arts and Sciences at Georgetown University provided financial support in the form of several grants and a subvention. The staff in the Special Collections Department at Georgetown University Library helped with some of the illustrations.

Because footnotes can never adequately articulate the personal and scholarly debts we accrue, I want to acknowledge here the support and influence of other sound-related scholars. Rick Altman, first my teacher and mentor and then my colleague and friend, shaped my thinking about sound media in countless ways. From shouting matches in his office to the more civilized but equally passionate quiet conversations over Gauloises in his living room, Rick and I argued about many of the issues raised in this book. Those conversations often extended beyond sound to include not only teaching, program development, graduate student unionization, a difficult job market, and a frankly odd series of academic appointments but also family, loss, perseverance, and joy. While I believe we learned from each other, by far the greater debt is mine. It was due to the encouragement of John Durham Peters that I began systematically to think beyond cinema and consider sound across multiple media forms. The work of fellow sound scholars James Lastra and Emily Thompson often inspired me to rethink familiar questions in fresh ways. My metro-D.C. colleague, Lisa Gitelman, helped me move forward through validation and her example when I became mired in self-doubt.

Years ago at Columbia University I had the good fortune of falling into a work-study job at the Center for American Culture Studies. It was at the Center, largely through the example of my fellow graduate student employees, that I learned firsthand what it meant to be a scholar. Quite literally, I would not be an academic today were it not for the mentoring, advice, and inspiration of Center director Jack Salzman. Thank you, my friend.

My students—in all the classes over all the years at four different schools—reminded me almost daily why I belong in a classroom. They have often been my reason to keep writing. My teaching colleagues at Georgetown, Bowdoin, Illinois State, and the University of Iowa provided friendship and inspiration that made my professional life better. I want to particularly thank Penny Martin, Carol Donelan, Matthew Tinkcom, Pam Fox, Jeffrey Shulman (the James Brown of higher education), and especially the men and women of the ill-fated COGS-SEIU unionization campaign at the University of Iowa. One of my colleagues at Georgetown, David Kadlec, died before this book saw the light of day. As a teacher, a media scholar, and a gentle and generous soul, David continues to inspire me.

As I worked on this book, in encounters varying in duration from several days to several years, a number of academics and normal citizens collided with my life in timely ways that ultimately proved both productive for the book project and comforting for my soul. I particularly thank Carolyn Bernstein, Jeffrey "The American Scholar" Hammond, Edgar at the Hawk and Dove, Bill Rock, Stephanie Brownell, John and Carol Walsh, Jenn Manno, Christina Heintze, and especially Barbara Bradford—each made my life better in some significant way.

My "nontraditional," geographically dispersed family, linked to me in ways beyond blood and marriage, provided unflagging support and en-

Acknowledgments

couragement. Norma Tilden, Shari Zeck, Steve Mathews, Cindy Stretch, Simon Mathews, Julie Monroe, Scott Curtis, Doug Loranger, and Don-E Larson all helped me to redefine my notion of family as a lived series of mutually fulfilling relationships. The Colorado contingent, Gail Wurtzler, Ross Buchan, and Katie Buchan, proved perpetually supportive and ever curious about when they might see this in a bookstore. Decades ago, I took my first undergraduate film class (David Bordwell's "Intro to Film") only after Gail's persistent prompting. Without that, I would be a very unhappy engineer right now.

My partner, Laura Baker, has been patient and accepting of disappointment as we sacrificed too much for our careers. Her encouragement, as this project and the larger academic two-step left me alternately preoccupied, anxious, obsessed, and disconsolate, helped to sustain me. It's over. We can go for a walk now.

This book is dedicated to my mother, gone for some time, but she seems to be perpetually at my elbow whenever I walk into a library or during each encounter with a student. Decades ago she took me to the old Menomonee Falls firehouse (converted into the public library) and started me on a path of reading, questioning, and exploration that continues today. She was my first and best teacher.

Electric Sounds

Introduction

In an essay on some of the historical and theoretical implications of the cinema's conversion to sound, Alan Williams poses a question rarely asked in histories of the period: "And this is one of the great mysteries of this part of film history. Why, with no previous indications of dissatisfaction, did audiences suddenly embrace the talkies, acting as if they had been dissatisfied with 'silent' cinema for a long time?"[1] One solution to this "mystery" requires reconceptualizing the question, broadening the issue from a media-specific approach that considers in isolation technological change in the cinema to a perspective that examines larger issues of sound technology during the late 1920s to early 1930s. Movie audiences' sudden enthusiasm for the talkies then becomes one small part of a general reaction to much larger technological shifts. Hollywood's conversion to sound and cinema audiences' enthusiasm are merely components of the larger pervasiveness and reaction to a new technological mediation of sound. This new technology applied electrical principles to sound recording, transmission, and reproduction in a number of different media forms (predominantly radio, the phonograph, and the cinema). Emerging sound media and the rhetoric surrounding each brought about new interest in and new auditors' relationships to sound. Technologically mediated sounds took on an increased importance, a new pervasiveness, as they fulfilled new social functions and embodied new and exacerbated preexisting social anxieties.

This book examines the process of technological change and the always contested emergence of new media forms by focusing on a crucial period in the history of U.S. mass media. Throughout the book I demonstrate that the ultimate form a technology takes is never predetermined, but instead results from struggles between conflicting visions of technological possibility in economic, cultural, and political realms. Corporations such as AT&T, General Electric, Westinghouse, and RCA (Radio Corporation of America) used extensive capital investment and strategic patent acquisition to control the innovation of sound technology and its application to various consumer products and services. But emerging sound media became sites around which coalesced competing notions of how media might function in U.S. society. Celebrants and critics of new media alternately framed sound technology either as an outlet for socially valorized "high culture" forms or as a purveyor of degraded popular culture, as a tool of education and cultural uplift or as a threat to standards of public taste. Innovators also hailed emerging acoustic media (particularly radio) as instruments that would fulfill the promise of participatory democracy through a better-informed citizenry, while others warned that the new media might function as tools to manipulate a gullible populace. Throughout the period, celebrations of a new acoustic "consumer democracy" uneasily coexisted with fears of the crowd readily transformed into a mob. Innovators thus sought not only to determine the most profitable method of deploying new media but also whether and how sound media might serve interests beyond commerce.

Ultimately, I argue that the innovation of electrical sound technology prompted a restructuring and consolidation of corporate mass media interests, shifts in both representational conventions and patterns of media consumption, and a renegotiation of the social functions assigned to mass media forms. In addition to these struggles to determine what roles sound media would play in U.S. culture, this time period includes some of the most important developments in American mass media history: the film industry's conversion to synchronous sound, the rise of radio networks and advertising-supported broadcasting, the establishment of a federal regulatory framework on which U.S. communications policy continues to be based, the development of several powerful media conglomerates with substantial interests in the film, radio, and recording industries, and the innovation of a new acoustic commodity (the product of so-called media "synergy") in which a single story, character, song, or other product is made available to consumers in multiple media forms and formats.

With sound technology as the primary object of study, traditional notions of media specificity begin to fray about the edges. Issues surrounding Hollywood's adoption of synchronous sound are conceptually linked at the outset with related developments in U.S. broadcasting and the phonograph industry. Within this conceptual framework, three of the dominant mass media forms before World War II (cinema, radio, and the phonograph) are less specific media than three related applications of the same technology. That technology, what I will refer to as *electrical acoustics*, involved the electrical and mechanical means through which sound was successfully encoded, amplified, transmitted, decoded and then reproduced. In what follows I make the terminological distinction between a technological *apparatus* (an organized system of applied scientific principles and hardware) and the various technological *applications* through which it is deployed (in commodity forms that address real, perceived, or created consumer needs). While the synchronous-sound film, the radio, and the phonograph are quite distinct media—that is, they involved specific technological *applications*—in terms of sound technology each is the selective arrangement of the same basic *apparatus*.[2]

Electrical acoustics consisted of three basic phases in the technological augmentation of sound: collecting/encoding, storage/transmission, decoding/reproduction. Collecting/encoding involved the collection of a version of an acoustic event and the transformation of the physical energy of sound waves into electrical energy by a microphone. The apparatus amplified and transmitted this electrical energy to a storage/transmission medium. At the second phase the apparatus moved the encoded signal through space and/or time using a radio carrier wave, a phonograph disk, or a film soundtrack (whether on a disk or a variable-area or variable-density optical film soundtrack). The third phase, decoding/reproduction, essentially repeated the first phase but in reverse. The electrical-acoustic apparatus transformed the encoded version of an acoustic event from electrical energy back into physical energy through a loudspeaker and dispersed the resulting sounds into space as sound waves.

The three primary applications of electrical acoustics differ most obviously in their respective media of storage/transmission. They differ as well in recording conditions (the absence of a visual component for the radio and the phonograph obviously allowed greater freedom in microphone placement) and the acoustic conditions of playback (in the late 1920s, the acoustically damped living room as the primary site of radio and phonograph reproduction contrasted sharply with the far more reverberant movie theater). However, as numerous engineers of the period frequently noted, the three applications of electrical acoustics shared the same basic technological hardware and applied scientific principles.

Much of the research and development work central to this technological change took place within the corporate boundaries of AT&T, its wholly owned subsidiary Western Electric, and Bell Laboratories. Research related to the amplification of telephone signals led to the development of the vacuum tube amplifier and, in 1915, AT&T used the device to inaugurate transcontinental telephony. The telephone amplifier resulted from innovations to the triode vacuum tube (the rights to which AT&T acquired from inventor Lee De Forest in 1913) undertaken by AT&T employee H. D. Arnold. Arnold's work was part of a systematic attempt by the telephone company to expand its research in the broad field of communications, a research effort that also led to the innovation of the condenser microphone, improvements in loudspeaker technology, the application of the vacuum tube to both radio transmission and reception and electrical systems of recording on both phonograph records and motion picture film. During the early 1920s, AT&T and Western Electric combined elements of this applied research to create public address systems and then electrical systems of phonograph recording and reproduction and synchronous-sound cinema. In 1925, Western Electric successfully marketed its electrical phonograph innovations by licensing two of the major U.S. phonograph companies, Columbia and Victor, and it found a market for a sound-on-disk film system (now called the Vitaphone) when Warner Bros. signed as a licensee in 1926. Through its wholly owned subsidiary Electrical Research Products Incorporated (ERPI), Western Electric eventually signed up virtually all the major studios in Hollywood by May of 1928.

AT&T was not alone in innovating electrical methods of augmenting sound. General Electric employee Irving Langmuir independently duplicated Arnold's development of the vacuum tube amplifier as part of a well-funded and sustained program of industrial research at GE. That research and development work also led to an electrical system of phonograph recording and reproduction and a competing system of synchronous-sound film. While Western Electric had been first to market successfully the application of electrical principles to the phonograph and film industries, GE garnered a share of the market by licensing the phonograph company Brunswick-Balke-Collender and, later, film company RKO and several other studios.

In addition to General Electric, a number of independent entrepreneurial inventors simultaneously developed competing applications of electrical principles to different media. Lee De Forest developed a sound-on-film system (called the Phonofilm) and sought with minimal success to market it to the motion picture industry from 1923 through the late 1920s. Theodore Case, an Auburn (New York)–based inventor, created a similar sound-on-film system and sold it to Warner Bros.' competitor, the Fox Film Corporation. Within several months of the August 1926 premiere of the Warner Bros.' Vitaphone films, Fox released synchronous-sound films under the Movietone brand. While De Forest had astutely retained rights to develop his triode vacuum tube when he sold the patent to AT&T, other inventors— such as Case—ultimately had to seek a license for this crucial component of electrical acoustics from either AT&T or General Electric, a development that gave these two large organizations control over a crucial component to any electrical sound system.

Corporate innovators like General Electric and AT&T were able to exercise significant control over technological innovation through their patent rights. At two key moments in the early history of electrical acoustics, they worked collaboratively with other patent holders to try to manage technological change. In 1919–20 the major patent holders in the field of wireless communication entered into a patent pool agreement that sought to develop radio in a manner that limited competition and maximized their profits. AT&T, General Electric, Westinghouse, the United Fruit Company, and several other smaller concerns licensed each other under their patents, negotiated a split of the anticipated profitable applications of wireless technology, and formed the corporation RCA as a patent-holding company. But this wireless patent pool failed to anticipate consumer enthusiasm for radio broadcasting, and by 1926 the dominant corporate patent holders renegotiated an agreement to carve up the field of radio. These patent pools affected all applications of electrical sound technology since they covered key patents for components (crucially, vacuum tube amplification) necessary for the various applications of electrical acoustics.

While major corporations sought to maintain control over the innovation of electrical sound technology, crucial institutional and representational transformations took place within all three of the primary sound-reproducing media. From the mid-1920s through the early 1930s, Hollywood became a consumer of the technological innovations at AT&T and GE and converted to synchronous-sound recording and reproduction. The film industry also underwent the difficult process through which aesthetic experimentation and debates among film critics, directors, sound engineers, and industry pundits led to the establishment of representational conventions for Hollywood recording and sound/image relationships. Simultaneously (and not unrelatedly), the U.S. film industry experienced an important period of consolidation and industrial realignments. While in 1925 two major companies significantly dominated the industry (Famous Players and Loew's/MGM), by arguably the end of Hollywood's conversion to sound the industry had realigned with power vested in five vertically integrated major film studios and, to a lesser extent, three smaller companies. Douglas Gomery concludes that "the greatest outcome of the coming of sound was the establishment of the studio classic system that dominated through the 1930s and 1940s."[3] Hollywood's conversion to sound also caused some significant shifts in the types of films produced and the theatrical filmgoing experience. Most obviously, Hollywood borrowed formats from other entertainment forms (Broadway, vaudeville, nightclubs) and began producing musicals. In general, the experience of theatrical filmgoing in the period was also marked by a shift from programs that combined live and recorded performances characterized by multiple modes of address to more unified representational patterns with relatively rare and highly conventionalized moments of direct address incorporated within the fully recorded.

As in motion pictures, the mid-1920s through the 1930s involved crucial developments within radio. The wireless patent pool had sought to fix the potential marketable identities of radio technology, legislating in effect each member organization's share of point-to-point wireless telegraphy and telephony, but it only barely and very ambiguously anticipated the ultimately dominant form radio technology would take: point-to-mass broadcasting. From 1922 on, radio broadcasting exploded in pervasiveness and popularity. A handful of transmitting stations in 1922 were joined within two years by over 500 licensed broadcasting stations. By one account, in the decade after 1922 the diffusion of radio receivers in U.S. households went from a market penetration of 0.2 percent to 55.2 percent before eventually reaching a total of approximately 81.5 percent of all households immediately before World War II.[4] Radio broadcasting's rapid diffusion throughout the country was accompanied by a series of struggles to define what broadcasting would become and who would pay for it. Conflicting spatial definitions of the medium (radio that offered a local service or a national mode of address) and conflicting economic models (government funding, a tax on receiving tubes and sets, commercial sponsorship, indirect or direct advertising) vied for dominance in the latter half of the 1920s. By 1927, Congress established the Federal Radio Commission (FRC) to oversee radio broadcasting and formally codified the regulatory assumptions that continue to inform much of U.S. broadcasting policy. But after 1927 and the formation of the FRC, the radio industry continued to exhibit dramatic changes. Radio network NBC officially began broadcasting in November 1926, but the RCA-owned network quickly expanded to two networks (NBC Red and NBC Blue) in January 1927. NBC was joined on the airwaves by competitor CBS in the fall of 1927. Radio advertising arguably began at AT&T's New York station WEAF in the early 1920s, but in a series of rhetorical, economic, and political struggles throughout the 1920s and into the early 1930s, broadcast advertising shifted from the indirect sponsorship of programming (only the corporate sponsor's name mentioned intermittently in the broadcast) to the full-scale, direct promotion of individual products and services. In a series of shifts related to such institutional and regulatory developments, programming on radio also underwent significant changes, including the development of announcing conventions, the gradual professionalization of radio performance, the development of scheduling conventions designed to target and call into being demographically specific audiences, and the increasing marginalization of some political voices and educational programming.

Within the phonograph industry, dramatic changes began prior to the shift in 1925 from acoustic to electrical methods of recording and reproducing sound as the industry faced slumping sales of phonographs and records and sought to navigate its relationship to the growing popularity of radio. During the period, the phonograph record's status as a commodity also underwent significant change in response to the potential offered by and limitations endemic to the new electrical systems (for example, a shift in the frequency range of phonographs), what some have pointed to as the influence of radio on popular tastes among the record-buying public, and the proliferation of jukeboxes throughout the 1930s. Electrical recording changed both recording practices and the types of music recorded as the act of recording was gradually reconceived from the preservation of an original event to the creation of calculated effects.[5] Because of the increased sensitivity of microphones, musicians in recording studios no longer had to cluster around acoustic sound-collection devices, and the spatial arrangement of musicians and their instruments could more closely approximate the conditions under which they normally performed. The extended frequency range (especially the more efficient reproduction of bass tones) reduced jazz musicians' reliance on strings and the cornet to provide a rhythmic structure. More sensitive sound collection allowed musicians to play less loudly and thereby better moderate their tone.[6] Electrical recording also, arguably, made possible new performance styles such as the crooner.

Significant recent work has engaged each of these media in turn, constituting a kind of second generation of historical scholarship on this period in U.S. media history. Recent book-length publications have revised and fine-tuned the dominant historical narratives that for years had functioned as the sources for our understanding of these media. Much can be learned from this work as it has carefully mined significant archival sources and brought to bear on the period principles of contemporary historiography. In each of the three media, a dominant historical work had functioned as a kind of master narrative, a map of sorts, against which subsequent efforts were positioned. For radio, Erik Barnouw's three-volume study fulfilled this function for more than a generation. First Hugh Aitken and Susan Douglas and subsequently Robert McChesney, Susan Smulyan, Michele Hilmes, Thomas Streeter, Douglas Craig, and then Douglas again provided essential book-length examinations of American radio particularly relevant to the 1920s and early 1930s. Shorter works, focusing on more specific historical issues, by Catherine Covert, Frank Biocca, Mary Mander, Leonard Reich, again Hugh Aitken, Hugh Slotten, and others productively augment our understanding of radio's origins and early development.[7] For the phonograph, Oliver Read and Walter L. Welch's *From Tin Foil to Stereo: Evolution of the Phonograph* (and to a lesser extent Roland Gelatt's *The Fabulous Phonograph: From Edison to Stereo*) mapped out histories of recorded sound. Recently, Andre Millard, Michael Chanan, William Kenney, and Lisa Gitelman have provided new book-length revisions to our understanding of this history. Exciting, essay-length work by Marsha Siefert, Emily Thompson, and Gitelman directly address the phonograph as a technological object, the social identity of which was in fact subject to economic and cultural determination.[8] For Hollywood's conversion to sound, Harry Geduld's book provided an early master narrative that was productively revised by Douglas Gomery's 1975 dissertation and the crucial essays it spawned. James Lastra, Donald Crafton, and the essays and editorial work of Rick Altman have provided a productive rethinking of this crucial period.[9]

The significant body of comparatively recent work on sound media has added breadth and depth to our understanding of the radio, the phonograph, and the film, but much of the existing literature retains a predominantly media-specific approach and hence fails to engage more fully the crucial links between these media forms. The first portion of Michele Hilmes's *Hollywood and Broadcasting* provides a useful overview of the industrial connections between film and radio in the United States.[10] While some of the phonograph histories include sections on film or radio, until quite recently little of the scholarly work on sound media has *systematically* explored sound technology as a transmedia phenomenon.[11]

James Lastra's *Sound Technology and the American Cinema* links issues surrounding Hollywood sound practices, particularly amid the industry's adoption of synchronous sound, to larger theories and debates about perception and representation as they were shaped by (and in turn shaped) both photography and phonography. Emily Thompson's *The Soundscape of Modernity* is less concerned with media than it is with a larger culture of sound and of listening that, although profoundly influenced by technological mediation, was not ultimately shaped or determined by any specific medium. Her work ranges across shifts in the study of acoustics, acoustic treatment in building materials and architectural design, methods of measuring and controlling noise, to a new culture of listening and the representational practices that produced the clear, direct, nonreverberant sound favored by this culture. Daunting in its scope and rigor, Thompson's book remaps how we might broadly conceive of acoustic culture and its practices even as her book points toward additional ways in which this period might be explored. Jonathan Sterne's *The Audible Past* is even more expansive in its scope than Thompson's work. Although concerned with technology and with media, Sterne's work is more focused on practices of listening and the ways in which they have been shaped by various conceptualizations of sound and hearing. While technology and media play a role in Sterne's presentation, to my mind he seems more concerned with their embodiment of both theories of sound/hearing and their expression of cultural contexts than with their material specificity as technology or as media. Ultimately, Lastra, Thompson, and Sterne all underscore the need for additional historically informed work on sound that transcends any particular medium.[12]

Media-specific approaches to the history of electrical acoustics result in part from observable material differences of media forms. Our contemporary notions about the specificity of the radio, the cinema, and the phonograph are additionally embodied in both substantial and trivial aspects of our professional lives. From the curriculum we often teach through various institutional alignments, to Library of Congress cataloging conventions, media-specific approaches provide a commonsense method to organize this admittedly chaotic period of technological change.[13]

In general, I see three overlapping reasons for revisiting the innovation of electrical acoustics with a broader conception of the object of study. First, the material technological base that provided synchronous recorded sound for the cinema, that improved the fidelity of the phonograph, and that allowed widely dispersed listeners to tune into the same radio programming all applied the same principles for technologically augmenting sound and relied upon many of the same component parts.

Second, a variety of social groups during the period viewed and accordingly acted as if these otherwise distinct media were crucially interrelated. Period commentators, especially engineers, were well aware of the mutual interdependence of various applications of the electrical-acoustic apparatus. Although recognizing such interdependence, accounts of sound technology in the late 1920s and early 1930s asserted multiple and often conflicting descriptions of causality. In 1931, public address engineer Gordon Mitchell claimed that, "Public address is the youngest offspring of the Sound family," while five years earlier, Western Electric engineers asserted that, "It was through applying the principles of sound amplification as they have been worked out in Public Address Systems to the phonograph that the new process of electrical recording of sound was developed."[14] Well-publicized public address performances from as early as 1921 undermine Mitchell's assertion that public address was the youngest child of the sound family. Rather than the result of sibling rivalry, simultaneous developments in the fields of radio, public address, phonography, and motion picture sound might best be seen as mutually reinforcing products of research and development accompanying a single apparatus that had multiple profitable identities. If we seek paternity within the sound family, then perhaps the telephone and its ability to electrically encode, transmit, and reproduce voices and music might serve as a progenitor.

Third, our contemporary historical moment resonates powerfully with developments in the 1920s and early 1930s. We increasingly live in an intermedial world. Ongoing technological change blurs distinctions between cinema, television, the Web, radio, and recorded sound. Various screens distributing various media content proliferate in public and private spaces. As media hardware grows increasingly mobile (SUV onboard digital video, iPod, each new "generation" of cell phone that promises to deliver both quantitatively more and greater variety of content), so too does media software. Portability propels media into new spaces, and content migrates with increasing apparent freedom from film to video game, from recording studio to the Web. Our contemporary experience of the intermedial might be productively informed by an understanding of a previous moment when the distinctions between media forms were not as clear as media-specific historical scholarship might suggest. While this book does not constitute a comparative study overtly positing links across seventy-five to eighty years of U.S. media history (that would most assuredly be a different book), it does provide a series of perspectives and frameworks, a series of questions and answers that might enliven our contemporary experience of media convergence with a historical perspective.

Work on historiography and technology from outside film and media studies provides a potential model for a non-media-specific approach to this period. In an essay outlining a "social constructivist" approach to the history of technological innovations, Trevor Pinch and Wiebe Bijker adapt a model drawn from the sociology of scientific knowledge to the study of technology.[15] They outline a three-stage program for research premised on a "multidirectional" model of historical innovations. That is, a model that takes into account and accounts for stages in the development of a technological artifact with varying degrees of success. This model thus avoids the retrospectively imposed linearity of many accounts that view through the filter of hindsight only those stages of a technological artifact that were successful. Their model also avoids the tendency of many economic-based analyses of innovation that fail to account for or even address the technology itself. "The Social Construction of Technology (SCOT)" describes the development of technological artifacts as an alternation between variation and selection, thereby allowing historians to account for why some variants of an artifact fail while others are selected.

The first step of SCOT involves demonstrating the "interpretative flexibility of a technological artifact."[16] Here, the historian of technological innovation demonstrates that within the development of a technology there is flexibility in both how people think or interpret a given artifact and the design of that artifact. The notion of such a flexibility hinges on the observation that a given technological artifact will have different identities and thereby present different problems to different social groups. The historian then traces how such flexibility was closed down and the identity and design of the artifact stabilized through various "closure mechanisms" through which consensus in terms of meaning and design were both achieved and enforced. The third stage of the research program involves "relat[ing] the content of a technological artifact to the wider sociopolitical milieu."[17] SCOT provides an alternative to linear accounts of technological "successes" by accounting for the various false starts and blind alleys often elided from such histories. Through the notions of interpretive flexibility and closure mechanisms, this historical approach explicitly acknowledges the contestation surrounding an artifact's meaning and acknowledges as well that such flexibility is closed down through discursive as well as economic or regulatory means.

SCOT provides a way of thinking about technology, a method of approaching the technological artifact, that helps to circumvent some significant historiographic and conceptual problems. By attending to the technology itself, such an approach avoids the dissimulation of technological processes and artifacts that focuses instead on texts, individuals, and/or institutions. Such an attention to the technology per se also directs the historian beyond media-specific approaches to this period of technological innovation. The notions of "variation and selection" point to the manner in which innovation in representational technologies results neither from a teleological evolution toward ever greater realism (idealist historiography) nor from a series of progressive design solutions responsive to consumer desires and/or market demands (economic determinism). Bijker's notion of "interpretative flexibility" illustrates that the identity of representational technologies (their relationships to and effects on aesthetic practices, the social meanings ascribed to them, their perceived potentials and limitations) are subject to discursive and practical contestation such that the eventual deployment of the technological artifact into its "social milieu" is less the product of qualities inherent to the technology itself than the result of a series of often historically elided "closure mechanisms." Finally, and perhaps most importantly, Pinch and Bijker's third category of inquiry, relating the artifact to the wider milieu, points to the necessity of viewing synchronous cinema sound, network radio broadcasting, or electrical phonographs as different solutions, different end results, of a redefinition of the identity of the larger electrical-acoustic artifact.

Rick Altman's book-length study of silent film sound illustrates how some of the principles and procedures of SCOT might be applied and adapted to media studies.[18] In *Silent Film Sound*, Altman outlines and puts into practice what he calls "crisis historiography." Setting himself against idealist historiography and its tendency to begin with an object (i.e., television) as it exists today as if it were a single consistent object of study, Altman suggests instead that, in their formative periods, representational technologies have plural, often conflicting identities. Coexisting with these plural identities, but not reducible to them, "jurisdictional conflicts" pit competing groups in struggles to control technological change. While these conflicts play out in the economic realm, they also inform representational practices as innovators of a new medium struggle to position it in relation to existing media conventions. Ultimately for Altman, "overdetermined solutions" (much like Pinch and Bijker's "closure mechanisms") resolve—at least temporarily—jurisdictional struggles, but through a process of social construction that is ongoing and multiple. "Rarely is the outcome of this process a winner-take-all affair," Altman notes. "Because the negotiation is complex and multisided, rather than an epic one-on-one clash, establishment of an acceptable, durable solution usually depends on discovering an arrangement that simultaneously satisfies a maximum number of users."[19] The history that results from this method is nonlinear in presentation and offers a multi-perspectival view that takes account of and accounts for the failures as well as successes that led to the incorporation of new technology into commodity and social relations.

A technological apparatus, such as electrical acoustics, and the processes surrounding and shaping technological change are inexorably interwoven into the larger fabric of social formations, institutional structures and practices, patterns of consumption, and often contradictory ideological formations. The existing literature on technological history, specifically the history of mass media and sound-related technologies, suggests several approaches to a technological object of study. Some studies take as their object the technology itself, often focusing specifically on a particular technological artifact and reading from the physical qualities of the artifact not only the application of technological principles but also the embodiment of certain practices of production and consumption. Shifts in the design of consumer objects are traced for the way they mirror and enforce (in a mutually codetermining manner) shifting conceptions of technological identity and individuals' relationships to it.[20] Other studies conceptualize technology as itself a social practice and emphasize its incorporation within existing systems of production and/or consumption.[21] Another approach to technological history of media explores the relationship between technology (envisioned as either object or practice) and mass media texts. Such studies have frequently theorized connections between aesthetic practice, the technology that supports it, and a larger ideological function underwritten and served by the nexus of technology/text/consumption.[22] Still other approaches view technology and its artifacts as a kind of symptom—historically meaningful for the manner in which they can embody, or their implication within, (often) hegemonic ideological systems and/or constellations of power.[23] In general, these approaches conceptualize technology in a number of different ways typically exploring the relationships between technology as object, practice, and/or discourse.

Much of what follows is informed by the observation that these three conceptions of technology (as object, as practice, as discourse) always work in concert. Technological objects and the practices they inspire are shaped by and always embedded within social relations. Technology is produced, reproduced and, importantly, *understood* through various discursive and representational practices that seek to discipline the ways individuals and society envision their possibilities, as well as talk and think about, use, and consume technological objects or technologically mediated texts. Public discursive constructions of and images depicting technology establish interpretive and practical frameworks through which their potential is understood and exercised, thereby constructing knowledge about and hence realities of their use. The material, experiential basis for our shared sense of media specificity thus arises from a series of historically preceding, historically determined, and historically erased decisions that shaped the development and deployment of mass media technologies.

The chapters that follow are not a direct application of either the SCOT program or Altman's "crisis historiography." Instead, what I share with both are a series of overlapping and related observations, indeed presumptions, about history, technological change, and/or mass media. These might be summarized as follows. Upon their innovation, media technologies have multiple, often conflicting identities. The ultimate meanings they take within social relations are the product of contestation and struggle. Various mechanisms—some economic or regulatory, others ideological or material—function to privilege one set of meanings over others as media technology is successfully secured within social habits and practices. While technology and media are indeed the product of social construction, they are also always sets of material objects and "the way things work" (or don't work) has much to tell us. The task of the historian is not to construct a grand narrative that traces the innovation of contemporary forms through a series of problems and solutions that lead inexorably to the present moment, but is instead to understand the myriad ways that the ultimate shape of a technology emerges from struggle and contestation. This necessitates addressing not only successful applications but also the many false starts, blind alleys, configurations, and agendas that are ultimately unpursued, for they have much to tell us about eventually "successful" forms.

Chapter 1 explores the economics of innovation, tracing the process through which corporate innovators sought to maximize their profits by shaping the introduction of new acoustic media. While the research and development surrounding electrical acoustics was undertaken by both entrepreneurial-minded inventors and large, corporate-based research laboratories, a small group of large corporations, often acting in concert, essentially controlled the commercial exploitation of acoustic innovation. Using the selective prosecution of patent rights, restrictive licensing agreements, and the power of capital investment, corporations involved in telephony, power engineering, and electrical manufacture dominated first the introduction of radio and then the application of electrical methods to the phonograph and the synchronous-sound cinema. The introduction and diffusion of electrical sound media emerged within and facilitated an increasing concentration of media ownership, so that by the early 1930s a handful of large and often interlocking concerns wielded enormous power in U.S. media. Ultimately, the innovation of electrical sound media was both profoundly shaped by and itself further exacerbated the growing power and pervasiveness of large, ubiquitous corporations in American life.

The economic efforts to control technological change described in chapter 1 were accompanied by rhetorical struggles to shape emerging sound technologies. Chapter 2 examines the attempts to influence public perceptions of electrical sound media by focusing on the ways in which technological change was presented to the public. I define such "announcements" broadly to include theatrical performances that demonstrated the new apparatus; highly publicized events that illustrated particular applications; journalistic accounts of innovation; and print advertisements promoting new media. These announcements most frequently fulfilled a dual pedagogical function, not only illustrating and explaining scientific principles but also seeking to establish specific uses and related social meanings for emerging media. Corporate innovators of electrical acoustics used these announcements to shift popular perceptions about the location of American innovation from out of the workshop of the lone inventor and into the well-funded, hierarchical, and "scientific" research labs of major corporations, thereby naturalizing and making seem inevitable and beneficent the economic arrangements described in chapter 1. In addition, emerging sound media were often hyperbolically hailed as the technological fulfillment of utopian goals such as universal uplift and the fulfillment of democracy. Locating public pronouncements about emerging media in terms of previous innovations, this chapter demonstrates that even after an apparatus is designed, its relationship to consumers and its position in a larger social fabric must also be engineered and constructed.

Having heralded technological change by, in part, calling consumer attention to applied electrical and mechanical principles, innovators of electrical acoustics then sought to redirect consumer attention away from the machines and to the sounds they reproduced. Chapter 3 examines three sites entered by emerging sound media and the manner in which both the physical design of technological artifacts and consumers' performances in reproducing sounds embodied struggles to establish a medium's identity. I begin with a survey of the phonograph industry's successful redefinition of sound reproduction before the innovation of electrical methods (during the first two decades of the twentieth century) from both a business device and an instrument of public performance to a musical instrument offering domestic entertainment. I then trace the links between the emerging medium of radio and the already-established phonograph and conclude with an analysis of early sound-film projection practices. Immediately preceding the conversion to electrical methods, the design of the mechanical phonograph and consumer interactions with it were characterized by two countervailing tendencies: on the one hand, discourses surrounding phonography and qualities of the machines themselves dissimulated the machinery of sound reproduction; but on the other hand, the materiality of the technology returned through consumer performances of the recorded repertoire. Radio's progress throughout the mid- to late 1920s involved multiple attempts to heal the apparent gap between efforts to disguise the machinery of sound reproduction and countervailing perspectives that viewed sound reproduction as an active performance. An identical process was enacted within the film industry in that while the conversion to sound had promised to fulfill a long-standing desire for standardized film accompaniment, in the short term at least, motion picture projectionists became active performers of reproduced sound.

Through the announcements described in chapter 2, corporate innovators of new sound technology proposed certain identities for the apparatus and its different applications, but they did not ultimately determine how emerging sound media became meaningful. Chapter 4 examines struggles to determine what social roles new media would perform. The spheres of commerce, culture, and politics provided a preexisting discursive grid onto which new media were mapped, a set of conceptual matrices through which the social functions of emerging acoustic media were understood and experienced. Running across and through these three spheres, neither wholly dependent on nor independent of them, a series of oppositions provided a further framework through which sound media were understood. The differences between local and national, between rural and urban, and between the immigrant and the native-born provided a focus both to how sound technology was imagined and how it was materially grafted onto social relations. I argue that a long-standing overdetermined desire, a perceived need, for media technology to serve as an instrument of national unification provided the governing logic that mapped emergent sound media onto preexisting institutions, relations, and practices.

Chapter 5 examines the struggle to establish conventions for representing with sounds. I begin by outlining two conflicting models for the manner in which the apparatus was to be used. The "transcription" paradigm, largely inherited from the mechanical phonograph, viewed electrical sound technology as mimetic in nature. A series of discourses and practices (as well as an implicit theory of sound representation) applied electrical sound technology as if its appropriate function was the accurate transcription of preexisting sounds. The second paradigm, "signification," viewed the apparatus as a medium of expression, without any responsibility to reproduce accurately, fully, or objectively an acoustic event. Instead, representational practices of the signification paradigm used the apparatus to construct new acoustic events. After tracing advocates of each approach, I focus specifically on film sound and illustrate how the conflict between these competing paradigms was successfully resolved with a new paradigm for technology's relationship to sounds. Called "signifying fidelity," this model for acoustic representation sought to use the creative potential of electrical sound technology to signify a mimetic relationship to an (often nonexistent) original sound event.

One

Technological Innovation
and the Consolidation of Corporate Power

"A Roaring Empire of Commerce"

In the fall of 1930, film studio RKO released *Check and Double Check*, a film featuring the popular blackface radio performers Amos 'n' Andy.[1] (See fig. 1.1.) In prerelease publicity, RKO announced that the film would open simultaneously in more than three hundred theaters during the week of October 24. Rather than initially releasing the film at a New York City and/ or Los Angeles movie palace or a series of urban centers and then promoting the film in subsequent venues based on these premieres, RKO opted for a massive, simultaneous national release of the film.[2] Through this strategy, RKO adopted a cinematic equivalent of network radio's national mode of address in which citizen-consumers located in both entertainment centers and the hinterlands simultaneously experienced the film, unifying "all" of America through this "event." The extensive promotional campaign for the film included a national radio broadcast over the NBC Red Network (on the "RKO Radio Hour"), originating from New York station WEAF and carried by thirty-eight other stations. The broadcast featured live performances by Amos 'n' Andy and the Duke Ellington Cotton Club Orchestra, also featured in the film. Advertisements in the motion picture trade press noted that RKO had developed tie-ups for the film with Pepsodent (a toothpaste, sponsor of the "Amos 'n' Andy" radio show), Williamson Candy

FIGURE I.I NBC radio stars Amos 'n' Andy in a promotional photo for the RKO film *Check and Double Check*, an example of the intermedia commodity facilitated by the concentration of media ownership. (Quigley Photographic Archive, Georgetown University Library, Special Collections Division, Washington, D.C.)

(makers of the Amos 'n' Andy candy bar), and Marx toys. This merchandising campaign promised to make available to exhibitors some 200,000 store windows with cooperative publicity underwritten by these other companies. Music publisher T. B. Harms also planned an advertising campaign in support of *Check and Double Check* and sheet music for songs from the film. Victor, a dominant force in the phonograph industry, coordinated the release to some 20,000 retailers nationwide of both Amos 'n' Andy and Duke Ellington phonograph recordings that were to be connected with the film. RKO announced plans to run single- and double-page advertisements in over 150 newspapers on October 25 and 26. And the film was used to open the new 2,300-seat RKO Mayfair theater on Broadway and 47th Street. In prerelease advertisements, RKO characterized this intensive promotion campaign as a "roaring empire of commerce geared to fighting pitch for show world's greatest demonstration of co-operative promotion."[3]

As a film, *Check and Double Check* capitalized on the fame Amos 'n' Andy had garnered in a different medium, radio, and to a lesser extent on the celebrity of Duke Ellington cultivated through radio and phonograph recordings as well as live performances. RKO's promotion of *Check and Double Check* linked the three dominant entertainment media of its day: cinema, radio, phonograph. The boundaries of the film as a commodity blurred, and radio broadcasts, the film's premiere, phonograph recordings, and sheet music became mutually reinforcing. Such tie-ins were certainly not originated by RKO; the production and marketing of sheet music overtly linked to motion pictures, for example, was a common practice throughout the 1920s. RKO also did not innovate the cross-promotional strategies that linked the film with apparently unrelated commodities (toothpaste, candy bars, and toys), although this practice would become increasingly common and pervasive as the decade progressed. What is significant about *Check and Double Check*, and the array of related media texts, is the extent to which this "roaring empire of commerce" originated within and was orchestrated by a single multiple-media conglomerate. RKO's parent company, radio giant RCA, controlled the NBC radio networks and station WEAF. It owned the Victor phonograph company and had a financial relationship with music publisher T. B. Harms. This concentration of media ownership allowed a single corporation to maximize its control of the profits generated from a variety of media forms.

While today such "synergy" is commonplace, in 1930 *Check and Double Check* and RKO's aggressive promotional strategies expressed in commodity form new economic relationships within U.S. media industries. This new media economy included both an intensified concentration of media ownership in which a few large firms controlled interests in what had been separate media (phonograph, radio, film) and a series of less formal relationships between media firms that promoted performers and performances simultaneously circulating in multiple media formats. The intermedia landscape in which Amos 'n' Andy emerged as a national sensation was the direct result (albeit not an *inevitable* result) of the advent of electrical acoustics. This chapter describes the process through which electrical methods of augmenting sound moved from corporate research laboratories to commodity forms. The innovation of electrical sound technology and its application to several media is largely a tale of corporate power and the methods used by large corporations (sometimes acting in concert) to limit competition and to manage technological change. Action (and indeed *inaction*) in the realm of public policy most often functioned to serve corporate agendas as corporate innovators successfully identified a market for new technology and both shaped that market and adapted to it. But the innovation of electrical sound technology is not simply a tale of success, for like all instances of technological change it was also characterized by a series of false starts and unanticipated consequences. Economic power provided large corporations with the flexibility to adapt to unanticipated developments comparatively better and more efficiently than those concerns that were less well capitalized. Even more than economic power, the key to corporate management and control of technological innovation resided in the acquisition and selective prosecution of patent rights.

Industrial Research, Patents, and Economic Closure Mechanisms

Corporate giant RCA, controllers of the "Amos 'n' Andy" brand, originated from a patent pool through which powerful corporate innovators of radio sought to manage the emerging medium. Events preceding and during World War I had powerfully illustrated some of the potential offered by wireless communications. But crucial wireless patents were distributed between a variety of corporations and individual inventors, largely preventing any one individual or organization from profitably and efficiently exploiting wireless innovation. In an effort to facilitate wireless progress through an orderly apportionment of the technology and its markets, AT&T and General Electric negotiated an agreement in 1919–20 that split the wireless field between the participating organizations and established a patent pool through generalized cross-licensing agreements. The patent pool created a new radio operating company, the Radio Corporation of America (RCA).[4]

The 1919–20 wireless radio patent pool that created RCA had precedents in previous attempts by corporations to rationally manage U.S. mass media innovations.[5] In 1879 the Bell Telephone interests entered into an agreement with Western Union that essentially split the field of distance communication into two distinct realms.[6] The agreement sought to circumvent pending and future patent litigation and to rationalize the business of distance communication. While the Bell interests maintained some financial and technological interest in wired telegraphy (their emerging "longlines" system was initially deployed at the service of telegraphic communication), the agreement essentially defined the telephone and the telegraph as distinct economic entities. This method of economic closure whereby real or potential competitors split a field into distinct and separate spheres of influence and exploitation would characterize future Bell System interactions with regional phone companies as well as the Bell System's approach to wireless telephony and broadcasting. General Electric and Westinghouse had used a similar agreement to pool their patents for the electrical power, lighting, and appliance industries in 1896.[7]

The process of solidifying the wireless patent pool began with General Electric's purchase of all American-based Marconi wireless interests and the formation of RCA as a holding company. On July 1, 1920, General Electric and RCA signed a patent cross-licensing arrangement with AT&T that provided for the sharing without royalties of all current and future radio patents for ten years.[8] Under the terms of the agreement, AT&T was to receive an exclusive license for applications in wired telegraphy, wired telephony, and some rights for the exploitation of wireless telephony in relation to its existing telephone network. General Electric was granted exclusive license to make, use, lease, and sell all wireless apparatus for amateurs. The new company, RCA, was granted exclusive rights to commercial applications of wireless. Although the initial provisions of the agreement separated potential development of wireless technology into apparently distinct fields, many of the provisions of the agreement were ambiguous, as later events would dramatically illustrate.

The patent pool was expanded in 1921 to include Westinghouse, which had aggressively increased its position in the field through strategic patent acquisition. While Westinghouse had manufactured radio apparatus throughout World War I, its comparatively weak patent position provided the corporation with minimal ability to shape postwar developments. In a 1920 agreement with the International Radio and Telegraph Company, Westinghouse gained Reginald Fessenden's patents covering continuous-wave transmission and heterodyne reception, and it then acquired exclusive rights to inventor Edwin Armstrong's patents for regenerative circuits and the superheterodyne receiver. These patents gave Westinghouse the influence needed to join the wireless patent pool, and on June 30, 1921, they entered into agreement to share patent licenses with the other participants.

RCA would purchase 40 percent of its radio apparatus from Westinghouse and 60 percent from General Electric.

In addition to codifying economic arrangements between the participating corporations, the wireless patent pool contained a series of different conceptions of what radio could become and the commercial possibilities it represented. A belief that wireless technology would augment already existing point-to-point communication services dominated these anticipated identities. AT&T, for example, initially conceptualized radio as an alternative to point-to-point wireless telegraphy and transoceanic cable communication.[9] This conservative approach imagined the technology according to the model established by wired telephony, and it coincided with AT&T's existing corporate focus. In 1915, AT&T executive John J. Carty examined the corporation's wireless tests and concluded that radio's future resided in the extension of wired telephony to previously unreachable places.[10]

The wireless patent pool only marginally codified the identity that ultimately proved to be the most influential application of radio technology, point-to-mass broadcasting. Ambiguously contained within provisions governing the manufacture and sale of amateur apparatus, radio's identity as broadcasting quickly became the primary site of consumer interest, research and development, and, consequently, corporate and regulatory concern. During 1920, a small number of experimental broadcasting stations had begun transmitting on limited schedules to amateurs who often listened with homemade receiving sets. Among the first was a station owned by the *Detroit News* that in 1920 broadcast news bulletins with a De Forest "radio telephone" transmitter. Dr. Frank Conrad of Westinghouse also broadcast from a homemade studio in his garage in 1920. His experimental transmitter and the local demand for radio receivers and equipment that it inspired became the basis for Westinghouse's support of station KDKA in Pittsburgh. KDKA inaugurated arguably the first regular broadcasting service on November 2, 1920, by reporting the results of the Harding versus Cox presidential election returns. From these initial experimental broadcasts, radio defined as broadcasting quickly expanded and eclipsed the ability of the patent pool and its participants to manage the emerging technology. By the end of 1922 there were thirty licensed broadcasting stations in the United States, but by 1924 that number had grown to over 500.[11]

With the construction and operation of an increasing number of wireless broadcasting stations after 1920, manufacturers encountered an unanticipated demand from consumers for wireless receiving apparatus. Initially taking the form of components for home-built receiving sets, growing consumer interest in radio-related items led to an eventual market for pre-built radio receivers, and consumer demand for wireless receiving equipment soon eclipsed the manufacturing capacity of participants in the patent pool. Radio broadcasting had grown so quickly that General Electric and Westinghouse were incapable of producing receiving sets and parts as fast as RCA could sell them, prompting RCA to run a national advertising campaign explaining the situation to anxious consumers. Simultaneously, a variety of organizations clamored to purchase broadcasting equipment. If AT&T was to be believed, there was a backlog in the manufacture of radio transmitters as well, although AT&T's response to customers (that they could simply purchase time on AT&T's toll stations instead of owning their own transmitters) raised some doubts about these production delays.

These early days of what we might retrospectively think of as the radio industry (actually several quite distinct industries at different time periods and as "radio" came to inhabit different identities) were characterized by an economic chaos echoed by a broadcast chaos on the airwaves. Broadcasting stations flooded the ether with an eclectic mix of programming all broadcast on the same frequency. Established wireless communications corporations and a variety of upstart concerns clashed amid a largely untested and increasingly confusing patent situation.[12] As radio rapidly developed into a consumer-based mass product, preexisting anitmonopoly sentiment became focused on the still-confused radio situation, and Secretary of Commerce Herbert Hoover, among others, loudly proclaimed that no one organization would control radio as a public resource. This antimonopoly sentiment exacerbated existing strains within the radio system established by the wireless patent pool and hastened its collapse.

The unanticipated prominence of radio broadcasting largely undermined the patent pool's ability to bestow economic order on technological innovation. But the 1919–20 pool successfully solidified the power and prominence of AT&T, GE, and Westinghouse and structurally guaranteed their continued dominance of subsequent acoustic innovations. RCA, itself a product of that patent pool, would also powerfully shape subsequent events. With the sudden importance of point-to-mass broadcasting, electrical sound technology's identity was not firmly and inevitably determined. A series of institutional and technical questions had yet to be navigated as part

26

26

Technological Innovation and Corporate Power

of broadcasting's development. These questions involved economic, regulatory, technical, and aesthetic issues, and not surprisingly, given the patent situation, a handful of large corporations largely dominated this process.

Although the wireless patent pool ultimately failed in its goal of efficiently managing technological change, its failure did not undermine the tremendous power patents provided to participating corporations.[13] For organizations like AT&T and General Electric, research and development, technological innovation, and the resulting patents formed crucial components of larger strategies designed to guarantee profits and to protect existing markets. Corporations exercised control over technological innovation largely through their patents, whether those rights were protected and exercised exclusively, pooled with other competing organizations possessing similar interests, or carefully distributed through licensing arrangements. Patents provided means to gain financial control of not only technological innovation but also the profit-making applications of emerging technology in the marketplace. This control also allowed some corporate entities to play a significant role in determining and defining the manner in which an apparatus or technological innovation was deployed.

In his study of the rise of corporations in the period from 1880 to 1930, historian David Noble argues that rather than protect the individual inventor, the U.S. patent system had come to reward science-based corporations. Noting the period also involved comparatively minimal judicial restraint of corporate patent monopoly, Noble outlines the ways in which large corporations used their patent positions essentially to circumvent antitrust laws. Noble cites as particularly significant "the agreements through which the efforts of individual companies were coordinated in the interest of all, and insulation from national and international competition in particular fields was secured."[14] Industrial research played an especially crucial role in perpetuating a corporation's economic power by generating new patents.

In an internal memorandum from 1927, John E. Otterson, the Bell System's executive in charge of sound motion picture activities, outlined the Bell System's approach to competition and its potential to impinge on the company's maximization of profits. Noting that the primary function of AT&T was to maintain and defend its position in wired telephone service, Otterson described some of the policies used to undertake that defense, stressing the importance of ongoing research efforts beyond those designed to improve wired telephony services:

The engineering in the major field extends beyond that field and over-laps upon the engineering of another major field and sets up a competi-tive condition in the "no-man's land" lying between.... . The regulation of the relationship between two such large interests [AT&T and General Electric] is accomplished by mutual adjustment within "no man's land" where the offensive of the parties as related to these competitive activities is recognized as a natural defense against invasion of the major fields. Licenses, rights, opportunities, and privileges in connection with these competitive activities are traded off against each other and interchanged in such a manner as to create a proper balance and satisfactory relation-ship between the parties in the major fields.... . It seems obvious that the best defense is to continue activities in "no man's land" and to main-tain such strong engineering, patent, and commercial situation in con-nection with these competitive activities as to always have something to trade against the accomplishments of other parties.[15]

Industrial research and development represented sorties into the "no man's land" of unexploited technological terrain, with patents fortifying newly established positions. Research could function preemptively, allow-ing a corporation to gain ground in as yet underexploited fields. Once such territory was gained, the position and patents could be strategically sur-rendered (as the Bell System eventually would do with many of its radio patents) or colonized and transformed into economically viable new tech-nologies (as in the Bell System's approach to sound motion pictures). The preemptive research strategy outlined in the Otterson memo had shaped the Bell System's expanded investment in research after 1909. That re-search program directly led to significant developments (and patents) cover-ing voice amplification, electrical recording, radio-signal transmission, and loudspeaker design—in short, the components of the electrical-acoustic apparatus. In its 1930s investigation of the telephone company, the Federal Communications Commission (FCC) concluded that the Bell System's re-search activities were designed "to obtain a degree of knowledge, influence, and control in adjacent industries sufficient to protect the return upon pri-vate capital invested in its wire telephone plant against diminution by the competition of emerging forms of communication."[16] This strategy sought to develop technologies and, importantly, *patents* in subsidiary fields not only for exploitation but also to preempt incursions into AT&T's domain.

These patents would also protect the Bell System's interests in future economic struggles.

New initiatives, when undertaken commercially, used exclusive licensing agreements that often mandated specific formats or prohibited interchangeability with competing systems. These license provisions could function as powerful tools in resolving or circumventing potential economic conflicts. In another memo later in 1927, Otterson described the intense competition between the Bell interests and RCA to license the major motion picture producers for sound films. Noting that such competition would "ultimately result in a situation highly favorable to the motion picture interests and opposed to our own," Otterson described the desirability of monopolizing the emerging application of sound films. "This is an extensive and highly profitable field and it is quite worth our while to go a long way toward making it practically an exclusive field. I believe that we could justify from a commercial standpoint, paying a large price for the liquidation of the Radio Corporation for this purpose alone."[17]

On the surface, licensing agreements that mandated industrywide formats by prohibiting the use of competing apparatus or services (so-called standardization and interchangeability clauses) would seem to function in restraint of trade. A 1937 U.S. District Court ruling, however, would legally affirm in principle the long-standing practice of economic protectionism through such interchangeability restrictions. Although finding that the noninterchangeability and the repair and replacement clauses of AT&T's contracts for sound motion picture apparatus (through their subsidiary Electrical Research Products, Inc. [ERPI]) were in violation of antitrust acts, Judge John P. Nields ruled that such restrictions during "the research and promotional period of the industry" were legally allowed.[18] The Bell System later successfully used similar licensing restrictions to eliminate potential competition from telegraph companies for wired transmission service (most notably in providing wired connections between broadcasters to create either temporary or permanent radio networks).[19]

The Bell System's deployment of its patent position illustrates four ways that the results of basic and applied industrial research could be used to a corporation's advantage. Bell's patent position allowed it to improve existing services to its customers and to eliminate competition within the telephone field. The patents also allowed the Bell System to preempt the development of fields that might potentially compete with wired telephony or with other commercial interests in which they had a financial stake (such as sound motion pictures). Finally, patent rights could be traded and exchanged in treaties with existing or potential competitors, thus providing corporate capital of a different order.

Corporate giants AT&T and General Electric used the strength of their well-funded research and development laboratories and their acquisition and prosecution of key patents to limit competition and to heavily influence technological innovation. Simultaneously, however, independent entrepreneurial-minded inventors struggled to develop competing, comparable systems that applied electrical principles to sound media. Susan J. Douglas and Hugh Aitken have chronicled the crucial role played by inventor-entrepreneurs in the paradigm shift that transformed radio from point-to-point wireless telegraphy to point-to-mass broadcasting, but with many of the resulting crucial patents acquired by GE, AT&T, and Westinghouse and subsequently managed through a radio patent pool, opportunities for independent inventors significantly diminished.[20] The patents surrounding the use of vacuum tubes for electrical-signal amplification jointly held by GE's Irving Langmuir and AT&T's H.D. Arnold proved crucial in constituting and maintaining this corporate hegemony in mid-1920s electrical-acoustic innovation. Significantly, when Fox Film adopted Theodore Case's independently developed Movietone sound-on-film system, they required a license from AT&T for vacuum tube amplification.

While engineers in the research laboratories of AT&T and General Electric applied electrical principles to sound recording and reproduction, other inventors also developed electrical-acoustic media, but their success was limited by the control the industrial giants exercised over key patents. Chicago's Orlando Marsh engineered a functioning, commercially viable process of electrical recording at his Marsh Laboratories, Inc., and successfully released electrically recorded phonograph disks a full year before Victor and Columbia's license with Western Electric. Initially releasing these recordings on its in-house Autograph label and then later on the Electra label, Marsh Laboratories also provided recording masters, predominantly of Chicago-based performers, to other lower-priced labels, including Silvertone, Paramount, and Gennett. During the 1920s, Marsh also offered recording and duplication services to clients who then distributed the recordings themselves.[21] Besides several jazz recordings by Jelly Roll Morton and King Oliver, Marsh is best known among today's record collectors for a

series of 1924 electrical recordings of movie theater organists, most prominently Jesse Crawford performing on the Wurlitzer organ at Balaban &
Katz's Chicago Theatre. In collaboration with Marsh, Crawford recorded at
least eighteen performances, released on both the Autograph label and others (Domino, Paramount, and Silvertone). While contemporary collectors
describe the results of Marsh's electrical process as varying tremendously
from one recording to another, they seem to agree that these electrical-
process recordings represent a significant advance over attempts to record
organ performances using prevailing mechanical methods.[22]

Crawford's theater-organ recordings demonstrate that Marsh successfully applied electrical principles to the phonograph independently of the
research and development efforts of Bell Laboratories/Western Electric or
General Electric. His small company released electrical recordings prior to
what is generally considered to be the phonograph industry's conversion
to the new electrical processes. But like other independent entrepreneurial inventors working with electrical sound technology, Marsh's recording
system relied upon the vacuum tube for amplification, and this inherently
limited his ability to capitalize on his innovations. Without a license from
either Western Electric or General Electric, Marsh could not himself license other companies for the use of his system, a crucial part of commercializing that system since he lacked the manufacturing facilities to
produce phonographs. Nor could he successfully expand his production
of electrical recordings without running afoul of corporate patent holders.
Apparently, Marsh never sought a patent for his electrical recording system
nor a license for the use of vacuum tube amplifiers from Western Electric
or General Electric.[23]

Well-capitalized, established firms used their economic power as well
as extensive research and development resources to control and to manage
the process of technological innovation. Patent rights, whether the product
of in-house research or purchased from independent inventors, proved to
be the linchpin in a larger system that successfully consolidated economic
power. While Theodore Case or Orlando Marsh could innovate devices that
competed technologically with the products of well-funded corporate research laboratories, their systems were ultimately not competitive in the
marketplace.

Patents or membership in a patent pool could guarantee a corporation's *participation* in an emerging media field, but the nature, source, and

amount of potential profits from a medium like radio, for example, would be directly linked to its market identity and its relationship to other preexisting media. While the 1919–20 wireless patent pool sought to control future radio developments, the unanticipated importance of broadcasting left the corporate participants struggling to position themselves profitably in relation to the new medium. In the early 1920s, representatives of a single corporation, AT&T, offered competing visions for the future of radio broadcasting. Two events involving AT&T in February 1923 illustrated radically different versions of radio's potential identity in American life. One, a well-publicized broadcast, asserted that radio broadcasting should appropriately function as a national mode of address. The other, an internal meeting of Bell System executives, outlined radio's future according to a local-service model.[24] Each vision for radio's future, and subsequent events, illustrate the ways in which AT&T sought to manage technological innovation.

Posing the Question: What Is Broadcasting?

During a February 14, 1923, broadcast over AT&T-owned radio station WEAF, audiences heard described two not entirely incompatible visions of the future of radio and other electrical sound technologies.[25] The broadcast event was a meeting of the American Institute of Electrical Engineers (AIEE), held simultaneously in both New York and Chicago. Through the auspices of AT&T long distance, wired voice transmission, and public address installations in both venues, members of the AIEE heard the presentation of scientific papers and engaged in discussion of those papers despite the distance between the two convention sites. The broadcast event itself, albeit not of compelling interest to audiences broadly conceived, was of interest to many of those who in 1923 were involved in radio and thereby concerned with electrical developments.[26]

In an address to the AIEE, Frank Jewett, vice president of Western Electric, celebrated sound technology's ability to transcend spatial limits and electrically unite dispersed members of a group, thereby reinforcing shared identity. "For the first time groups of men and women, separated by hundreds of miles, are gathered together in a common meeting under a single presiding officer to listen to papers presented in cities separated by half the span of a continent and to take part in the discussion of these papers with the ease characteristic of discussions in small and intimate gatherings."[27]

According to Jewett, sound technology could (re)constitute community, an intimate gathering characterized by the copresence (albeit electrically achieved) of all participants. Jewett's remarks emphasized sound technology's ability to foster a type of discourse characterized by rational discussion and exchange. In short, Jewett framed technology as an instrument that facilitated a utopian public sphere, although a public sphere populated not by society at large but by a scientific community.

Jewett acknowledged in his remarks the implications of broadcasting the proceedings as well. "At the same time unnumbered thousands in their homes are auditors of our deliberations through radio broadcasting."[28] The participatory public sphere of scientists and electrical engineers had an audience as well, a "public" called into being not through the auspices of membership, copresence, or even the electrically augmented pseudo-presence of reciprocal communication, but instead this public was constituted by overhearing the rational deliberation and debate. The "scientific public sphere" of the AIEE became an aural spectacle to be consumed by all those absently overhearing rather than directly participating in the proceedings.

Later in the same conference, chief engineer of Western Electric, Edward B. Craft, described a similar vision of radio. Craft noted to his New York City audience that the first mayor of New York in 1685 could have easily addressed his entire constituency of 1,100 without raising his voice. He continued by contrasting this earlier moment of direct, unmediated address with the possibilities now available through electrical means. "The ideas on which this nation was founded were spread abroad largely by word-of-mouth. With the gradual growth of the population, of the newspaper and of the mail service, the printed word gradually displaced the spoken word. The loud speaking telephone system will tend to re-establish oratory by making it worth while for the nation's leaders to sway the emotions of tremendous audiences."[29] Craft invoked the model of a public addressed by its leaders, a vision far from the public sphere of Enlightenment ideology. Here, emotional sway replaced rational debate, and the model for political modes of address was not a dialogue but precisely the one-way, theatrical performance that eventually would be codified through the rise of radio networks, commercially sponsored broadcasting, equal-time provisions, and other exclusionary methods of disciplining electrically augmented speech. Craft's remarks foreshadowed technological possibility as the perpetuating of power relations according to a preexisting status quo, all the while evoking an earlier "utopian" moment of copresence and unmediated address.

Taken together, the remarks of Jewett and Craft point toward many of the constitutive features of what American commercial radio would eventually become. The AIEE conference featured two-way, reciprocal communication that fostered participation and technologically augmented the possibilities of the rational discourse of scientific exchange, arguably a model of greater inclusion. The broadcast obviously did not involve such reciprocal communication but instead, like membership in the AIEE, this one-way communication entailed a necessary structural exclusion. Some spoke, others listened. In keeping with the celebratory rhetoric and the event's emphasis on the utopian possibilities of technology, "participation" was rhetorically redefined as listening in on distant events. Not coincidently, broadcast advocates made the same argument when supporting proposals to transmit congressional proceedings in the early and mid-1920s and in the aftermath of the broadcast of the 1924 national conventions of the Democratic and Republican parties. Audiences were rhetorically constructed as participants in these events.[30] Participation so defined elides the structural exclusion of one-way communication. While spatial absence was compensated for by broadcast, by dispersion into the ether, the "unnumbered thousands" of auditors in their homes mutely consumed an aural spectacle rather than genuinely or meaningfully participating in it. We can trace in these remarks by AT&T officials some of the underwriting assumptions embodied within the eventual U.S. commercial broadcasting system. Radio as a technology could be grafted onto and further solidify existing configurations of power, maintaining existing structural exclusions all the while rhetorically celebrating an illusory enhanced participation in American political and cultural life carefully contained through the act of consumption. But neither the ultimate form of American broadcasting nor the assumptions implicit in Jewett's and Craft's remarks were necessarily inevitable in February of 1923.

In the same month, other executives within the combined corporate boundaries of the Bell System advocated a different model for the diffusion and use of American radio. This model combined both a conception of radio as a one-way medium of communication to an audience with a view of that audience as neither national nor essentially anonymous. Instead the audience within this competing model for radio, like the members of the

AIEE, was linked by shared components of an identity, namely residence in a community. On February 26, 1923, AT&T held an internal conference attended by representatives of associated Bell System companies to discuss and to develop a strategy for maintaining and consolidating their interest in radio. While radio as a medium in early 1923 had been more or less successfully defined as predominantly broadcasting (as opposed to a two-way medium of, for example, wireless telephony), the ultimate shape through which the technology would be diffused was far from predetermined. Various corporate interests (including Westinghouse, General Electric, the Radio Corporation of America as well as AT&T) vied with entrepreneurial-minded inventors and loosely aligned interest groups (including educators and the Amateur Radio Relay League) to develop an identity for the emerging media form. At the 1923 conference, the representatives of the Bell System developed a coherent vision for radio's dissemination according to a local-service model.

AT&T proposed developing U.S. radio as a dispersed, largely decentralized array of noncompetitive local broadcast stations operating at transmission-power levels designed only to reach members within a specific community.[31] According to the plan, AT&T would construct, own, and operate a single broadcasting station in virtually any town of consequence in the United States. Within this model, programming would be predominantly local in origin with the community itself determining the content of broadcast material. Programming policies and decisions were to be vested in the hands of a local broadcasting association consisting of community leaders and the general public. When the local broadcast association wished it, AT&T would provide through a modified version of its long-distance telephone system interconnections between local stations for access to programming of abiding national or regional interest. Under the plan, competition from private stations would be minimized through the efforts of the local broadcasting associations to function inclusively, encouraging all in the community who were interested in radio to join in the shared collective effort. Such a vision for American radio poses a striking contrast to the eventual (predominantly) for-profit, advertising-supported, competitive oligopoly of powerful radio networks and their affiliates that would come to characterize American broadcasting well beyond the transition to television. By some standards, the AT&T proposal sounds almost utopian in nature; this model for radio reproduced and reinforced utopian notions of a locally defined public freely and collectively determining what would be in their own community's best interest. That public could have been truly participatory, not simply the passive consumers of radio programming (an audience), but active participants both in shaping broadcast policy and in developing the resulting programming.

Whether AT&T genuinely intended for this system of radio to constitute a broadcasting version of an ever-illusory participatory public sphere (an issue I will address below), such a model for radio would appear to have created the *structural* conditions within which a participatory deployment of radio could have developed. One can imagine an alternative history of American broadcasting in which the technology provided voice to individuals, organizations, and perspectives largely excluded from a for-profit medium that in the first instance addressed its audience as consumers (commercial programming) and in the second instance (community affairs, political commentary, political broadcasts) addressed its audience as consumers of hegemonic discourse.

Beyond the structural conditions that could have created such a decentralized community-controlled system of radio, some evidence from the period surrounding AT&T's proposal suggests what such a local-service model of radio might have sounded like. In a community with a strong organized labor presence, ordinary working people might have been addressed directly as a social class, perhaps by fellow working folks speaking precisely about labor issues. Such a mode of address was in fact adopted a few years later by Chicago radio station WCFL before the pressures of competitive, commercial broadcasting undermined the station's initial mission. In a community with a strong ethnic population, programming could have cultivated that identity rather than fostering an ethos of assimilation. Early broadcasting history included examples of stations and programming that addressed ethnicity as a shared cultural identity rather than either as merely a market segment or an identity strategically erased by attempts to constitute a national consuming audience.[32] Radio did not inevitably have to adopt the assimilationist impulse described by Michele Hilmes as "a common mode of address, an engagement with the interests of the audience that at once recognized their diversity and attempted to delineate and address common problems that that diversity produced (while, arguably, directing them to solutions compatible with radio's commercial base)."[33] Within such an alternative history of broadcasting, women might have played a far more active role both as listeners and as broadcasters. Susan Smulyan has tracked the extent to which the rise of commercial broadcasting hailed women as consumers and homemakers, thereby reinscribing them within patriarchal domination.[34] However, widely circulated radio journals in the 1923–24 period like *Wireless Age* offered countermodels for feminine, indeed proto-feminist uses of radio that incorporated women (as listeners and as broadcasters) into active roles within a political rather than consumption-driven public sphere. While a journal like *Wireless Age* did include articles on radio as a tool in engineering a more efficient home economics, other models for women in radio existed before radio's full-scale commercialization.[35] Within AT&T's model, when the corporation's desire to maintain profits by adding more stations to this broadcasting system combined with the increased diffusion of radio receivers to minority populations, a crucial point would have been reached when AT&T might have constructed stations in predominantly African-American communities like Harlem or Chicago south of the Loop. Here, programming policies of individual stations might have overtly addressed African-American audiences in the way that vaudeville circuits (TOBA), record companies (race labels), and newspapers (*Chicago Defender* et al.) had already done.

Such retrospective speculations about a different radio history views technological innovation as potentially utopian in nature, and perceives media developments as following a declension narrative in which radical or even reformist potential is progressively foreclosed through the almost inertial weight of capitalist investment and regulatory initiatives that structurally follow the logic of capital. But before invoking "economics in the last instance" as the ultimate arbiter of the social shaping of U.S. broadcasting, it is important to examine AT&T's rationale for this local-service model of radio. Despite its carefully managed rhetoric of "universal service" throughout the 1910s and 1920s, the Bell System was not philanthropically proposing this approach to radio solely as a contribution to an actual or virtual communications democracy. In this historical instance at least, the interests of a large corporation briefly coincided with those who sought a noncommercial, decentralized system of U.S. broadcasting.

AT&T's 1923 plan for the diffusion of radio according to a local-service model resulted from the interaction of multiple determinants that shaped the organization's perspective on radio. Foremost among these was AT&T's desire to maximize profits by exercising its patent position and the "exclusive" rights granted to the organization by the wireless patent pool agreement of 1919–20. AT&T maintained, in a position later determined to be erroneous, that the patent pool granted it the exclusive rights to build, lease, sell, and/or license radio transmitters.[36] Until those patent rights expired, profits could be maximized through the construction and operation of a large number of radio stations. A model for radio featuring an abundance of broadcasters operating at lower power levels rather than fewer high-powered stations financially appeared to be in the Bell System's short-term interest. Such a plan offered continuing long-term profits through service contracts with local community broadcasting associations and through the introduction of improvements in transmission technology developed within the Bell System's established research initiatives. Innovations in the "state of the art" could guarantee continued sources of profit through new patents as the old patents expired, thereby foreclosing potential competition from future organizations.[37] AT&T's 1923 radio plan also would have successfully secured its role as the sole provider of interconnections between local radio stations or for wired transmissions between broadcasters and remote events. Radio, from this perspective, provided yet another potentially profitable use for the Bell System's existing long-distance facilities.[38]

AT&T's desire to maximize profits through the execution of its patent position and the potential for radio to supplement the income from its existing long-distance lines shaped this 1923 radio proposal. But other forces shaped the plan as well. A decentralized, local deployment of radio like that of the plan was consistent with ongoing public relations efforts within the Bell System that stressed the public service provided by this large, increasingly ubiquitous corporation. A local-service model for radio complimented both the Bell System's rhetoric of "universal service" and its related attempts to craft a benevolent public service identity, thereby forestalling public antimonopoly sentiment and potential regulatory intervention.[39] Through a local-service model of radio, AT&T could nurture an additional local identity that was structurally linked to community leaders and a definition of radio broadcasting as a public service provided by the Bell System. In addition, AT&T's plan for the proliferation of low-power, community-centered broadcasting facilities conservatively applied the successful model for the diffusion of the telephone to a new and emerging media form. AT&T would be a communications *provider*, constructing, owning, and operating the instrumentality of communication while individual local consumers determined how best to use it. Finally, AT&T's 1923 radio proposal should be understood in light of the organization's ongoing experience as the operator of a broadcasting station, WEAF (New York City). AT&T was attempting, with mixed success, to operate WEAF in a manner analogous to existing telephone service. The company provided a broadcasting facility, the content of which was solely to be determined by whoever purchased the use of that facility. The 1923 radio proposal expanded to a national scale the experiences and approach of WEAF.[40]

AT&T's patent position for broadcast transmission equipment, the proposal's consistency with AT&T's public presentation of the Bell System as precisely a public service, and the conceptual conservatism that made logical a view that radio might be best diffused in a manner analogous to the telephone, all underwrote the approach to radio outlined internally by AT&T officials in February 1923. Rather than any utopian, participatory vision of wireless technology, AT&T's plan for a decentralized, noncommercial, publicly controlled model for broadcasting not surprisingly arose out of a desire to maximize profits. A large corporation sought to exploit commercially its patent position in the "no-man's land" adjacent to its major field. Despite its for-profit origins, however, this model for radio would have structurally enabled a drastically different approach to broadcasting than what developed in subsequent years.

The conflicting visions of radio articulated within the corporate boundaries of AT&T during 1923 attest to the presence of multiple, competing identities for radio broadcasting as an application of electrical acoustics. Despite AT&T's proposed local-service model for radio and complementary spatial conceptions of radio's potential that emphasized local programming and participatory values, ultimately a stable system of U.S. broadcasting emerged in the late 1920s that more closely resembled the model articulated by Craft and Jewett and the values that model embodied.

As radio receivers successfully entered American homes, the experience of listening to radio broadcasts shifted from attention to the apparatus itself to the service it provided. Simultaneous with this shift in auditors' relations to electrical-acoustic technology, mutually reinforcing developments in radio industry economics, broadcast programming, and federal regulation coalesced to make progressively dominant one of broadcasting's competing identities. The competing models for broadcasting articulated within AT&T in 1923 shifted toward a hierarchy of identities that subsumed notions of radio as a local service within a larger political economy of broadcasting that privileged radio as offering a national mode of address.

AT&T led the way in conceptualizing radio in this form with dramatic demonstrations of the potentials offered by wired radio networks. The simultaneous broadcast of presidential speeches by several stations or the interconnection of multiple stations during national defense preparedness exercises demonstrated radio broadcasting's potential service in moments of national emergency, offering an instrument of mass public address that promised, in principle at least, eventually to reach all citizens. Having failed to control radio's potential profits by licensing all broadcasting stations, AT&T sought to consolidate its interest in radio by promoting wired transmissions between broadcasting stations. AT&T was also instrumental in innovating corporate sponsorship of radio programming, instituting through its radio station WEAF "toll broadcasting," in which advertiser sponsorship of programming first appeared on American radio. This identity for radio broadcasting inspired a particular type of economic arrangement—the establishment of radio networks and their contractual relationships with local affiliates as well as the eventual incorporation of advertising agencies into the political economy of American broadcasting. Radio's identity defined as a national mode of address also drove regulatory efforts as U.S. radio regulation followed developments elsewhere in the industry, conservatively reinforcing emerging economic relations and a specific identity for broadcasting.

The ultimate shape of U.S. broadcasting was overdetermined, that is, U.S. broadcasting was shaped by multiple, overlapping, and mutually reinforcing determinants. These included the conservative manner in which the radio was constructed as an entertainment device, a series of economic arrangements largely orchestrated by powerful electrical and communications corporations, and federal radio regulation that ultimately served the interests of corporate broadcasting control and radio spatially conceived as a national mode of address.

Corporate giant AT&T, despite its ubiquity and powerful patent position, was able neither to predict nor to control the development of radio. However, that patent position and the corporation's wealth allowed it to successfully adapt to broadcasting developments. Ultimately, in a series of moves that no doubt shaped Otterson's 1927 perspective on patents, AT&T

in a revised patent pool ceded the business of constructing radio stations to its rivals, sold its radio broadcasting stations to RCA, and largely withdrew from the field of radio to instead focus its radio-related efforts on the highly profitable business of providing the wired transmission facilities connecting broadcasting stations that became necessary for radio networks.

Grafting New Technology Onto Old Media: Phonograph and Cinema

During the initial struggles to determine the ultimate shape of radio broadcasting, some of the corporate members of the radio patent pool successfully applied electrical principles to two preexisting media—the phonograph and the motion picture. While radio was in a state of flux with multiple, competing identities vying for commercial, regulatory, and rhetorical dominance, preexisting media provided relatively stable identities, established markets, to which electrical-acoustic innovations might be applied. Research and development within both AT&T's affiliates and General Electric approached electrical acoustics as a set of related principles and devices that might be adapted to multiple applications. Although technically situated in what Otterson in his 1927 memo labeled the "no man's land" between the two organizations' major fields, the phonograph and cinema industries proved to be fertile ground on which the two corporate giants might cultivate a brand identity and reap significant profits. Existing phonograph and cinema companies ultimately became consumers of emerging technology developed by these large corporations.

During the early to mid-1920s, major phonograph corporations like Victor and Columbia lacked well-financed research and development capabilities even as they sought to increase the volume and fidelity of phonograph recordings. Research at these firms tended to proceed in an empirical, often haphazard manner. In contrast, the Bell System's coordinated program of industrial research developed methods that could provide phonograph recordings with greater fidelity through innovations related to several components of the phonograph as a technological system. The first innovation dramatically increased frequency response in the *reproduction* of recorded music, although it did not involve the application of electrical principles. Building on previously existing insights, engineers at Bell Laboratories developed the "theory of matched impedance," which applied theories of electrical systems to mechanical systems. In short, mechanical systems (such as that which translated the encoded sound waves on a phonograph disk into acoustic energy and subsequently amplified the resulting sound) could be *systematically* designed and developed to insure minimal distortion and loss in (acoustic) energy.[41] Such an insight, when practically applied, moved phonograph development far in advance of the empirical "cut and try" methods that, for example, led Victor's Eldridge Johnson to invent the "gooseneck" tone arm in 1903. The theory of matched impedance led, in practice, to the development of the folded acoustic horn, which provided a substantial increase in both the volume produced by and the frequency response of mechanically reproduced sounds while simultaneously maintaining a small enough size to fit into existing cabinet model phonographs.

Unlike the folded acoustic horn, the second fidelity-based innovation Bell Laboratories brought to the phonograph industry involved the direct application of electrical acoustics to phonography, an electrical system for phonograph recording. When used in combination with the folded acoustic horn, recordings produced through the electrical process demonstrated a substantial improvement in terms of apparent fidelity. Mechanical phonographs reproduced a frequency range of approximately three octaves, but when electrically recorded disks were reproduced through the folded horn, the frequency range increased to five and one half octaves.[42]

In addition to the theory of matched impedance and electrical recording, Western Electric's engineers also developed an electrical system for reproducing phonograph recordings. This system essentially applied radio principles to phonograph reproduction in that the encoded sound waves on a disk were first translated into an electrical signal that was then amplified and reconverted into acoustic energy through a cone-type radio loudspeaker. Western Electric claimed that a combined electrical recording and reproducing system increased recording of overtones, added "atmosphere" to recordings, and dramatically increased frequency range and volume.[43] (See fig. 1.2.)

The engineers at Bell Laboratories and Western Electric were by no means alone in attempting to apply electrical principles to the phonograph.[44] Besides Orlando Marsh's work in Chicago, Victor was in the midst of collaborations with engineers from Westinghouse on the theory of matched impedance when Western Electric demonstrated its innovations to Victor's executives.[45] Rather than viewing the Western Electric demonstrations as

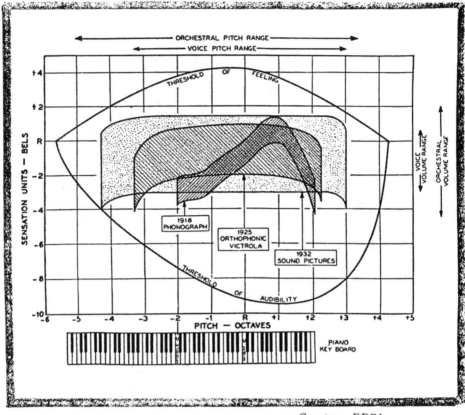

Courtesy *ERPI*

Graph showing how frequency range and volume range have been extended since 1918.

FIGURE 1.2 Electrical-acoustic innovations allowed the reproduction of a dramatically expanded range of sound frequencies. Here, the sound film of 1932 is favorably compared with both the mechanical phonograph of 1918 and the Orthophonic Victrola of 1925. (Author's collection)

affirmation of their own research and as encouragement to continue, Victor's engineers and executives instead used their independently developed insights to evaluate Western Electric's advances and signed a license with the AT&T subsidiary.[46]

Western Electric used both its patent position and its reputation, as well as the engineering strengths of its system, to secure in 1925 licensing agreements with Victor and Columbia, the two largest and most powerful U.S. phonograph corporations. The licenses with industry leaders established Western Electric's dominance as a provider of electrical equipment to the phonograph industry and reinforced its broader reputation within the emerging field of electrical sound devices. But these initial licensing agreements did not guarantee Western Electric a monopolistic position. As would be true for film sound, General Electric independently developed its own system of electrical recording and reproduction that would compete with Western Electric's. By licensing the phonograph company Brunswick, General Electric insured a market share for its competing system. The Panatrope, advertised as a product of the combined research of RCA, General Electric, Westinghouse, and Brunswick-Balke-Collender, most significantly differed from the Western Electric system in microphone design.[47] Brunswick advertised a frequency response of 16 to 21,000 cycles, far in advance of mechanical recording methods (128 to 2,000 cycles) and mechanical reproduction (250 to 6,000 cycles), although these claims probably exceeded the actual performance of the Panatrope.[48]

Unlike the emerging medium of radio broadcasting, the phonograph industry supplied a preexisting market for electrical-acoustic technology— that is to say, an identity for the medium did not have to be engineered and developed. The application of electrical principles essentially grafted new methods of augmenting sound onto already established technological artifacts, industry structures, consumption and listening habits, and representational practices. The acoustic results of electrical methods fulfilled previously stalled areas of research and development that had been empirically pursued by phonograph innovators for some time (namely the desire for greater fidelity and more consumer control over reproduced sound). Nonetheless, the industry's conversion to electrical methods was neither immediate nor simple. After signing a licensing agreement with Western Electric, Victor began the task of converting its existing recording studios, developing new recording techniques, and designing, engineering, and manufacturing new phonograph models that incorporated these technological changes. To facilitate that conversion, Bell Laboratories assigned J. P. Maxfield to Victor's Camden (New Jersey) studios.[49] Maxfield set up an experimental electrical recording studio on January 27, 1925, but Victor's first studio wasn't successfully converted to electrical recording until March 11, and the organization continued to produce some phonograph recordings using the now-outmoded mechanical process until the end of July 1925.[50]

Because of slower-than-anticipated sales throughout 1924 and into 1925, Victor had a substantial backlog of inventory at its various dealers, distributors, and factories.[51] Before publicly introducing newly designed phonographs and recordings, Victor sought to dispose of this existing stock. In June of 1925, it notified its dealers of the impending technological change and with national dealer cooperation cut sales prices in half for all existing phonographs. This sales program proved successful, and by the fall of 1925 when Victor prepared to announce to the public the introduction of new machines and new-process recordings with the ballyhoo of a nationally declared "Victor Day," the backlog of existing mechanical phonographs had been reduced. This announcement of the new devices and recordings had, of course, to demonstrate to consumers the desirability, if not the necessity, of the compulsory repurchasing that would eventually become a standard procedure of technological change within the American recording industry.[52]

As the phonograph industry adopted electrical technology, Victor, still the most powerful organization in the industry, released an important hybrid device, one that applied electrical principles to only *portions* of the phonograph as an artifact. Ironically, Victor recordings manufactured according to the new electrical process were successfully introduced to the public through a heavily and successfully promoted line of new *mechanical* phonographs. Orthophonic Victrolas, redesigned all-mechanical home phonographs, included no electrical components for sound reproduction. Instead a redesigned mechanical pickup device translated the grooves of recordings into acoustic energy that was then amplified through a folded horn conforming to Western Electric's design. The Orthophonic Victrolas competed successfully with Columbia's Vivatonal and Brunswick's Vivatonal and Panatrope phonographs, which relied upon electrical methods of reproduction and amplification and projected sound through cone loudspeakers. By many accounts, the Orthophonic Victrolas reproduced the new electrical-process recordings better than an electrical phonograph with either the cone or horn loudspeakers available to radio listeners in 1925 and 1926.[53] Since, the Orthophonic Victrola allowed the reproduction of consumers' existing records as well as the new and improved electrical recordings, the device arguably facilitated technical change by maintaining consumers' confidence in the value of their existing record collections and briefly delaying the compulsory repurchasing encouraged by a format change while simultaneously introducing consumers to the new fidelity standards embodied in electrical-process recordings.[54]

Soon after the initial introduction of electrical sound technology, the phonograph industry began releasing to the public a variety of machines that combined phonographs with radios. Phonograph companies took out licenses from RCA for radio equipment manufactured by Westinghouse and General Electric to be installed in combined phonograph-radios or for the manufacture and sale of domestic radio receivers.[55] Some of these machines used the folded (Orthophonic) horn as an amplifier for radio signals reproduced through an electro-dynamic speaker and included a mechanical switching device that enabled the listener to choose between projecting a radio show or a phonograph recording through the folded horn.[56] Some of the more expensive combination phonograph-radios provided for AC and or universal (AC/DC) attachments that allowed the radio receiver to be operated without batteries, or that used electrical current to drive the turntable (replacing spring-wound turntables).

By 1927, Victor's customers could select from a series of acoustic devices in a variety of cabinet styles that offered various combinations of the following: mechanical phonograph reproduction, electrical phonograph pickup and amplification projected either through a cone loudspeaker or through the Orthophonic horn, tuned radio frequency or superheterodyne radio receivers with programming projected through a cone loudspeaker or the Orthophonic horn, battery or electrically powered radio receivers, and electrical or spring-driven turntables. From 1926 through 1929, sales of Victrola/Radiola consoles (combined phonograph-radio receivers) steadily increased while sales of both the Orthophonics and Electrolas peaked and then fell off (1926 for the Orthophonics, 1927 for the Electrolas). By 1929 the two most popular sales lines for Victor were the combined radio and phonograph consoles and portable Victrola devices.[57] As the 1930s began, the phonograph increasingly became an adjunct to the domestic radio receiver rather than a separate and distinct domestic entertainment device.

The phonograph as an industry and as a consumer product provided a preexisting commercial identity onto which electrical acoustics could be grafted. Phonograph industry rhetoric of the period framed the conversion to electrical methods as the fulfillment of long-standing consumer desire for greater fidelity, "more sound," and greater control in performing sound reproduction. The absence of systematic and sustained research and development efforts within phonograph companies necessitated that they become consumers of electrical technologies designed, developed, and manufactured by organizations outside the industry. When Victor and Columbia signed licenses with Western Electric, and Brunswick adopted the system developed by General Electric, the major corporate innovators of radio expanded both the identity of the electrical-acoustic apparatus and their commercial control of that identity. The technological convergence of radio and phonograph embodied in combined machines was literalized in the economic realm when, in 1929, RCA acquired the Victor Company. Victor, for decades the leader in disk-model phonographs, became a subsidiary of the radio industry giant, itself formed only a decade earlier as a patent-holding company.

When AT&T's engineers adapted its system of electrical recording to motion picture applications, this was neither the first attempt to create a synchronous-sound film system nor the first such system based on applied electrical principles. Throughout the first two decades of the century, a number of entrepreneurial-minded inventors around the world had devised practical methods of producing synchronous-sound films, and several sound-film systems reached a commercial stage of development. Some of these efforts, like Edison's Kinetophone, the Cinephone, the Cameraphone, and the Gaumont Chronophone, used mechanical phonograph technology to record and reproduce film soundtracks.[58] Attempts to engineer synchronous-sound cinema by combining film with the mechanical phonograph had failed for several reasons. Perhaps most important were the absence of adequate means of sound amplification and difficulty designing a reliable method of maintaining synchronization between film projector and sound reproducer. Amplification proved to be a problem not only in the reproduction of recorded sound to a large audience but also in recording. The reliance on an acoustic horn as a sound-collection device made difficult, if not impossible, simultaneous recording of sound and image. The acoustic horn could not, after all, appear on screen, yet to adequately record voices, it needed to be close to the actors.

Eugene Augustin Lauste had by 1904 resolved the problem of maintaining synchronization when he developed an experimental apparatus that recorded a soundtrack on the same film as the motion picture image. Although Lauste's system relied on electrical principles (a microphone encoded sound waves and a selenium cell translated the electrically encoded sound back into acoustic energy), the absence of an efficient source of amplification for the electrical signals or the reproduced sound limited its commercial viability.[59] Joseph Tykociner, a professor at the University of Illinois with a background in wireless technology, publicly demonstrated a sound-on-film system in June of 1922. Tykociner applied many of the principles of electrical acoustics and adapted components from other electrical communications devices. Using a carbon telephone transmitter as a microphone and adapting an early Magnavox radio loudspeaker for reproduction, Tykociner relied on several components that would eventually characterize GE's and Western Electric's commercially marketed sound-film systems (a lamp modulated by electrical signals encoded sound waves onto the film, a photoelectric cell translated the visually encoded soundtrack back into electrical impulses, and perhaps most importantly, vacuum tubes provided signal amplification).[60]

While innovators like Lauste and Tykociner solved the problem of synchronization by engineering combined cinema-sound devices that encoded the soundtrack on the same strip of celluloid as the image, amplification remained an obstacle to the innovation of a commercially viable system of film sound. Lee De Forest's Audion triode vacuum tube, with the appropriate modifications independently developed by H. D. Arnold at AT&T and Irving Langmuir at GE, provided a reliable, distortionless source of signal amplification, but the wireless patent pool's cross-licensing agreement between GE and AT&T effectively prevented entrepreneurial inventors from using vacuum tube amplification.

Lee De Forest's attempts to introduce his Phonofilm sound-on-film technology in 1923–1927 illustrate the myriad obstacles even successful inventors faced when competing with the participants in the radio patent pool. De Forest sold the patent rights to his triode vacuum tube to AT&T in 1913, but he had astutely retained personal rights to continue commercially developing his invention. Unlike electrical phonograph inventor Orlando Marsh, De Forest did have access to vacuum tube amplification. In 1923, after designing and demonstrating a functioning method of producing synchronous-sound films called the Phonofilm, De Forest sought without success to generate commercial interest in his sound-on-film system. (See fig. 1.3.) De Forest's lack of both capital resources and the extensive manufacturing capacity necessary to supply equipment for the film industry seriously hampered his efforts. Further, the extent to which the film industry

FIGURE 1.3 In this 1925 photo, inventor Lee De Forest holds the attachment used to convert film projectors for the reproduction of Phonofilm sync-sound films. (Quigley Photographic Archive, Georgetown University Library, Special Collections Division, Washington, D.C.)

was vertically integrated simultaneously damaged the prospects of small entrepreneurial electronics firms while working in favor of large, established electronics manufacturers.

The film industry underwent a period of consolidation during the 1920s, as did much of U.S. business, and the emerging, dominant vertically integrated concerns were hesitant to embrace wholesale technological change. This industrywide consolidation was conservative in nature, offering economies of scale as producers/exhibitors acquired smaller competing producers and existing or emerging chains of movie theaters. This consolidation maximized large corporations' profits within an existing industrial status quo. Having identified and cultivated a product (the feature-length film, or in Richard Koszarski's words "an evening's entertainment") and in the midst of applying more economically efficient methods of producing and delivering that product, Hollywood producers were justifiably hesitant to reconceptualize the nature of the entertainment commodity they produced.[61] In addition, as the emerging major producers/exhibitors assembled ever larger and widely dispersed holdings, they also accumulated ever larger debt. This economic conservatism both driving and resulting from film industry consolidation had, no doubt, more to do with producers/exhibitors' disinterest in embracing technological change than did the often-cited hesitancy caused by previous failures to innovate sound film successfully.[62]

When Lee De Forest was unable to interest established film producers/exhibitors in his Phonofilm sound-on-film system in 1923, he faced the difficult task of not only selling his system to independent theater owners but also providing them with synchronous-sound films. For De Forest to market successfully his Phonofilm system, he had to promote the device to independent-minded and, in many cases, already financially embattled owners of individual theaters, engage in research and development to improve his system, establish and maintain manufacturing facilities, and not only produce sound films but also establish representational practices for those films. Having innovated what he perceived to be a new representational technology, De Forest had then to establish protocols for representing with that technology. While a contract for sound-film projection equipment with a chain of vaudeville theaters provided De Forest with an initial market for the Phonofilm, he still had to supply those venues with a constantly changing series of films. De Forest's Phonofilm system successfully solved problems of both amplification and synchronization, but its commercial failure resulted from a range of factors well beyond strictly technological issues.

Before Warner Bros.' introduction of Vitaphone, attempts to innovate synchronous sound within the American cinema failed owing to the absence of one or more necessary conditions for technological change. While none of these conditions was individually sufficient to provoke such change, each proved necessary to the successful application of electro-acoustics to the cinema. These conditions included adequate technology (especially amplification and a reliable method of ensuring synchronization between recorded sound and image during reproduction), a large, well-capitalized organization capable of manufacturing and installing sound-reproduction equipment for theaters in a relatively rapid manner (thereby guaranteeing producers a return on the investment necessary to convert to the production of sound films), some degree of vertical integration within the industry or other industrial conditions that would encourage producers of films and film exhibitors to act in concert in instituting technological change (similarly guaranteeing an economic return on the capital investment necessary for wholesale or even partial conversion of the industry), and a conception of recorded sound as appropriate for the cinema as a mode of representation and as an entertainment event. The desire among powerful film producers/exhibitors to increase standardization at the level of exhibition, the growing popularity of radio, and the increasing pervasiveness of electrically augmented sounds helped to fulfill this final condition. Ongoing shifts in the film industry's economic structure during the 1920s and the innovation of electrical acoustics by large, well-capitalized firms fulfilled the remaining conditions.

In an elaborately staged public premiere, Warner Bros. "introduced" electrically recorded, synchronous-sound film to the American cinema on August 6, 1926.[63] The program, at New York City's Warners' Theatre, featured the Vitaphone presentation of the John Barrymore film *Don Juan* with recorded orchestral accompaniment and selected synchronized sound effects. A series of sync-sound short subjects accompanied the feature, including performances by the New York Philharmonic, Giovanni Martinelli, Efrem Zimbalist and Harold Bauer, Mischa Elman and Josef Bonime, Anna Case, and Roy Smeck as well as a recorded speech by Will Hays, president of the Motion Picture Producers and Distributors of America. Warner Bros. rhetorically framed the event as the inauguration of a new epoch in mass-produced entertainment, a landmark in the inevitable teleological advance of cinema as an art form toward ever greater realism. Favorable initial press accounts lauded the realism of the films and echoed Warner's rhetoric about the cultural and artistic potentials offered by this technological advance.

The Western Electric–designed Vitaphone system presented to New York audiences in August 1926 was conservative both in its design and in Warner Bros.' use of the technology. It essentially relied on enlarged phonograph recordings as a medium for storing sounds, and like an earlier generation of devices that combined the phonograph with the cinema (the Kinetophone, the Cameraphone, etc.), the Vitaphone was less than perfect. Without proper care in operation, the Vitaphone system could slip out of synchronization during projection. Soundtrack disks were fragile and designed to be played a maximum of twenty times, and when a film print broke and was respliced (necessitating the removal of some individual frames), that print could never routinely be projected in a way that consistently retained synchronization.

Warner Bros.' initial use of Vitaphone also conservatively conceptualized the possibilities offered to cinema by recorded sound. Recordings of orchestral accompaniment (even when supplemented by selective sound effects) simply repeated the current conventions of silent film's live accompaniment.[64] As Charles Wolfe has observed, Vitaphone's early short subjects essentially provided filmed phonograph recordings.[65] Although conservative in exploring the aesthetic possibilities of recorded sound, both uses of the Vitaphone offered economic benefits beyond differentiating Warner Bros.' films from competing products. Recorded orchestral scores promised to reduce a movie theater's operating expenses by eliminating musicians, and performance shorts could simulate live prologues at significantly lower costs in the long run.[66] This conservative grafting of electrical sound technology onto the existing institutions and practices of cinema also potentially increased the standardization of film presentation.[67]

Despite the apparent economic benefits of sync-sound film, the other major vertically integrated U.S. film companies responded cautiously to Warner Bros.' introduction of the Vitaphone. Like the phonograph industry, the major concerns within the U.S. film industry lacked established research and development programs. The Fox Film Corporation provided an important exception: during the summer of 1926, Fox Film had already begun innovating a competing synchronous film-sound system, the power of which hinged, not surprisingly, on its patent position. Fox Film commercially developed Movietone, a sound-on-film system designed by independent engineer Theodore Case and his assistant Earl Sponable during the early 1920s.[68] Encoding the soundtrack on the same strip of film as a motion picture's images, Movietone had obvious benefits over the Vitaphone sound-on-disk system. During projection, the Movietone soundtrack could not slip out of synchronization with accompanying images, and breaks and splices in a film, while they would generally introduce momentary noise into the system, did not throw the remainder of a reel of film out of sync. Further, Case's single-system method of recording sound and image proved much more portable than Western Electric's competing disk-based sound recording.[69]

William Fox purchased exclusive rights to Case's system during the summer of 1926 and then invested substantial capital into further research and development within the newly organized Fox-Case Corporation. The Fox Film Corp. introduced Movietone to New York audiences in 1927 with a sound newsreel, and Fox's newsreel unit became the primary site for initial research and development of Case's sound-on-film system.[70] Fox also personally gained a potentially important patent position by acquiring the U.S. rights to the so-called Tri-Ergon patents (which formed the basis for a German sound-on-film system), but the Fox-Case Movietone system still had to rely on a license from Western Electric for use of vacuum tube amplifiers and loudspeakers.[71] This Western Electric license and a large loan guaranteed by AT&T eventually provided the leverage needed when hostile forces caused Fox to lose personal control in 1929–30 of the company that bore his name. Fox blamed the collaboration of the phone company with New York City banking interests for his financial troubles and suggested that their motive involved his patents.[72]

While Fox Film sought internally to innovate a synchronous-sound film system, undertaking research and development within its corporate boundaries, Warner Bros., like phonograph industry leaders Victor and Columbia, became consumers of applied electrical-acoustic innovation developed within AT&T and its wholly owned subsidiary Western Electric. Research and development work within AT&T that led to the Vitaphone had begun as early as 1915, with an effort to develop improved methods of sound recording using such components as the condenser microphone, the vacuum tube amplifier, and the balanced armature loudspeaker. Initially intended as an instrument to study and reduce the introduction of noise into AT&T's long-distance telephone lines, the resulting system of electrical recording was then adapted for commercial applications under the insistence of Edward B. Craft, an assistant to the chief engineer at Western Electric.[73]

While Warner Bros. and AT&T/ Western Electric brought a commercially successful system of electrical sound technology to the American cinema in 1926, the telephone interests were not the only established communications corporations applying the principles and components of electrical acoustics to motion pictures. As it had done with the electrical phonograph, General Electric developed its own sound-on-film system, eventually marketing it under the name RCA Photophone. While Vitaphone originated in telephone noise research, GE's competing system similarly developed as an unintended consequence of communications research. Seeking a method of recording transoceanic wireless telegraph signals, General Electric engineer Charles Hoxie developed a photographic system of sound recording. Like the Fox-Case Movietone system, sounds translated into electrical signals were photographed onto celluloid only to be later retranslated into electrical signals and then sounds.[74]

General Electric initially used Hoxie's system (then called the Pallophotophone) to record speeches by President Coolidge and the Secretary of War that were then broadcast in 1922 over radio station WGY (Schenectady).[75] Hoxie eventually discovered that only a narrow width of his photographed sound recording was necessary for efficient reproduction, and this realization made possible the recording of both sound and image on the same strip of celluloid. With access to vacuum tube amplifiers and the innovation elsewhere within GE of the dynamic, paper-cone loudspeaker by Chester Rice and Edward Kellogg (offering a significant improvement in the quality of reproduction available for domestic radio receivers), GE had the components for a commercially viable sound-on-film system. Like AT&T, GE initially applied the principles of Hoxie's Pallophotophone to phonograph recording and reproduction, and this was the system licensed to Brunswick-Balke-Collender in 1925.

Having successfully applied the principles of the Pallophotophone to the phonograph industry, General Electric continued to develop Hoxie's system as a sound-on-film device. Hoxie and other GE engineers created a "soundhead" attachment that could adapt a silent-film projector to sound

projection, refined GE's loudspeakers to compensate for highly reverberant movie theaters (and thereby improved the reproduction of speech),
designed a compatible system of sound-on-disk reproduction for those theaters that desired one, and developed increasingly portable sound-on-film
recording equipment. In cooperation with the motion picture division of
General Electric's publicity department, GE's engineers produced a series
of demonstration films during 1926–27. Although major film producers did not show immediate interest in the system, General Electric collaborated with Paramount to produce a synchronous soundtrack for the
1927 aviation drama *Wings*. Produced after the film was completed, the
soundtrack consisted of recorded music and selected sound effects. Paramount distributed *Wings*, with the GE sound-film system now renamed
the Kinegraphone, as a "roadshow" feature, a special attraction booked in
selected theaters with prints of the film and the necessary equipment traveling from venue to venue.

The application of electrical acoustics to the cinema took place in two
overlapping phases. In the first, major innovating corporations introduced
their respective systems to the film industry, creating subsidiaries or licensing individuals to promote the introduction of the emerging technology.
Having successfully gained a foothold in the film industry, the major firms
then sought to maximize their share of the potential market. On January
1, 1927, Western Electric formed Electrical Research Products, Inc. (ERPI)
as a wholly owned subsidiary designed to manage all film-related business
within the Bell System. In April of 1927, Fox became the first studio to sign
with ERPI, renegotiating its license for those components of the Movietone
system that relied upon Western Electric patents and equipment.

With Warner Bros. now a licensee of Western Electric's Vitaphone system and the Fox Film Corporation introducing and promoting its competing and incompatible Movietone sound-on-film system, the other major
vertically integrated motion picture firms collectively agreed to refrain from
adopting sound-film technology until further study determined whether
technological change was warranted and, if so, how best to accomplish this
change without undue disruption to existing economic arrangements. Famous Players, Loew's, Universal, First National, and Producers Distributing
Corporation sought to make a collective decision as to technological format
in order to guarantee that sound films produced by each would be compatible with equipment in the others' theaters. The so-called "Big Five" also

sought to establish, to whatever extent feasible, an industrywide standard that could be adopted by the less economically important but numerically significant independently owned movie theaters and small local chains. The major film producers/exhibitors sought a system that was technically sound and was manufactured by a well-capitalized firm that controlled the necessary patents and had both technological expertise and an established manufacturing capacity to supply producers and exhibitors. Although the film industry's ad hoc committee considered a variety of devices (including the Fox-Case Movietone, De Forest's Phonofilm, and even small upstarts like Vocafilm), Western Electric and General Electric far exceeded potential competitors in their ability to meet these criteria.[76]

The wireless patent pool provided AT&T/Western Electric and RCA/General Electric with equivalent rights to develop commercial applications of electrical acoustics, but Western Electric, through ERPI's licenses with Warner Bros. and Fox, had acquired an early competitive advantage in the sound motion picture field. ERPI developed business tactics to maintain and to protect this advantage by creating the impression of unequaled patent protection in order to limit RCA's participation in the sound-film market. The "Big Five" agreement, executed on February 17, 1927, effectively froze further licensing of film producers and their theater chains for the time being. The conversion of independent theaters and those owned by Fox and Warner Bros. did, however, continue during the yearlong period that the Big Five's committee examined various competing sync-sound-film systems. Ultimately, the major film producers signed contracts with ERPI in May and June of 1928.[77] In addition to adopting Western Electric equipment in their production facilities, the producers/exhibitors also signed licenses for Western Electric sound-projection equipment in the theaters in which they controlled an interest.

Having initially gained an advantage over RCA through the licensing of Fox and Warner Bros., Western Electric extended that advantage by licensing the other major producers and the theaters under their control. At the end of 1928, ERPI installations outnumbered those of all competitors combined by a ratio of ten to one. ERPI's success in licensing the major U.S. film organizations and various restrictive clauses of its contracts did not, however, prevent either General Electric or competing equipment offered by less-established firms from finding a market share.[78] In 1928, General Electric combined its sound-film efforts with similar developments from

Westinghouse and RCA, and the firms created a new subsidiary, RCA Photophone, Inc., designed to exploit commercially the sound-on-film system. This arrangement repeated the method used by these corporations to develop radio. RCA provided sales and promotional expertise for applied research and development that was undertaken within GE and Westinghouse. General Electric and Westinghouse would provide both research support and manufacturing expertise and facilities. RCA then constructed a new vertically integrated motion picture company by strategically acquiring and combining several existing entertainment firms. First, it formed the Keith-Albee-Orpheum Corporation on January 28, 1928, to acquire the stocks of the B. F. Keith Corp., the Orpheum Circuit, Inc., the Vaudeville Collecting Agency, the B. F. Keith-Albee Vaudeville Exchange, and the Greater New York Vaudeville Theatres Corporation. This assembled in one company control of vaudeville booking offices and, more importantly, a chain of vaudeville and motion picture theaters in the United States and Canada with a total seating capacity of over one million. In October of 1928, RCA then combined the Keith-Albee-Orpheum vaudeville distribution network and theater chain with the Film Booking Office and Pathé (two film-producing organizations) to produce a vertically integrated motion picture production, distribution, and exhibition corporation.[79] Called RKO Pictures (for Radio-Keith-Orpheum), by 1930 it was one of only six such vertically integrated companies in the motion picture industry. Having been defeated in their attempts to license the film industry's leading firms, the radio interests created a new motion picture organization that could rival the powerful ERPI licensees.

Putting Broadcasting on a Commercial Basis

The 1919–20 patent pool agreement that created RCA failed in its goal of smoothly and predictably incorporating electrical acoustics into existing commodity relations. While the agreement successfully anticipated and, to a large extent, managed the potential for wireless to augment existing point-to-point reciprocal communication, it failed to anticipate the power and prominence of radio's identity as an instrument of point-to-mass broadcasting. As work progressed on applying electrical principles to the phonograph and film industries, and as various forces struggled to determine and to shape the identity of radio broadcasting, the dominant patent-holding corporations entered into private negotiations in order that they might forge a new agreement that could profitably control emerging developments in the broadcasting field.

In a revised radio patent pool in 1926, AT&T withdrew from all broadcasting-related activities and sold its transmission facilities (stations WEAF and WBAY) to RCA. As a provision of the new agreement, AT&T obtained exclusive rights to link broadcast stations through a modified system of its existing long-distance telephone lines. Following established corporate policy, AT&T had used patent acquisition and research and development in a field arguably adjacent to its main business (point-to-point voiced communication) and then strategically surrendered that position in an agreement designed to maximize telephone-related profits. AT&T would henceforth cultivate a monopoly position in the profitable service of connecting radio stations into radio networks, arguably a business practice more consistent with its established expertise than managing broadcasting stations or selling time on those stations. The 1926 radio patent pool also allowed RCA to consolidate further its ownership of existing broadcasting transmitters and to establish the radio network NBC. The formalization of network-affiliate relations through NBC further solidified radio's identity, in the first instance, as a medium offering a national mode of address, and that identity would be regularly reiterated with each well-publicized NBC national broadcast.

The revised radio patent pool of 1926 sought to enforce a new economic model on the emerging industry of radio broadcasting, further consolidating control of the medium in large corporations. Within the realm of public policy, an emerging regulatory framework for radio had essentially defined the medium (the broadcast spectrum) as a public resource while ultimately serving the interests of emerging media corporations. In January of 1921, Secretary of Commerce Herbert Hoover had taken an important step in professionalizing (and indeed commercializing) radio broadcasting by prohibiting amateurs from "broadcast[ing] weather report, market reports, music, concerts, speeches, news or similar information."[80] This mandate effectively imposed a boundary between amateur wireless enthusiasts, now limited to point-to-point reciprocal communication, and point-to-mass commercial broadcasters. In September of 1921, the Department of Commerce began issuing licenses to commercial stations on a designated wavelength.

As interest in radio soared, limitations to the useable frequency spectrum for broadcasting and the proliferation of stations created a perceived need for some method to control access to what was increasingly viewed as a valuable national resource. Consumers could demand more sensitive radio receivers, and broadcasters could transmit at ever-increasing power levels, but while increased transmission power achieved broad geographic distribution of signals, overlapping signals disrupted the ability to hear individual stations. While some advocated a view that free-market competition would eventually and "naturally" establish a system of broadcasting that could best exploit the frequency spectrum, consensus relatively quickly developed that some form of external regulatory control was necessary.[81]

A series of four annual radio conferences, called by Hoover between 1922 and 1925, provided the technical and administrative bases for further radio regulation. Significantly, the Radio Conferences also provided an aura of legitimacy for subsequent public policy decisions. The agendas of these conferences and the participants invited inevitably shaped the nature and potential outcomes of deliberations. Historian Thomas Streeter concludes that the general purpose of Hoover's Radio Conferences conformed to a corporate liberal agenda in that they ultimately set out "to encourage the development of broadcasting on a for-profit basis with maximum autonomy for capital and minimal governmental interference."[82] Perhaps the most powerful policy initiative to come out of these conferences was the use of spectrum allocation to bestow order on a radio situation perceived to be inefficient and chaotic. Spectrum allocation ultimately had the effect of minimizing the power of nonprofit broadcasters while consolidating and legitimizing the increased power and control of corporate broadcasting interests.

Spectrum allocation proceeded through a series of steps, each of which ultimately functioned increasingly to consolidate this corporate power. On December 1, 1921, the Department of Commerce distributed broadcasting stations on two frequencies—360 meters for music and entertainment, and 485 meters for broadcasting of agricultural and weather reports. In 1922, Commerce established two classes of radio licenses and two spectrum allocations: while most stations continued to broadcast on 360 meters, more powerful stations (over 500 watts) were assigned 400 meters under the stipulation that programming content must be original (i.e., broadcasting of phonograph recordings was prohibited). In the spring of 1923, Hoover's

Commerce Department established a third, more marginalized, class of stations. These "Class C" stations were limited to broadcasting under 500 watts of power, and all were assigned the same transmission frequency. At the same time, the Department of Commerce began distributing the existing Class B and Class A stations across the radio spectrum, maintaining a 10-kilocycle separation between broadcasters and giving primary consideration to the higher-powered stations that avoided prerecorded programming. The Class C designation hindered the economic viability of nonprofit educational and religious broadcasters, requiring them to share broadcasting time, thereby limiting their potential audience and predisposing them to struggle in an increasingly competitive radio environment that was quickly emerging as a marketplace.[83]

The Radio Conferences and limited public debates about broadcasting successfully framed policy decisions like spectrum allocation as *technical* decisions. The "neutrality" of engineering and the complexities of radio technology provided a rhetorical screen, of sorts, that disguised highly political policy initiatives as merely functional decisions.[84] In large part through Hoover's Radio Conferences, policy decisions that might have focused on issues like democratic control of media as a public resource or debated the potential clash between private property and public interest, were instead successfully framed as the product of technical necessity. These public policy deliberations not only functioned conservatively to foreclose other potential models for radio broadcasting, but they also successfully established a legitimating framework that ultimately underwrote corporate dominance of radio broadcasting. Streeter notes, "A political and legal rhetoric was also refined to legitimate these activities, the rhetoric of broadcasting in the public interest, where the public interest was associated with an orderly, managed, corporate capitalism that worked hand in hand with a modest and cooperative government agency for the good of all."[85] By 1926 and the emergence of NBC, Hoover's Department of Commerce had established a spectrum allocation plan, and radio regulation both served and legitimized a prevailing system of power vested in corporate broadcasting interests.

The regulatory control established through Hoover's licensing and spectrum allocation plan unexpectedly unraveled, however, during 1926. In what some historians have suggested was an intentional attempt to undermine the regulatory authority of Hoover's Department of Commerce, Chicago radio station WJAZ in early January 1926 began broadcasting on a frequency to which it was not assigned and at hours well beyond the two hours per week it had been allowed when initially licensed.[86] The frequency WJAZ chose (329.5 meters or 910 kHz) had been allotted by informal agreement to Canadian broadcasting stations. The U.S. attorney prosecuted WJAZ and its owners, but ultimately a federal judge found in favor of the defendant in *U.S. v. Zenith Radio Corporation et al.* The ruling, which Hoover and the U.S. government never appealed, effectively undermined the Department of Commerce's ability to allocate and to enforce radio transmission frequency assignments and limitations on when a radio station broadcast.

The *Zenith* decision essentially gave all radio stations the freedom to transmit at whatever hours and whatever frequency they chose regardless of the limitations that had been imposed on them by the Commerce Department. While various broadcasting industry voices encouraged stations not to "jump frequency" and to maintain informally the time provisions and frequency assignments given them, various broadcasters took advantage of this temporary freedom from regulation. The "chaos" often fearfully invoked in radio journals never quite came to pass, but the sudden legal blow to federal radio regulation provided added momentum to the efforts of those who supported congressional intervention in the broadcasting situation. Although several local radio-regulation initiatives sought to restore order to broadcasting, congressional action ultimately reestablished federal authority to regulate broadcasting stations and transmissions.[87] The 1927 Radio Act formed the Federal Radio Commission (FRC) to function as the instrument of that federal authority. While the Radio Act rhetorically established "the public interest, convenience, and necessity" as the guiding principle upon which the FRC was to act, ultimately it provided the legal basis for, and thus reaffirmed, the emerging corporate control of radio broadcasting.

Once established, the FRC sought to exert control over the broadcasting spectrum through licensing and allocation. The FRC's "General Order 40" of 1928 divided the broadcast spectrum into ninety separate channels that were then allocated to individual stations based on a specific, although largely unarticulated, interpretation of "the public interest, convenience and necessity." Forty of the available ninety channels were designated "clear channels" and allocated to individual broadcasting stations while the remaining fifty channels were split up and shared by some six hundred additional broadcasters. This initial instance of spectrum reallocation and the principles by which "the public interest, convenience and necessity" were defined in practice dramatically reinforced the specific political economy of radio codified in the 1926 patent pool.

So-called clear-channel stations (with individually designated frequencies and permission to operate at higher power levels) codified through regulation radio's identity as an instrument of national address. Many of the prestigious clear-channel stations were either owned by RCA and its partners in the "radio group," or were affiliates of NBC. Still others would soon form the basis of the CBS network. Even those clear-channel stations unaffiliated with RCA/NBC or CBS could reasonably claim to broadcast "nationally" because of their frequency allocation and transmission power. A definition of radio as a local service was also legally codified through regulation in the form of stations licensed to operate at lower transmission power levels. While such a regulation scheme did not in and of itself lead directly and inevitably to the marginalization of radio's potential identity as a local service, the institution of time-sharing provisions, in which such "lesser," that is local, stations were required to split a broadcasting day with several competing stations, did effectively marginalize the efforts of non-commercial and, especially, educational broadcasters. Ultimately, federal regulation efforts and the allocation of frequency assignments to broadcasting stations legally codified what had already largely been achieved economically, technologically, and rhetorically, that is, the construction of an identity for U.S. broadcasting that privileged in the first instance broadcasting that offered a national mode of address. As Robert McChesney has pointed out, the battle for radio's identity in American life had not ultimately been determined by the late 1920s.[88] Alternative visions of radio that had been foreclosed were never fully eclipsed. Ultimately, the Radio Act of 1927 functioned not to create but to reaffirm and to provide the legal basis for an emergent system of broadcasting in which power was largely vested in the hands of corporate interests. In 1928, with the formation of CBS, the shape of broadcasting in the United States was established for decades to come. Radio was dominated by (marginally) competing networks that functioned as national disseminators of professionally produced programming distributed locally by affiliate stations.

By the time of the 1926 patent pool and its new attempt at achieving economic closure, electrical-acoustic technology was in the midst of becoming pervasively diffused albeit not universally present in American society. Although not every household owned a radio set or an electrical phonograph, these devices had become increasingly commonplace in Americans' homes. The motion picture industry was in the earliest stages of its conversion to sound, and the phonograph industry was on its way to the transition to electrical methods. U.S. radio in the period between 1926 and 1930 developed so as to marginalize broadcasting's potential identity as a local service. Radio's identity moved almost inexorably toward the increasing prominence of a national mode of address and accompanying assumptions about radio's social function. Through first the rise of indirect advertising and corporate sponsorship of programming and then the progressive intensification of that sponsorship culminating in direct advertising, radio shifted from a consumer product to also a medium that served consumer society. Radio content also became increasingly professionalized, and as the production of network programming grew more centralized, radio reasserted a preexisting spatial definition of entertainment in which production centers addressed geographically dispersed consumers. Invoking many of the same rhetorical tropes that underwrote the phonograph's construction as a domestic entertainment device, both program sponsors and radio networks presented themselves as collapsing the distinction between urban entertainment centers and the geographic hinterlands of consumers. Accompanying these shifts, a cautionary rhetoric that darkly invoked the potential for noise and disorder made these developments seem if not inevitable then the most desirable of available alternatives. National networks presented themselves and commercial sponsorship to the public as a solution to the potential disorder of too many stations broadcasting too much "substandard" programming.

A national mode of address presenting popular, predominantly urban, entertainment originating in entertainment centers became the acoustic manifestation of a specific, corporate-based interpretation of the public interest, convenience, and necessity. For-profit, commercially sponsored radio claimed the status of a public service through the successful invocation of rhetorical constructions like "education" and "uplift," "musical democracy" and the ever-promised and perpetually deferred electrically augmented political public sphere.[89]

Federal regulation and allocation of the radio spectrum responded to, reinforced, and essentially served these developments. The 1927–28 reassignment of radio frequencies by the Federal Radio Commission (ostensibly to bring order to the air) in fact reinforced an order of a particular type. The same principles that provided legal support for broadcasting defined, in the first instance, as a national mode of address ultimately underwrote an acquiescence to commercial sponsorship of programming. Radio advertising rode a rhetorical wave that supported specific definitions of quality programming, and in the emerging political economy of broadcasting during the mid- to late-1920s, the oft-asked question "Who will pay for radio?" was answered by default through commercial sponsorship. While various answers to the question were intermittently posed (such as a financial subsidy provided by radio equipment manufacturers, a tax on the sales of domestic radio receivers, etc.), commercial advertising, almost by default, provided a solution consistent with radio defined as a national mode of address. Commercial sponsorship provided an instrument through which programming perceived to be in "the public interest, convenience and necessity" could be provided by a system of broadcasting that defined itself as a national service.

Multiple-Media Conglomerates and Cross-Media Links

The innovation of electrical acoustics exacerbated an increasing concentration of ownership and power within the U.S. mass media. The establishment of radio networks, the institutionalization of network-affiliate relations, and the increasing role played by national advertisers in programming content centralized the economic control of U.S. broadcasting. American cinema took important steps toward the "mature oligopoly" of a few powerful vertically integrated film organizations during the 1920s. Immediately before and simultaneous with Hollywood's conversion to sound, film producers/distributors made a concerted effort to expand their theater holdings through the acquisition of existing theater chains and, in some cases, a program of building new urban theaters. The aftermath of technological change involved still further consolidation within the film industry. Warner Bros. expanded its power by acquiring its competitor First National in 1928 and the Stanley Company's chain of 250 theaters. Fox Film, after innovating Movietone, radically expanded its theater chain and announced a takeover of Loew's/MGM. This latter acquisition prompted merger talks between Paramount and Warner Bros., a combination that would have exceeded the power and scope of both a combined Fox/Loew's and even RCA. Although Department of Justice disapproval (on antitrust grounds) prevented Fox's takeover of Loew's and squelched merger talks between Warner Bros. and Paramount, the film industry emerged from the conversion to sound with a handful of vertically integrated firms dominating the industry.[90]

This increasing concentration of power within a relatively small number of media firms accompanied a related process through which individual corporations gained interest in multiple media forms. Corporate giant RCA held substantial interests in consumer sound equipment sales, but it also controlled several important radio stations, NBC's two radio networks, the film studio RKO and its chain of movie theaters, and Victor's phonograph recording and manufacturing facilities as well as its exclusive contracts with performers.[91] RCA was not alone in developing financial interests across multiple applications of electrical acoustics. As part of the larger mid- to late-1920s consolidation of corporate interests within the film industry, most of the major vertically integrated film studios forged institutional relationships with various entertainment interests across media lines, extending their financial interests and control well beyond the nation's movie studios and theaters.

With a Los Angeles radio station, the acquisition of several large music publishing companies (M. Witmark and Sons, Harms Music Publishing Company, and partial interests in others) consolidated within its Music Publishers Holding Company, and even a short-lived chain of music stores, Warner Bros. established multiple media connections simultaneous with its conversion to synchronous sound. Loew's/MGM owned radio station WHN in New York City as well as the Robbins Music Publishing Company. Paramount came perhaps closest to mimicking the horizontal integration of RCA. It established a music publishing company, began broadcasting from its Hollywood studio in 1928, and in 1929 temporarily acquired half interest in radio network CBS through an exchange of stock.[92]

The increased concentration of ownership in U.S. media that accompanied and facilitated the introduction of electrical sound technology made possible, logical, and profitable a variety of cross-media links. Performers, having achieved success in one medium, could easily migrate to different media. While also true of an earlier generation of stars (i.e., Vernon and Irene Castle's use of Broadway, nightclubs, and record sales), the period surrounding Hollywood's conversion to sound featured the previously unheard-of proliferation of the multimedia star largely through the auspices of technologically augmented sound.[93] Eddie Cantor, Paul Whiteman, Bing Crosby, Burns and Allen, and Ruth Etting, among others, paved the way for later performers (most notably the Big Bands) whose success was directly linked to their simultaneous "presence" on records, radio, and film.

Content might be simultaneously exploited in several forms and formats with various media commodities mutually reinforcing each other through combined and integrated publicity. Although originating before cinema's conversion to sound, the "movie theme song" proliferated subsequent to Hollywood's adoption of electrical acoustics. Theme songs allowed not only promotion of a film through sheet music sales, phonograph releases, and radio airplay but also provided multimedia conglomerates several related sources of income. By 1929 the practice had become so widespread that some pointed to the talking film's creation of a "new national ballad."[94]

A single song, for example, could be introduced in a motion picture, heavily and pervasively promoted via radio airplay, and sold as sheet music and phonograph records. Trade publications like *Motion Picture Herald* ran lists of the hit songs from current films and identified the songs' publishers, to facilitate exhibitors' tie-ins for promotion.[95] In 1930, theater managers in the Loew's circuit (MGM) were explicitly instructed to actively promote music published by Robbins. "The entire exploitation and publicity machinery of the Loew Theatre circuit should be placed squarely and enthusiastically behind Robbins songs and music. The Robbins Music Corporation is an important unit in the Loew-Metro family and Loew managers and publicity men should bear in mind that what they do for Robbins music they do for their own company."[96] Douglas Gomery notes that by 1929 the film industry's control over music publishing was so pervasive that "over 90 percent of the popular music in America was being generated by music publishers owned by Warners, Paramount, RCA, or Loew's."[97]

Warner Bros. noted in on-screen credits for some early 1930s films that all radios depicted in the motion picture were manufactured by Brunswick. Warner's Vitaphone short subjects frequently featured exclusive Brunswick recording artists performing songs released on Brunswick records and available on sheet music from Warner-owned music publishers. Further, Warner Bros.' animated films often featured popular songs that originated in its feature-length musicals. While such product tie-ins would throughout the 1930s and beyond become a frequent component of U.S. film and its relationship with a larger culture of consumption, crucial here is the extent to which Warner Bros. was able to maintain within its expanded corporate boundaries control of the profits generated from multiple, related commodity forms released in a number of media formats. As owners of Brunswick-Balke-Collender, a number of music publishing houses, for a time a chain of music stores, and even, briefly, a printing facility, Warner Bros. as a diversified multiple-media conglomerate was well suited to benefit from the profits generated by intermedia commodities.

The introduction of the sound film and the rapid expansion of radio broadcasting accelerated the commodification of entertainment on a national scale. The economic consolidation of the entertainment industry provided powerful financial incentives and the structural wherewithal for an expansion of cross-media links. An intermedia commodity—like Amos 'n' Andy in *Check and Double Check*—conformed to and reinforced the structural logic of media concentration. Such "synergy," in today's terms, often linked otherwise apparently competing media firms, but in its most mature and efficient form all of the accrued profits would remain within the control of a single large media corporation.

Even absent structural links between national media firms, cross-media collaboration and cooperation often took place on a local level. By May of 1930, Loew's informed managers of theaters in the circuit that "virtually every city in which there is a Loew theatre has a Loew radio program." Broadcasts ranged from daily presentations of organ music to providing local stations performers from the theater's stage show for a "Loew State Half Hour" (New Orleans) or "Loew's Variety Hour" (Boston), for example. Some theater managers had established cooperative arrangements in which local broadcast orchestras performed movie-related music and mentioned the film and theater. Other theater managers provided stations with the latest "movie theme song" records that then played as part of a regularly scheduled movie music program. Loew's theaters in cities as large as Boston and St. Louis or as small as Reading (Pennsylvania) actively cultivated relationships with local radio stations. Cautioning managers to "handle radio with kid gloves," the New York office of the Loew's chain reminded managers that theater acts should never perform the same material in both a stage presentation and a radio broadcast and to "bear in mind your reason for broadcasting is to advertise your theatre—not merely to entertain the radio listeners."[98]

When faced with the potential competition posed by an important national broadcast, some local theater owners incorporated radio into their motion picture program. In the summer of 1930, Russell Bovin, manager of the Loew's theater in Canton (Ohio), countered the potential effect of the national broadcast of the Jack Sharkey versus Max Schmeling fight by broadcasting the fight to his theater audience. In an elaborate tie-in with the local Philco radio dealer, Bovin borrowed a 17-foot-high prop Philco from the local dealer and, tuning a radio in the manager's office to the fight, placed the microphone for the house public address system in front of the radio to broadcast the fight to his theater patrons. Marquee banners, newspaper stories, radio announcements, and an extensive advertising campaign, much of it paid for by the local Philco dealer, promoted the "event," which featured both the blow-by-blow national broadcast and the theater's regular feature, Gary Cooper in *The Texan*. In its national inhouse publication to theater managers, Loew's highlighted Bovin's stunt as a particularly showman-like adaptation to the potential competition of radio and encouraged other local theater managers to orchestrate similar broadcast performances.[99] Whether through one-time collaborative relationships with other media organizations (like Bovin's) or more systematic cross-media exploitation (like *Check and Double Check*), U.S. media organizations, texts, and experiences took on an increasing intermedia character in the aftermath of technological change. These developments expressed in commodity and experiential form the increased concentration of media ownership of the late 1920s and early 1930s.

Conclusion

The innovation of new electrical methods of augmenting sound in the aftermath of World War I is largely a tale of corporate power. Dominant organizations in communications (AT&T) and electrical power (General Electric and Westinghouse) used the strength of their research laboratories and resulting patent positions as well as their enormous capital to shape and to manage technological innovation in what ERPI's John Otterson referred to as the "no man's land" between their established interests. Acquisition and prosecution of patent rights allowed large corporations to limit competition and largely to control technological innovation. Although entrepreneurial inventors successfully created competing technological systems, the large corporations controlling electrical-acoustic innovations limited the ability of inventors to successfully market their own innovations. The power of GE's and AT&T's vacuum tube amplifier patents could limit the ability of an entrepreneurial inventor like Orlando Marsh to introduce his system of electrical recording, and their capital and manufacturing capacity could make their sound-film systems more attractive to the film industry than Lee De Forest's Phonofilm. Ultimately, however, the process of technological change was neither seamless nor inevitable.

The technological processes and artifacts of electrical acoustics exhibited what Pinich and Bijker refer to as "interpretive flexibility," the ability to take on different, multiple, or overlapping meanings for different social groups. The failure of the wireless patent pool to anticipate the importance of point-to-mass broadcasting as a commercial identity for wireless illustrates how this interpretive flexibility could undermine a "closure mechanism" designed to anticipate and manage profitably technological development. It further underscores that despite their attempted patent monopoly and economic power, the major innovating corporations had to *respond* to developments in addition to initiating them.

The conflicting visions of radio broadcasting articulated within AT&T in 1923 demonstrate that even after technological systems are successfully engineered and introduced, such interpretive flexibility can linger. In the movement from technology to media, a medium must be defined discursively and economically in addition to being successfully engineered. While the innovating organizations powerfully shaped subsequent events— through RCA and later NBC, the patent pool of 1926, AT&T's introduction of "toll broadcasting" and advocacy of networked broadcasts—the development of radio in the 1920s illustrates that the system of broadcasting taken for granted by the late 1930s might well have developed otherwise.

While forces in the public and private sectors struggled to define radio broadcasting as a medium and as a profitable enterprise, the major corporations innovating electrical acoustics adapted the apparatus for use in the phonograph and film industries. These entertainment industries provided preexisting commodities and established commercial relations onto which electrical sound technology could be profitably grafted. While the established organizations in film and phonography became consumers of technology developed elsewhere, institutional realignments related to technological change exacerbated ongoing consolidation within each industry.

Amid the conversion of phonograph and film industries, struggles to shape radio broadcasting continued. The process of technological change and the means through which radio's identity was shaped led to and ultimately reinforced economic consolidation within and across U.S. sound media. The 1920s and early 1930s saw an increasing concentration of media ownership in the hands of progressively fewer, larger horizontally and vertically integrated firms. The most advanced articulation of this process was the "intermedia commodity," a series of interrelated products that embodied in commodity form the economic relations within emerging multimedia conglomerates.

But these *material* efforts to fix the identity of broadcasting and other applications of electrical acoustics accompanied equally important *rhetorical* struggles to shape the technology's identity. Utopian predictions surrounding radio's and the sound film's future roles in cultural uplift and even political emancipation uneasily coexisted with dire warnings of the emerging media's potential to debase public culture and even to threaten democracy by addressing a mass audience all too easily transformed into a mob. Rhetorical and material struggles over emerging acoustic media thus became sites in which previously established positions on U.S. cultural and political life were again rearticulated and frequently clashed. Technological change did not cease with the licensing of U.S. movie studios and phonograph companies, nor with codification of a regulatory process for radio or the increasing prominence of broadcasting considered as a national mode of address. In subsequent chapters I take up and explore further the rhetorical and practical struggles surrounding efforts to secure emerging acoustic media within social relations.

Two

Announcing Technological Change

In a February 1887 report to *Scientific American,* a correspondent recounted his trip behind the scenes of the central office of the telephone company. He framed his account in terms of a conceptual contradiction that quite literally haunted late nineteenth- and early twentieth-century acoustic technologies: the simultaneous materiality and immateriality offered by the telephone, phonograph, and radio. The technology itself was materially present, visible, and audible as it performed its representational task, yet the origin and source of technologically augmented voices and other sounds were inevitably absent from the scene of their reproduction. Immaterial voices unhinged from their corporeal origins traversed space and/ or time and emerged from technological objects. The *Scientific American* reporter noted "the mysteriousness, the sense of material non-existence, of that part of the machine and its belongings that lies beyond one's own instrument and that of the person at the other end." The communications channel, the means and method of transmission, maintained a wondrous immateriality. "Between us two there is an airy nowhere, inhabited by voices and nothing else—Helloland, I should call it."[1]

Although an immaterial nonexistence, an uncanny space between, an enigmatic nowhere, characterized telephonic communication, the materiality of the technology itself could suddenly return when the connection was broken. The reporter noted that when a connection was unexpectedly severed, "if you become impatient, and raise your voice in earnest demand or protest, the more you bellow, the more you become aware that you are idiotically shouting yourself black in the face against a mere inanimate box stuck against the wall."[2] This quality of the telephone (and like it later acoustic devices) to be both simultaneously material yet immaterial, physically present as an artifact yet producing that which was physically absent, utilitarian yet wondrous, resided at the heart of both their appeal and some of the anxieties acoustic technologies provoked.

After invoking the uncanny qualities of reciprocal telephonic communication, the reporter described his trip to the central office, explaining switchboard and telephone company procedure, importantly lingering in his account on the "healthy young women" and "girls of good physique" who calmly and politely assisted customers and made connections. The account of the women operators and their cooperative male superintendent successfully embodied the telephone company, populated the "airy nowhere of Helloland," and made tangible, material, modern, and efficient the "mysterious material nonexistence" of voice transmission. As the reporter concluded, "Helloland is not so ghostly afterall."[3]

"The telephone," the reporter wrote, "seems to have no visible agency."[4] This immateriality of the telephone as a communications apparatus and the ubiquity of the telephone company as a communications provider caused AT&T and the associated Bell System companies to enlist leading figures in the burgeoning field of public relations during the 1910s and 1920s. This public relations effort sought, in part, to provide the telephone and the telephone company with human agency. By strategically embodying AT&T as a corporation and striving to render coherent and material telephonic processes and technology, AT&T's public relations campaign sought to engineer consumers' perceptions and experiences of acoustic technology. That campaign provided part of the context within which later acoustic innovations were presented to the public. It provided as well precursors to some of the means through which electrical acoustics was made socially meaningful to consumers in the 1920s.[5]

The *Scientific American* article is predictive in that it displays traces of both the need to make technology meaningful to consumers and some of the rhetorical strategies used to introduce later technological change. The 1887 account begins by invoking discomfort with a new technology, acknowledging the potential confusion and alienation provoked by encounters with a wondrous new apparatus. Through the figure of the reporter, the reader journeys "behind the scenes" to discover the material processes and importantly the people that make the wondrous device function. The "ghostly" process is literally embodied in the "comely young women" and their helpful superintendent. "Helloland" is made material and comprehensible. Beyond this pedagogical function that explains technology, the account also seeks to discipline consumers' encounters with the device, training readers in appropriate interactions with the apparatus by seeking to limit their impatience, frustration, or anger. In the phone company superintendent's words, "We like to have people who have telephones come up here. It gives them an idea of how the thing is done, and we notice that they seldom get impatient in the use of their telephone afterward."[6] Appropriate consumption is here defined as knowledgeable and more patient. Glimpses behind the scenes of new technologies functioned as a kind of consumer pedagogy, not only providing consumers with insight into how an apparatus actually worked but also instructing them in its use, or more importantly shaping their attitudes toward it.

The demystification of technology (and technology providers) offered by behind-the-scenes accounts such as this sought to contain the potential anxieties provoked by disembodied voices. Widely circulated descriptions of new acoustic technologies sought to make material the otherwise mysterious processes that disembodied voices and moved them through space or across time and thereby to firmly locate new acoustic devices within consumer society. Accounts like that of the reporter's trip to the telephone exchange were common throughout the late nineteenth and early twentieth centuries. Similar glimpses behind the scenes accompanied later acoustic innovations, especially those involving the various applications of electrical sound technology. A variety of discursive practices introduced new acoustic technologies to the public, explaining the principles behind their operation and seeking to police consumers' encounters with new means of transmitting or reproducing sounds.

In an essay exploring the circuit of reception of new technologies around the turn of the twentieth century, Tom Gunning suggests that "newness and amazement became a mode of reception for technology" at international expositions, but that such "heightened astonishment, gradually [faded] into understanding as the dazzle of the first encounter yields to knowledge" (40–41). He continues, noting that as the period progressed "the cycle between amazement and explanation may have become shorter, but one could also claim that the increased pace of modernity supplies a constant stream of environmental changes, sufficient to renew wonder even in shorter cycles."[7] While Gunning focuses on the reception of new technologies as individuals encountered and engaged with technological change, in what follows I am more interested in rhetorical attempts to address consumers, attempts to shape precisely that reception. My work in this chapter suggests that innovators of electrical-acoustic devices built on established methods of presenting technology and simultaneously cultivated both relationships to innovations, both amazement and explanation.

Elsewhere in the essay, Gunning notes that, "A discourse of wonder draws attention to new technology, not simply as a tool, but precisely as a spectacle, less as something that performs a useful task than as something that astounds us by performing in a way that seemed unlikely or magical before. The discourse highlights and defines this magical nature. This wonder intrigues and attracts us, allowing curiosity to give way to investigation and education, usually carefully channeled by social discourses. However, habituation dulls our attention to technology."[8] I would suggest that by the mid-1920s, habituation had rendered commonplace both recorded sound (via the mechanical phonograph) and the transmission of disembodied voices (via the telephone), necessitating that innovators of electrical acoustics, while often using the familiar format of spectacle, relied rather more heavily on a pedagogical mode of address.

In what follows, I use the terms "announcing" and "announcements" to refer to public presentations of technological change in a variety of forms. Specifically, such announcements of electrical acoustics included public performances designed to demonstrate new devices, the pedagogical rhetoric adopted by informational magazine articles (especially those offering to consumers a glimpse "behind the scenes" of the culture industry), and advertising campaigns (particularly their inclusion of graphic attempts to explain applied scientific principles). The commonsense distinctions between such diverse practices as advertising, pedagogy, and public performance frequently blurred in these announcements. A given presentation of technological change might offer consumers a rudimentary lesson in acoustic principles while promoting a particular product line, or while making broad claims about the innovation's utopian potential to bring about social change. This chapter examines both a series of discursive, performance, and textual practices that introduced electrical sound technology to American consumers and the explicit and implicit messages about technology, innovation, progress, and media that these announcements conveyed. I locate these discourses within a larger historical tradition of earlier acoustic innovations that similarly combined often spectacular public performances with a pedagogical mode of address. Combining amazement with explanation, corporate innovators positioned technological change within broader narratives of progress, corporate identity, the shifting site of technological innovation, and the scientific fulfillment of utopian promises. Amid ongoing struggles in economic and regulatory realms to develop and to manage profitable identities for emerging sound media, innovators sought to shape public perceptions of this process. More than promotions for a new device and product line, these announcements offered to consumers ways of making sense of technological change, and they sought to make an emerging political economy of media seem natural and inevitable.[9]

Public Performances of Technology

Throughout 1927, the Victor Talking Machine Company used an enlarged version of the Orthophonic Victrola as part of a broader corporate strategy to introduce to consumers the potential offered by electrical recording and reproduction. Flanked by an eight-foot in diameter reproduction of a Victor disk and an even taller reproduction of "Nipper," the ubiquitous Victor canine trademark, the Auditorium Orthophonic Victrola appeared as a regular part of the program in higher-class vaudeville houses like the Loew's State Theatre in Boston. With a loudspeaker horn measuring eighty by ninety inches at its opening, the theatrical Victrola was presented as a magnified version of the Credenza Orthophonic available for consumers' home use. However, unlike the Credenza Orthophonic, which still relied upon mechanical methods of amplification, the Auditorium Orthophonic was fully electrical in its methods of reproducing sounds. Besides the Loew's State, the device also publicly performed in the leading Paramount chain of theaters in New England while also making appearances at recital halls and music stores.[10]

These public performances of an enlarged electrical phonograph emphasized, not surprisingly, the device's fidelity in reproduction and also both its increased volume and the operator's ability to adjust the "tone quality and shading" of reproduced sounds. Carefully stressing that the onstage demonstrations featured precisely the same Orthophonic-process records then available to consumers, these public performances and the press accounts accompanying them presented the conversion to electrical methods as an aural spectacle. Reporters claimed that the phonograph performances conformed to and indeed improved upon the standards of live entertainment to which vaudeville audiences were accustomed. In the words of one press account: "The results showed conclusively that the public accepts the talking machine as an outstanding form of entertainment, since even when in competition with star performers, the instrument has proved the hit of the bill."[11] Besides such theatrical demonstrations, the Auditorium Orthophonic Victrola also performed in other public settings as well. Between Christmas and New Year's Day, it entertained passersby on Boston Common and, by one account, could be heard up to a mile and a half away at Bunker Hill. Whether at the Philadelphia Sesqui-Centennial Stadium entertaining a crowd estimated at 150,000 or reproducing electrical recordings for Atlantic City boardwalk crowds from a yacht anchored a half mile off shore during a beauty pageant, the enlarged Victor device publicly demonstrated technological change to large crowds.

Performances by the Auditorium Orthophonic were not limited to New England. Versions of the enlarged electrical phonograph played in such venues as urban theaters (New Orleans and Buffalo), fairgrounds and festivals (the Indiana State Fair, Indianapolis fairground, Portland Oregon's Rose Festival), atop buildings (the Wead Building in Omaha), and in public parks (Philadelphia).[12] These presentations of electrical phonography were a more dramatic (indeed louder) continuation of local demonstrations of the new process records and reproducers inaugurated nationwide on November 2, 1925. Designated Victor Day by the company through a national advertising campaign, that morning Victor retailers across the United States simultaneously began highly publicized demonstrations using nine recordings selected by Victor to best demonstrate the new electrical process.[13]

Other phonograph companies besides Victor announced the new electrical processes through phonographic performances. Columbia's demonstration phonograph performed, for example, in France during 1928. At a Paris concert in December of that year, Columbia mixed live performance with recorded music (including a ballet performed to recorded accompaniment), screened a film demonstrating manufacturing processes entitled "The Birth of a Record," and publicly performed a recording session on the stage of the theater.[14] Even the less well-capitalized Gennett Record Company of Richmond, Indiana (a licensee of General Electric), entertained Cincinnati audiences with a demonstration climaxed by an onstage electrical recording session followed by the reproduction of the resulting record.[15]

These public performances of the electrical phonograph built upon a long tradition of similar spectacular demonstrations of new acoustic devices. Innovators throughout the late nineteenth century and first three decades of the twentieth century introduced emerging sound technologies to potential consumers through such public demonstrations. Often framed as educational events, these performances also functioned as a type of spectacle, presenting for their audiences the latest marvel in what became a grand, evolutionary narrative toward a scientifically engineered future. Wrapped in a rhetoric of edification, public performances of new sound technologies sought to align acoustic devices with preexisting values concerning cultural uplift and modernity.[16]

Alexander Graham Bell initially introduced the telephone to the public by demonstrating the device's ability to function as a medium of one-way communication—in effect, early broadcasting. Throughout 1877, Bell's "lecture-performances" featured the wired transmission of music across distances, explained how the telephone worked, and used the device to illustrate larger electrical principles. Early public demonstrations of the phonograph also contained such an educational mode of address, but regional roadshow performers who brought Edison's latest marvel to cities and small towns often emphasized showmanship rather than science in their public performances of the "wondrous" new apparatus. Before crowds variously estimated between a few hundred to up to a thousand, dozens of regional, touring phonograph exhibitors like New England's Sullivan Brothers and Pennsylvania's Lyman Howe presented entertaining and edifying exhibitions of sound recording and reproduction. Often these programs began with introductory remarks delivered by the apparatus itself. Rather than an object operated for an audience's benefit, the phonograph through a frequently rereleased cylinder recording, directly addressed audiences about the possibilities it provided. "Mr. Phonograph" literally sang his own praises:

I am the Edison Phonograph, created by the Great Wizard of the New World to delight those who would have melody or be amused. I can sing you tender songs of love. I can give you merry tales and joyous laughter. I can transport you to the realms of music. I can call you to join in the rhythmic dance. I can lull the babe to sweet repose or awaken in the aged heart soft memories of youthful days. No matter what your mood, I am always ready to entertain you. When your day's work is done I can bring the theater or the opera to your home. I can give you grand opera, comic opera, or vaudeville. I can give you sacred or popular music; band, orchestra or instrumental music. I can render solos, duets, trios, quartets. I can aid in entertaining your guests. When your wife is worried after the cares of the day and the children are boisterous, I give rest to the one and quiet the other. I never get tired and you will never tire of me for I always have something new to offer. I give pleasure to all, young and old. I will go wherever you want me, in the parlor, in the sick room, on the porch, in the camp, or to your summer home. If you sing or talk to me, I will retain your songs or words and repeat them to you at your pleasure. I can enable you to always hear the voices of your loved ones even though they are far away. I talk in every language. I can help you to learn other languages. I am made with the highest degree of mechanical skill. My voice is the clearest, smoothest, and most natural of any talking machine. The name of my famous master is on my body and tells you that I am a genuine Edison phonograph. The more you become acquainted with me, the better you will like me. Ask the dealer.[17]

Here, as in most such announcements, the technological apparatus itself, in addition to the sounds it reproduced, became the object of consumers' attention. The artifact was materially present competing for attention with the services it could provide.

In *The Phonogram*, a 1890s industry publication, M. C. Sullivan advised those planning to publicly exhibit the phonograph to structure their presentations with attention to the demands of showmanship and dramatic spectacle. Sullivan suggested that phonograph exhibitors generate suspense by alluding to upcoming selections, or limit the duration and frequency of serious recordings so as to make their effect more powerful when they did appear in the demonstration program. He also advocated careful staging of both the machine and the recorded repertoire to conform to existing standards of dramatic showmanship, all in a manner that would attract audience interest to the machine and away from its operator.[18] Photographs of the Sullivans indicate that their performances prominently featured a framed photograph of Edison as part of the theatrical mise-en-scène, thereby also drawing attention to, in the words of "Mr. Phonograph," "the Great Wizard of the New World." Much like Gennett's demonstration of electrical recording to Cincinnati audiences in the late 1920s, Lyman Howe's turn-of-the-century phonograph performances built to a dramatic climax when he illustrated the potential of the apparatus by recording onstage the live performance of a local band or prominent community figure and then immediately reproducing the resulting record.

Public performances of new acoustic technologies surrounded technological spectacle with a pedagogical discourse that both presented scientific principles and offered implicit or explicit assertions about the larger social significance of technological change. A typical program for the Sullivans or Howe included musical, comic, and spoken recordings, all of which were recorded by either well-known performers or performers from distant cities. In Scranton, Pennsylvania, or Bath, Maine, for example, audiences were introduced to the phonograph's ability to eclipse spatial limits by bringing even to diverse geographic locations the well-known and expensive performers from urban centers. This explicit demonstration that the phonograph both transcended time (through the preservation of sound) and transcended space, linking the hamlet with the urban center, would become crucial to subsequent introductory performances of sound technologies.

In 1913, Thomas Edison used a series of public demonstrations to announce his ill-fated attempt to wed his phonograph with motion pictures, the Kinetophone. Films produced for these performances not only dramatically enacted for the audience the *fact* of synchronized sight and sound but also illustrated a range of possible sounds in synchronization with their sources, thereby making implicit claims about the Kinetophone's potential contributions to the recorded repertoire and beyond. The first Kinetophone film released and usually shown at all introductory performances was called, appropriately, "The Lecture." Designed to demonstrate the successful synchronization of recorded sound and image and to explain the relevant scientific principles applied by the Kinetophone, "The Lecture," according to one popular account, consisted of "a man [who] discusses with many oratorical flourishes of voice and limb the history and the future of the invention, and then illustrates the possibilities of the Kinetophone by breaking plates, blowing horns and whistles, and by the introduction of a singer, a pianist, a violinist, and a barking dog."[19] A more sympathetic account of Kinetophone demonstrations detailed sound films subsequent to "The Lecture" that included a portion of a dramatic performance, a comic act lampooning a politician's speech, the opening of a blackface minstrel show, and a performance of Verdi, all of which concluded with the entire cast reassembling for a rendition of "The Star Spangled Banner," described as "so real, so vivid, and so stirring," according to this second account, "that the audience rose to its feet [in] a spontaneous tribute to the actuality of an event that had a genuine scientific importance."[20]

Edison's Kinetophone "The Lecture" publicly announced the successful combination of the phonograph and motion pictures through various demonstrations of synchronized sound and action. By all accounts, the short film also included an educational mode of address that introduced audiences to acoustic and mechanical principles. But not content to demonstrate that the apparatus indeed worked, or even *how* it worked, Edison also sought to align the Kinetophone with preexisting cultural values. According to another account, "The Lecture" concluded with the following oratory:

> Consider the historic value of a kinetophonic production of George Washington, if it were possible to show it now, and you will realize the splendid opportunities of future generations to study the great men of today. The political orator can appeal to thousands while remaining at his own fireside; the world's greatest statesmen, actors and singers can be seen and heard in even the smallest hamlet, not only today, but a hundred years hence. In fact there seems to be no end to the possibilities of this greatest invention of the wizard of sound and sight, Thomas A. Edison.[21]

Edison thus sought to shape the social uses to which the combined audiovisual apparatus would be put.

Reports of the Kinetophone demonstrations echoed this familiar Edison-originated rhetoric about the social, educational, and scientific contributions of representational technologies and their ability to preserve the present for posterity. Edison himself was frequently quoted asserting the immediate democratizing benefits of the Kinetophone: "You can see how this is going to put the best operas and dramas within the reach of poor people. The poor man will be able to see for a nickel performances which the rich man now sees for a high price. It is going to help educate the poor."[22] Newspaper and popular magazine accounts dutifully presented Edison's grandiose and utopian predictions of the Kinetophone's future impact while also glorifying the apparatus as yet another instance of the great man's genius and commitment to the greater social good.

Although the Kinetophone failed to revolutionize motion picture presentation, Edison again turned to public performances of technology between 1915 and 1920 to announce technological innovation. In close collaboration with local phonograph dealers, the Edison Company supported a series of performances throughout the country designed to demonstrate the merits of the Edison Diamond Disc phonograph system. Called Edison's "Tone Tests," they became one of the most pervasive and important methods used to construct an identity for the phonograph. Several thousand "Tone Tests" were conducted with an audience that may well have numbered over two million.[23] In theaters and concert halls, Edison recording artists performed onstage accompanied by Edison Diamond Disc technology to illustrate the fidelity of Edison's "re-creations" to original performers and performances. One period account claimed that "customers looked forward to the Edison Tone-Tests just as they did the Rexall 1 cent sale."[24] A typical program would include the performer singing accompanied by the phonograph as well as several dramatic moments when the lights were extinguished so that the audience would be incapable of confirming whether they heard a live performance or a recording.[25] The content and structure of the "Tone Test" demonstrations were highly standardized, and they enacted these theatrical comparisons between Edison performers and recordings of their voices within a carefully constructed set that mimicked an idealized version of a wealthy home.[26] The "Tone Tests" were thus designed not only to advocate the merits of Edison technology but also to link sound recording and reproduction to elite, high-culture values.

In her account of the "Tone Tests," Emily Thompson describes how these Edison promotions and their accompanying publicity were multiply illusory. Rather than providing conventional live performances, the Edison singers consciously sought to imitate the sounds of their recorded voices.

Further, local press coverage of these events often simply reproduced without editing the promotional copy supplied through Edison's press agents. Thompson concludes that the "Tone Tests" "provided listeners with a tool, a resource, that enabled them to transform their conception of what constituted 'real music' to include phonographic reproductions."[27] Edison's public performances of technology stressed the inability to distinguish between a live performance and its "re-creation." But in addition to promoting a product line, they also sought to shape consumer attitudes about the social functions and meanings of sound technology. The Tone Tests implicitly claimed that the phonograph was in fact a musical instrument.

Public performances of new acoustic technologies prior to the introduction of electrical methods served as something more than merely promotional presentations of new consumer devices. They combined entertainment with the "scientifically important," showmanship with education, a spectacular display emphasizing wondrous innovations with the explanatory rhetoric of applied science. Adopting a pedagogical tone, the performances themselves simplified scientific principles, essentially demystifying emerging technologies. They also went one step further, struggling rhetorically to position new devices in terms of preexisting cultural values. Utopian rhetorical claims emphasized the contributions acoustic technologies would make to the democratization of culture and cultural uplift, stressed their ability to eclipse spatial limits by bringing the greatest orators and performers to small towns, and invoked the historical value of recordings for subsequent generations. Each of these assertions would be consistently reinvoked in announcements of subsequent sound technologies, specifically with cinema's 1926–1929 conversion to synchronous sound.

The innovators of electrical sound technology followed the patterns set by Bell, Edison, and other phonograph companies when they presented new technologies to the public. In 1915, AT&T inaugurated transcontinental telephone service with an elaborate public performance. Widely reported in newspapers and through AT&T's own magazine advertisements, transcontinental telephony, or in the words of Bell System rhetoric this "Wonder of Wonders," was publicly demonstrated to audiences at the Panama-Pacific Exposition in San Francisco. In the words of AT&T's advertising copy: "Here, in a theatre de luxe, the welcome visitors sit at ease while the marvel of speech transmission is pictorially revealed and told in story. They listen to talk in New York, three thousand miles away; they hear the roar of the surf on the far-off Atlantic Coast; they witness a demonstration of Transcontinental telephony which has been awarded the Grand Prize of Electrical Methods of Communication."[28] Audiences entered the Bell Telephone Exhibit and, once seated in a large theater, heard an illustrated scientific talk on the history of and technological principles responsible for the telephonic transmission of speech. Then, through headphones attached to each of their seats, they listened in on a transcontinental transmission originating on the East Coast. Like previous technological spectacles, this presentation of transcontinental telephony made the technological artifact the object of a theatrical display and, through a synthesis of science and showmanship, dramatically culminated with the "wondrous" eclipse of previously impossible distances. But the performance also adopted a pedagogical mode of address that provided both a history lesson celebrating technological progress and an edifying presentation of scientific principles.

AT&T and its affiliated companies announced innovations in radio broadcasting, public address systems, and the sound film with similar public performances of technology accompanied by various rhetorical framing mechanisms. The corporation's initial attempt to dominate radio through its New York City station WEAF frequently included highly publicized demonstrations of carefully selected versions of broadcasting's potential. One series of such performances linked multiple broadcasting stations through the Bell System's long-distance telephone lines so as to broadcast simultaneously the same program, thereby suggesting that networked radio should characterize the medium's future. Overseen by AT&T executive John J. Carty, head of the Development and Research Department, these broadcasts typically began with a roll call of the participating stations. Carty, acting as a master of ceremonies, would call to a participating station and it would answer, thereby powerfully demonstrating to any listener the geographic scope of wired, networked broadcasts.[29]

In one of the most widely publicized of these performances, the September 1924 National Defense Test Day, interconnected broadcasting stations dramatized network radio's potential to respond to national emergencies. The appropriately redubbed *General* Carty undertook his roll call and the responses from dispersed broadcasting and repeating stations mobilized the wonders of AT&T's applied technology to serve national preparedness. Speeches by the Secretary of War and Gen. John Pershing, as well as reports from four regional military commanders, asserted the need for a transcontinental system of public address in a time of emergency.[30] In this instance the public performance of technological possibility, including General Carty's transcontinental roll call, illustrated a specific political use for sound media; broadcasting combined with wired telephony could enable a national mode of address. National Defense Test Day dramatized one possible future for radio, a future in which AT&T's maximization of profits through its monopoly control of long-distance telephone lines patriotically served the national interest.

AT&T also used well-publicized patriotic demonstrations to announce the public address system developed by its subsidiary Western Electric. During the most notable of these, on Armistice Day in 1921, Western Electric's new public address equipment conveyed the speeches, prayers, and musical performances commemorating the internment of the Unknown Soldier at Arlington National Cemetery to an estimated crowd of 100,000. Additional crowds estimated at 35,000 at New York City's Madison Square Garden and some 20,000 in the San Francisco Civic Auditorium listened to the ceremony transmitted over AT&T's long-distance telephone lines to public address systems in each venue. Besides stressing the potential to unify the nation through commemorative rhetoric addressed to a technologically created national audience, accounts of the demonstration again frequently referred to the potential civil defense and preparedness applications made possible by such a combination of acoustic innovations.[31]

AT&T continued to use similar public performances of technology to introduce Vitaphone sound-on-disk film technology. Among the most notable and widely circulated of these was a 34-minute film entitled "The Voice from the Screen," featuring Edward B. Craft, executive vice president of Bell Laboratories, who demonstrated and explained the Vitaphone. In a lecture directly addressed to audiences, Craft began with a brief survey of the origins of the Vitaphone, locating the precursors to individual components in Bell System's innovations of public address, electrical recording, and electrical sound reproduction. Craft introduced a version of a technological history for the Vitaphone that would be echoed throughout later AT&T and Western Electric announcements—namely, that synchronous-sound films resulted from the combined corporate research and development efforts undertaken exclusively within AT&T and its associated companies.

"The Voice from the Screen" next moved into a Vitaphone-equipped studio in which Craft stood before and then described both a camera booth and a sound recording room. This section of the film culminated with a demonstration of the sound-film recording process. In a static long shot, the male vocal duo Witt and Berg performed a song while a cameraman filmed and an audio engineer recorded the performance. The three tasks (original performance, image recording, sound recording) were arranged laterally across the screen, with the action spatially compressed so as to provide a film equivalent of a cross-section diagram. This staging functioned rhetorically to demonstrate the *simultaneous* recording of sound and image and thereby to distinguish the Vitaphone from several earlier sound-film processes that relied upon post-synchronous strategies of recording.

"The Voice from the Screen" next moved to a projection booth with two walls removed so as to reveal the machinery. But before he produced a Vitaphone disk and cued it on the turntable, Craft pointed out and briefly described the functions of amplifiers and volume controls, a cone speaker to monitor the volume of reproduced sound, the projector and attached turntable. After a projectionist stepped in for Craft, "The Voice from the Screen" then showed a Witt and Berg Vitaphone short subject. The duo performed in a static, uninterrupted long shot, directly addressing the camera and sound recording technology. This interpolated Witt and Berg Vitaphone short subject began with the song audiences just saw recorded in the studio demonstration, but the performance continued uninterrupted through two additional songs, each of which was punctuated by Witt and Berg's pausing and nodding to an imaginary audience and, in one instance, silently mouthing "Thank you."

After the presentation of the Witt and Berg short subject, "The Voice from the Screen" concluded, like many such public performances of technology, with a series of assertions about the device's wider social significance. In carefully scripted concluding remarks, Craft paid tribute to the largely anonymous labors of many engineers, cautioned the audience against taking for granted the achievements of applied science, and acknowledged the entertainment applications of the Vitaphone synchronous-sound film system.

But in his concluding remarks, Craft (and with him the associated Bell System companies) went further in outlining future applications for the device. Using a rhetorical trope at least as old as Edison's announcements of the phonograph, Craft noted, "Events, complete in actions and in sounds, can now be reproduced by these methods regardless of time or place. We of the immediate present and the generations to come will be the gainers by this possibility."[32] But while Edison and those who echoed his pronouncements most often invoked the possibility of preserving voices in the realms of politics and entertainment, Craft asserted the social value of preserving the voices and views of men of science: "The Faraday of the future, the Pasteur and the Galileo, may by this method make available to students in any place at any subsequent time a demonstration of their scientific researches and synchronize therewith their own comments, discussions and even their personalities."[33] Invoking the possibility of previously unheard-of communication and demonstration of science, Craft predicted changes to the educational process that would free teachers from the more mundane tasks of teaching and allow them more time for research or individual instruction.

Besides this concluding assertion of the Vitaphone's social value, Craft's rhetorical tribute to the "painstaking labors of many scientific workers over long periods" participated in a larger framing of acoustic innovation as not the accomplishment of a single inventor but as the result of the carefully orchestrated labor of anonymous engineers working together in the modern corporate laboratory. In stark contrast to the framed portrait of Thomas Edison that accompanied each Tone Test (and that had also figured prominently in the Sullivan Brothers' nineteenth-century phonograph demonstrations), Craft relocated the site of technological innovation from the heroic lone inventor (the "Great Wizard of the New World") to the modern corporation. When he paid tribute "to the many workers, some of whose names you will never hear, the many workers in the various fields of science and engineering, who, through their enthusiastic labors, have made possible such results," Craft was also paying tribute to a particularly modern method of technological innovation, that of the corporate-sponsored research and development laboratory.[34] This rhetorical tribute to corporate innovators effectively rewrote one of the dominant narratives of American innovation and would become increasingly pervasive throughout the introduction of electrical acoustics.

Although a filmed presentation rather than a live performance, "The Voice from the Screen" built upon the preexisting tradition of public performances that announced technological change. Like Alexander Graham Bell's lecture-performances or the Victor Auditorium Orthophonic Phonograph, "The Voice from the Screen" presented technological innovation as a kind of edifying spectacle. The film's pedagogical approach and simplistic presentation of scientific principles addressed general audiences, but "The Voice from the Screen" was initially presented at meetings of various technical societies. During October of 1926, the film accompanied Craft's presentation to the New York Electrical Society, and it was screened for the thirteenth annual meeting of the Telephone Pioneers. In the latter instance, the Craft film was accompanied by several Vitaphone specialties produced by Warner Bros.[35] Public presentations of the Vitaphone system before scientific bodies continued into 1927 when, for example, Craft addressed the metropolitan New York Section of the National Electric Light Association on "Some Electrical and Mechanical Problems in the Development of the Vitaphone." Craft's talk was augmented with both a screening of "The Voice from the Screen" and several Vitaphone shorts, including the *Tannhäuser Overture* and selections featuring tenor Giovanni Martinelli and concert violinist Mischa Elman.[36] In these contexts, Warner Bros.' entertainment short subjects functioned as public demonstrations of the possibilities of synchronous-sound recording. Rather than components of a larger entertainment program, the Vitaphone shorts became scientific evidence testifying to the engineering accomplishments of the associated Bell System companies. The recontextualization of Warner Bros.' Vitaphone films illustrates how public performances of new sound technologies strategically combined demonstrations of scientific principles and a pedagogical mode of address with mass entertainment that often bore the patina of high culture.

The simplistic pedagogical mode of address made the demonstration of Vitaphone technology within "The Voice from the Screen" accessible to audiences beyond members of scientific societies, and the film was shown more widely. Throughout the spring and early summer of 1927, Bell Laboratories employee R. E. Kuebler traveled throughout the Midwest and East publicly demonstrating talking motion pictures on behalf of AT&T. Staging demonstrations in such cities as Tulsa, Oklahoma City, Kansas City, Pittsburgh, and Norfolk, Kuebler traveled with the Vitaphone apparatus screening both "The Voice from the Screen" and selected Vitaphone short subjects.[37] Perhaps the largest audience for these demonstrations came at the Sesqui-centennial celebrations where AT&T, Bell Laboratories, and the Vitaphone had as audible a presence as the Victor Auditorium Orthophonic demonstrations.[38] For some American audiences at least, the initial direct encounter with the Vitaphone sound-on-disk system came in nontheatrical venues amid performance events distinctly framed as educational demonstrations.

These nontheatrical encounters with Vitaphone technology also included demonstrations in which the focus of the performance went arguably beyond the talking pictures themselves. In early 1926, Bell Laboratories constructed three portable sound-film projectors for AT&T and this equipment was placed at the service of associated Bell System companies throughout the United States for local celebrations of the fiftieth anniversary of "the birth of the telephone." Two sync-sound films were prepared for the occasion. The first featured Thomas A. Watson, Bell's assistant at the time of the development of the telephone, and the second illustrated the evolution of the telephone switchboard to the present day. During 1926, the portable sync-sound Vitaphone system performed in twenty-six different cities to audiences estimated at over 206,000, including a 23-week stint as part of the Bell System's Sesquicentennial exhibit. While these performances demonstrated to the public the possibilities of the synchronous-sound film and the application of electrical principles to motion pictures, they also connected this latest innovation in acoustic technology to previous developments (the telephone, the switchboard) and to the prestige and public service identity carefully cultivated by Bell System companies.[39]

Announcements of technological change that relied upon spectacular public presentations of new acoustic devices did not originate with electrical acoustics. Instead, when announcing innovations to consumers in the 1920s, corporations built upon a long tradition of previous acoustic demonstrations. These performances of technological change involved a consistent approach. First, the technology itself, as well as the electrical, mechanical, and acoustic principles behind it were put on public display. Rather than opaque instruments of wonder presented to awe and astound audiences, the technology was rendered visible and audible as an artifact. An educational rhetoric that purported to explain scientific principles addressed audiences and, in effect, demystified the machine. Second, these public performances often strategically assigned to the apparatus specific social uses and located those uses within larger, preexisting cultural traditions. Although the phonograph and other techniques of storing sounds obviously were assigned the task of transcending time and preserving for posterity, they were also credited with eclipsing spatial distances as well. As part of a larger rhetoric of technologically facilitated cultural uplift, these public performances credited sound technologies from the outset with collapsing the spatial distinction between urban centers and dispersed citizens, offering a democratization of culture that promised the best high-culture performers, educators, or orators to even the most widely dispersed consumers. Finally, these public performances of technological change framed the apparatus in terms of utopian promises that they would eclipse class distinctions so that high-culture values would become equally accessible to all social groups. Public performances of technological change presented mechanical and later electrical artifacts as instruments of a new scientifically engineered enlightenment.

Becoming "Sound-Wise": Consumer Pedagogy

Edward Craft's caution in "The Voice from the Screen" against taking for granted the wonders of applied science became a recurring theme throughout announcements of electrical-acoustic devices. While innovators presented emerging sound technologies as part of an evolutionary advance of applied science and engineering (the inevitable progression of modernity toward a more perfect, more convenient society), they also encouraged consumers to view technological artifacts as something more than opaque objects of wonder. Announcements of technological change frequently addressed consumers as interested in the means through which such technological marvels performed. Consumer pedagogy, a rhetorical strategy that presented to audiences often cursory (but sometimes quite detailed) views of how an artifact actually worked, both trained audiences in how devices operated and suggested how consumers might listen to the new electrically reproduced sounds.

A similar pedagogical mode of address had often rhetorically framed public performances of technological change dating at least as far back as Bell's initial telephone demonstrations and continuing with subsequent announcements, but presentations of the scientific and acoustic principles underwriting electrical sound technology also publicly circulated in a variety of other forms. Newspaper and popular magazine articles, as well as several advertising campaigns during the period, provided consumers with varying degrees of insight into applied science. Rendering the new technology familiar, accessible, and comparatively transparent, rudimentary scientific explanation of emerging sound media connected innovations with devices already familiar to consumers.

In general interest publications such as the *Saturday Evening Post, Liberty, Collier's,* or *Time,* as well as more specialized periodicals like *Phonograph Monthly Review, Scientific American,* or *Photoplay,* 1920s readers encountered detailed descriptions, diagrams, and photographs depicting the application of electrical principles to both the phonograph and the film. These accounts pervasively presented to the lay reader and especially to the potential moviegoer not only the apparatus but also the electrical, mechanical, and acoustic principles on which it was based. An interested reader in the period from 1927 to 1930 could find detailed descriptions of the comparative benefits of sound-on-disk systems versus sound-on-film systems. She could find simplified accounts of condenser microphones, amplifier tubes, alkaline-earth-oxide lights, photoelectric cells, or hairfelt and other acoustic-dampening materials. An interested filmgoer could read detailed, albeit not always accurate, accounts of shifts in Hollywood production practices. Consistently in magazines of the period, well-illustrated feature articles adopted an educational tone to inform readers about how this new application of sound technology actually worked.[40]

Coinciding with the Vitaphone's New York City premiere, the movie fan magazine *Photoplay* informed readers of the methods through which synchronous sound was brought to the screen.[41] Including descriptions of both the device's operation and the production methods used to create a Vitaphone short, the article's rudimentary explanation of the relevant electrical principles also located the Vitaphone in relation to previous acoustic innovations. Claiming that the Vitaphone was "miles ahead of the famous early Edison talking pictures," thereby reassuring readers who remembered Edison's ill-fated Kinetophone device, the author invoked no less of an electrical authority than Columbia University's Michael I. Pupin to attest to the success of the Vitaphone. Pupin cited the mimetic promise of earlier acoustic innovations: "No closer approach to resurrection has ever been made by science." This labeling of the Vitaphone as an innovation both new, yet familiar, continued in the article through the assertion that Vitaphone was not a new invention but rather a "combination of old and new ideas, an application of telephone, phonograph and radio principles."

Photoplay's early attempt to describe the Vitaphone system included many of the rhetorical methods that would subsequently educate consumers about new electrical sound devices. The account combined simplified scientific explanation with a historical perspective containing implicit arguments about technological innovation. New methods of handling sound were located within an ongoing tradition of innovation and compared to already familiar devices. Behind-the-scenes filmmaking details provided readers with privileged insight into the production of entertainment and accompanied implicit or even explicit claims about the social value of that entertainment.[42]

Two years later, *Photoplay* again presented an account of the technology involved in synchronous-sound film, this time a bit more thoroughly. In a highly visual presentation called "How Talkies Are Made," *Photoplay* illustrated for its readers both sound-on-disk and sound-on-film systems through simplified diagrams depicting microphone placement on a film set, a single-system camera (with sound recording accomplished inside the camera), the location of the soundtrack on a release print, a cross-section of a projector identifying features like the exciting lamp and the photoelectric cell, and a cross-section of a theater illustrating the location and shape of loudspeaker horns.[43] Once again, a written description accompanying these diagrams asserted that no single invention had caused the cinema's conversion to sound, but that the apparatus resulted from the successful combination of developments in radio, the telephone, and the phonograph.

While *Photoplay* introduced to many American movie fans the technology involved in cinema's conversion to synchronous sound, magazines like *Scientific American* presented announcements of the new technology to more technically minded readers. In a series of 1927 articles, the magazine described first the Vitaphone sound-on-disk and then the Movietone sound-on-film systems.[44] During the summer of 1927, it chronicled the technology and technique behind coast-to-coast radio broadcasts and as early as February 1926 described the application of electrical sound technology to the phonograph industry.[45] Readers of *Scientific American* in 1926–27 encountered the Vitaphone and Movietone systems as two specific examples of a wider deployment of a new approach to sound that was itself subsumed within an ongoing narrative celebrating applied science.

Specialized periodicals throughout the late 1920s included pedagogical presentations of emerging technologies that explained acoustic, mechanical, and electrical principles while simultaneously accomplishing a variety of other rhetorical tasks. An article by "Experts of the Western Electric Company" published in *Phonograph Monthly Review* presented phonograph enthusiasts with a fairly direct introduction to applied science and the shifts in recording practices caused by the new electrical processes of recording.[46] Like many such announcements, comparisons to more familiar apparatus helped to introduce the technology. The microphone and the recorder were explained in relation to the telephone, and principles of amplification were explained through comparisons to public address systems, thereby implicitly asserting that acoustic innovations arose out of telephone research. In another familiar rhetorical trope, the article characterized "old," outmoded mechanical recording practices as highly artificial in comparison to the new, more "natural" electrical methods that not only allowed the use of instruments identical to those used in live performances but also provided for greater freedom in spatially locating performers in a recording studio. Western Electric's "experts" even gave readers implicit suggestions on how to listen to the electrically recorded, identifying the increased recording of overtones, "the elimination of tinny effects," and the addition of atmosphere as well as an increased range of tones and volume as sources of an electrically enhanced fidelity to original sounds.

Later articles in *Phonograph Monthly Review* took readers behind the scenes and into the Okeh recording studio and provided updates on developing techniques since the initial attempts to use electrical recording technology.[47] Other articles presented and explained to consumers electrical recording systems competing with the Western Electric design such as the Brunswick Palatrope light-ray system of recording and electrical reproduction by the Brunswick Panatrope method.[48] Like the Western Electric experts, Brunswick's representatives announced these innovations through graphic and descriptive comparisons with both outmoded recording techniques and their resulting reproduced sounds. Illustrations dramatically embodied technological change by contrasting a recording studio containing musicians clustered in an unusual arrangement around an acoustic horn with depictions of the more "natural" arrangement of performers around studio microphones. Graphic renditions of increased frequency response through electrical means testified to the greater fidelity now available. Readers encountered simplified descriptions of applied science rhetorically invoked at the service of claims for greater fidelity. The scientific discourse of these pedagogical accounts served as apparently impartial rhetorical support for the fidelity claims made in advertisements often appearing in the same journals.[49]

Periodicals like *Photoplay*, *Scientific American*, and the *Phonograph Monthly Review* announced electrical sound technology to their particular constituencies by incorporating consumer pedagogy into their prevailing editorial practices. Other general interest periodicals like the *Saturday Evening Post* and *Collier's* also oriented consumers to the ongoing technological change. Graphic illustrations of synchronous-sound film technology accompanied descriptions of both the electrical and mechanical principles behind the technology and behind-the-scenes accounts of sound-film production. (See fig. 2.1.) Popular magazines often depicted an American film industry in the midst of overcoming an initial state of turmoil but quickly developing new techniques of production, distribution, and exhibition suited to changing technological circumstances. As early as the spring of 1929, articles used the past tense to describe panic-stricken producers and actors or even the unintentional accentuation of unwanted sounds by the "unforgiving" microphone. Such obstacles were the product of a now surpassed technological infancy. Magazine depictions of the film industry in the midst of technological change often included detailed descriptions of the technology itself, providing written equivalents to the graphic illustrations and labeling of technology circulating elsewhere in the larger culture.[50] Audiences could read how studios provided already completed silent films with a synchronous-sound track of special effects and music.[51] Actress Peggy Wood provided readers of the *Saturday Evening Post* with an insider's behind-the-scenes view of the sound-film production process that included not only comparisons from an actor's point of view of the differences between stage and screen acting but also accounts of the acoustic treatment of sets, innovations in lighting, makeup, and set design, illustrative anecdotes of shifts in film production practice from silent to sound filmmaking, and even a glossary of "talkie" terms.[52] Other articles clarified for consumers the various labels Hollywood currently used to identify different types of sound practices, explaining, for example, the differences between "sound," "sound and effects," "talking picture," and "all-talking picture."[53] (See fig. 2.2.)

FIGURE 2.1 (*Top, right*) Two Hollywood technicians prepare to enclose a camera operator (wearing a gas mask) in a soundproof camera booth. Such behind-the-scenes glimpses of Hollywood adapting to technological change shaped consumers' understandings of emerging technologies. (Quigley Photographic Archive, Georgetown University Library, Special Collections Division, Washington, D.C.)

FIGURE 2.2 (*Bottom, right*) Behind-the-scenes accounts of Hollywood's conversion to sound often focused on the new power of recording engineers to shape production practices. (Quigley Photographic Archive, Georgetown University Library, Special Collections Division, Washington, D.C.)

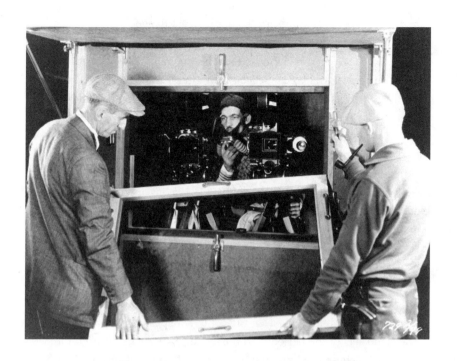

A series of Western Electric advertisements augmented the educational rhetoric of feature magazine articles that announced electrical sound-film technology. Advertisements such as "What Makes the Picture Talk?" overtly addressed consumer interest in applied science and sought to link the quality of sound reproduction to recognizable and familiar manufacturers of electrical appliances.[54] Informing consumers that "it pays to go to theaters equipped by the makers of your telephone—the Western Electric Company," the advertisement also included brief history lessons about microphone and loudspeaker technologies as well as an implicit assertion about the nature of technological progress. At the lower right of the advertisement, Western Electric printed drawings of, from top to bottom, a telephone transmitter, the familiar broadcasting-style microphone, and the condenser microphone used to make sound films. On the opposite side of the page appeared a brief graphic history of sound-reproduction equipment depicting from top to bottom the telephone receiver, the cone loudspeaker popular with many home radio sets of the period, and the "loud-speaking horn" used for sync-sound films. These brief graphic histories both made electrical film sound technology more familiar through comparisons to devices already located in the consumer's home *and* implicitly argued for a linear, evolutionary history of technological change: from the telephone through the radio to the sound film. In addition to promoting the Western Electric sound-film system and Western Electric as a larger contributor to contemporary life, "What Makes the Picture Talk?" provided consumers with models for thinking about technological change as an evolutionary progression.

The pedagogical mode of address continued elsewhere in the same advertisement as the top portion of the ad included a graphic depiction of the technology used to reproduce sound films in movie theaters. Echoing the descriptions and depictions appearing in feature magazine articles, the drawing illustrated the projector (equipped for both sound-on-disk and sound-on-film formats) with a detailed schematic of the sound-on-film system's exciting lamp and photoelectric cell and depicted the location of sound-reproducing horns positioned behind the movie screen. Rather than an invisible, mysterious, or incomprehensible object of wonder, the cinema's acoustic apparatus was here an object for the consumer's knowledgeable gaze.

"What Makes the Picture Talk?" was just one in a series of Western Electric advertisements that provided rudimentary technological explanation and urged consumers to attend theaters equipped with Western Electric equipment while simultaneously presenting a particular vision of the corporation. "Making a Sound Picture," another advertisement in the series, provided a brief written and visual demonstration of the technology and Hollywood production practices.[55] The advertisement's copy referred to the new need for silence on the film set, the placement of cameras within soundproof booths, and the role of the sound monitor. Illustrations depicted the filming of a musical production number and, through a series of photographs, various components in the Western Electric system (microphones, monitor, amplification rack, theater loudspeaker, and projector). These depictions of technology assumed some consumer familiarity with the relationships between the various components, a familiarity provided by the numerous other accounts and depictions of the technology circulating in the period. The second rhetorical function of the advertisement, its encouragement to patronize "appropriately" equipped theaters, was accomplished through text alone, linking "sound equipment that assures clear and natural tone" to the "half century's experience in making telephones and other apparatus for reproducing sound."[56]

This series of Western Electric advertisements depicted for the consumer technology and technological practices. Addressing and thereby partially constituting readers as interested in the apparatus behind the scenes, the advertising campaign labeled representational technology a site of consumer interest and encouraged a discerning approach to consumption that identified quality with the manufacturer of specific sound devices. Consumer choice was linked to consumer pedagogy, and the knowledgeable filmgoer should select his or her entertainment based on technological knowledge. This process of what we might today call the "branding" of sound motion pictures as a product of the telephone company reassuringly linked an emerging medium to an already well-known communications provider. (See fig. 2.3.) Moviegoers could trust the quality of Western Electric's new acoustic innovations in the way that they had come to take for granted the reliability of their telephones. Sound films rearticulated in a new form the public service identity of the Bell System and functioned as audible testimony to the social benefits of corporate-based research and development.[57] (See fig. 2.4.)

Advertisements for various products echoed the educational rhetoric of Western Electric's announcements about sound technology. Sometimes

Quiet in the Studio! Not even an undesired whisper may enter the sensitive microphones!

SOUND PICTURES
... a product of the telephone

OUT of a half century's experience, the Bell Telephone Laboratories developed for Western Electric the first successful system of sound pictures.

This system (embracing Vitaphone and Movietone) makes possible a great new art in entertainment. Now, in theatres all over the country—Western Electric equipped—you can hear stars of motion pictures, opera and stage in lifelike renderings from the screen.

Producers who use the Western Electric sound system exclusively:
WARNER BROS.
FOX FILMS
VICTOR TALKING MACHINE
PARAMOUNT FAMOUS LASKY
METRO-GOLDWYN-MAYER
UNITED ARTISTS
FIRST NATIONAL
UNIVERSAL
HAL ROACH
CHRISTIE
COLUMBIA PICTURES
HAROLD LLOYD

Hear and see the world's greatest personalities as they talk from the screen.

Hear orchestral accompaniment played from the screen . . . the actual roar of an airplane . . . the thunder of galloping hoofs!

Yesterday's dream is today's fact. And tomorrow? Here is an art now in the early stages of its development which is revolutionizing the field of motion picture entertainment.

Watch—and *listen!*

Western Electric
SOUND — The VOICE of ACTION — SYSTEM

GREAT NAMES
Pioneering in this great new art

AND *the world's leading makers of . . .*

Telephones

Switchboards

Cable

Trans-Atlantic Telephone Equipment

Telephoto Machines

Public Address Systems

BELL Telephone Laboratories—Western Electric—leading producers—progressive exhibitors—together these bring to you Sound Pictures!

Calling upon fifty years' experience in the telephone art, Western Electric produced the first practical system (used by Vitaphone and Movietone) for recording and reproducing Sound Pictures.

Producers have standardized on Western Electric equipment and are successfully meeting the technical difficulties natural to a new and revolutionary art.

Discriminating exhibitors, eager to provide the best in entertainment, have installed the Western Electric Sound System.

The success of Sound Pictures is history now. Continuing progress is certain. Make sure of enjoying it. Go to the theatres showing these great producers' pictures with the sound equipment recognized as the world's standard.

Western Electric
SOUND — The VOICE of ACTION — SYSTEM

these ads connected the innovation of the synchronous-sound film to other technological developments in apparently far-flung industries. For example, Masonite Presdwood insulation products successfully linked acoustic treatment for the home to Hollywood's conversion to synchronous sound. In "Presdwood Stars Again," advertising copy read, in part, "In *The Desert Song*, the latest Warner Brothers' Vitaphone talking and singing production, there stars a single, silent actor, Masonite Presdwood—a grainless wood board that is used in set after set to build gorgeous scenic effects."[58] The advertisement explained the miraculous acoustic properties of Presdwood and their necessity to the new, revolutionary technique of the talkies (as well as the wealth of other construction industry and consumer applications of Masonite wood products). Masonite "played a leading role in *The Desert Song*, clarifying the production of voice and music—most essential when a chorus of one hundred trained voices is featured." An example of the common practice through which product tie-ins linked Hollywood films to a larger culture of consumption, "Presdwood Stars Again" also illustrates the educational rhetoric surrounding Hollywood's conversion to sound, in this case education on the new acoustic requirements shaping Hollywood set design.

Other advertising campaigns even further afield than Presdwood also invoked the conversion to sound and audience knowledge of the consequent changes in film production practices. In a national campaign from the spring of 1929, Old Gold cigarettes linked claims for the quality of their tobacco ("they're as smooth as the polished manner of Adolphe Menjou") to the new production requirements of the film industry amidst the conversion to sound. (See fig. 2.5.) In signed testimonials, screen stars like Madge Bellamy and Richard Barthelmess linked the claims of Old Gold ("not a cough in a carload") to the new requirements for "quiet on the set." "He coughed . . . the Villain! and the love scene had to be taken all over!" exclaims Bellamy, while Barthelmess begins, "Please pardon my frown . . . but some one in the studio just coughed . . . *and spoiled our love scene*."[59] The

FIGURE 2.3 (*Top, left*) In 1929, Western Electric aligned its "brand" with the introduction of sound films through national advertising campaigns that claimed acoustic innovations directly resulted from telephone research. (Author's collection)

FIGURE 2.4 (*Bottom, left*) Western Electric's "branding" of sound-film technology presented "the talkies" as another in a long line of the Bell System's communication innovations. (Author's collection)

FIGURE 2.5 Manufacturers and advertisers of various products invoked new sound technologies, addressing a newly "sound-wise" public. (Author's collection)

ad campaign built upon consumer awareness of technological change by underscoring selected aspects of how the conversion to sound had affected production practices. Barthelmess' testimonial includes, "It's in the making of a 'talkie' that studio silence is required. No one is allowed to make any noise, but the players. For the slightest sound registers on the sensitive recording mechanism. And an uninvited cough is a calamity."[60] Like the Masonite advertisement, the Old Gold campaign both attests to the larger practice of advertising tie-ins linking the film industry with other types of consumption and, in the context of ongoing technological change, directly addresses a consumer educated and interested in the technology behind cinema's conversion to sound.[61] Advertising campaigns like those for Old Gold cigarettes or Presdwood suggest that consumer pedagogy was working; the public was becoming more knowledgeable about technological change. Manufacturers of other commodities could invoke this "insider's knowledge" about the film production process to link their products to a broader interest in sound technology as a particularly modern innovation.

As movie theaters around the country converted to synchronous-sound technology, consumers often encountered an even more vivid example of this pedagogical mode of address, the Fleischer Studio's 1929 cartoon "Finding His Voice." The cartoon, sponsored by Western Electric, illustrated for moviegoers the processes that made talking pictures possible by depicting the addition of a soundtrack to an animated representation of a silent film, named "Mutie." Praised in the film industry's trade publications directed to exhibitors, "Finding His Voice" provided a film version of the hundreds of articles in newspapers and magazines describing technological change.[62]

The film begins with a hand drawing a strip of movie film that soon comes to life. This figure, named "Talkie," is soon joined by a second animated figure named "Mutie," who is characterized with a gag tightly binding his face and an unhappy expression. He speaks through muffled sounds that appear onscreen in the form of printed titles. Talkie takes Mutie to visit his good friend Dr. Western, the scientist who provided Talkie with his voice. Mutie and with him the audience of the cartoon are then given a tour of the various steps in the production, processing, and exhibition of synchronous-sound films. As Talkie croons into a microphone, Dr. Western first shows Mutie the camera enclosed in its booth, describes how sound is picked up by a microphone, and then transports him to the mixer room. After illustrating volume control, Dr. Western takes Mutie to the amplifier room, to the machine that records sound on a separate strip of film, to a demonstration of printing a completed sync-sound print from an image negative and a sound negative, and finally to the projection booth and the loudspeakers behind a movie screen. Mutie, now minus his gag and fully voiced, upstages Talkie onscreen, interrupting his performance by producing a series of apparently random noises. The film concludes with Talkie and Mutie getting together and singing in harmony.

"Finding His Voice" adopts the pedagogical mode of address common to announcements of electrical acoustics, offering its audience a rudimentary overview of the principles behind the technology and a glance behind the scenes into the production process of a synchronous-sound film. The cartoon graphically depicts Dr. Western's explanations of technological processes through a series of closer schematic views of a microphone, a light valve, a photoelectric cell, the small holes in the movie screen allowing the sound to be transmitted, and so on. Technological components are made more familiar through the common practice of comparing them to devices already familiar to the audience. Dr. Western describes the microphone, for example, as "really a glorified telephone transmitter," and he characterizes the conversion of electrical impulses back into sound as follows: "the receivers connected to the horns convert the electrical vibrations back into sound waves exactly the way the telephone receiver operates." These comparisons both illustrate scientific principles and implicitly continue the historical argument familiar to Western Electric advertisements of the period, namely, that the sound film directly evolves from telephone research and development. Following his descriptive presentation of how a light-valve functions to translate electrical impulses from a microphone into variable density lines on a strip of celluloid, Dr. Western turns to directly address the film's audience and with a wink says, "Simple, isn't it?" This instance of direct address acknowledges and validates potential audience confusion, reassuring those who do not fully understand the technological details that they too can participate in technological change.

"Finding His Voice," like the magazine articles describing Hollywood's conversion, sought to make emerging technology understandable to consumers. Building on the long tradition of public performances of acoustic technology, consumer pedagogy constructed the apparatus itself as an object of consumer interest and attention, and it presented technological change to audiences in an educational way. This educational rhetoric addressed two different types of consumers of sound during the period. The first group included the portion of the U.S. population who were already amateur sound enthusiasts. They had actively embraced new acoustic technologies in the 1910s and 1920s, engaging in such activities as building their own radio sets, and then rebuilding those sets in an attempt to achieve greater clarity of reproduction and far-flung distance listening, or so-called DX listening. This group of consumers actively followed radio's development in the various technical and semitechnical radio journals that were published in the 1920s. As audiences for announcements of technological change, they were already well versed in the functioning of vacuum tube amplifiers, contemporary microphones, and sound-reproduction devices. But consumer pedagogy also addressed an even larger audience, those who only were to embrace radio as a medium after the radio industry redefined its market from *hobbyists* requiring tubes, insulators, and antennas to *audiences* requiring pre-built sets that included such labor-saving refinements as single-control tuning or receiving sets operated on household current rather than batteries.

Hailing consumers as knowledgeable *participants* in a period of technological change, offering readers and film audiences both education and reassurance, consumer pedagogy presented technology as accessible and understandable to the average consumer. One *Photoplay* article overtly addressed filmgoers as knowledgeable participants in technological change: "After all, the public is sound-wise. You know a good tone when you get it on your radio. You've been educated up to recording, whereas with the exception of a few amateur photographers, the public does not pay a lot of attention to good and bad photography."[63] Consumer pedagogy surrounding this instance of technological change went a long way in making audiences precisely "sound-wise."

History Is Progress: Corporate Embodiment and the End of "Cut and Try"

While announcements of electrical acoustics drew upon established traditions for presenting technological change, they also brought something new to the introduction of acoustic devices. Popular accounts of emerging media stressed not only an "epochal" advance in the methods of disseminating sound but also a dramatic shift in the techniques and processes of innovation. Simultaneous with the earliest announcements of cinema's conversion to sound, Western Electric ran an advertisement in several large circulation magazines proclaiming "*The* 'rule of thumb' *is over* . . . King Thumb rules no more. The rule of thumb, with all its costly guesswork, has no place in Western Electric."[64] (See fig. 2.6.) The end of the "rule of thumb," elsewhere in this period pervasively described as the eclipse of outmoded "cut and try" engineering, provided a rhetorical framework that distinguished contemporary methods of scientific investigation from earlier less precise (albeit successful) methods of innovation. If, as received history told consumers, Bell's breakthrough in telephony was literally the product of an accident ("Watson, come here. I want you."), then the new media of sound were the products of precise and controlled scientific investigation. Moreover, announcements of electrical-acoustic technology relocated the *site* of the labor of innovation from the lone inventor to the corporate-sponsored research and development laboratory.

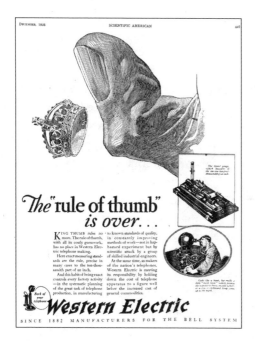

FIGURE 2.6 Coincident with its innovation of electrical acoustics, Western Electric proclaimed the end of haphazard "cut and try" engineering and a new era of corporate-based scientific innovation. (Author's collection)

During the redefinition of radio from wireless telephony to broadcasting, invention and technological innovation took place on essentially two fronts. Individual inventors working in small laboratories provided important contributions to carrier-wave technology, vacuum tube development, and other inventions that enabled the shift to a broadcasting model. Men like Lee De Forest, Reginald Fessenden, and Edwin Armstrong were often rhetorically linked to earlier lone, "heroic" inventors like Edison and Marconi in public presentations of the process of invention. Simultaneously, and far less prominently, a different group of engineers worked on radio and related technologies in well-financed and well-equipped research and development units within large corporations. Engineers like J. P. Maxfield and William Harrison at Bell Laboratories, H. D. Arnold at Western Electric, Irving Langmuir and later Harry Olson at General Electric, although less publicly prominent, were equally important to the development of electrical acoustics. As major corporations consolidated their control over acoustic innovations, the site of invention shifted from the workbench of the individual inventor to the well-capitalized research and development laboratory characterized by a division of labor and rigorous attention to scientific principles. In effect, the principles of scientific management that revolutionized manufacturing were being successfully applied to the tasks of invention and innovation. Or, at least, announcements of electrical sound technology actively presented to the public such a shift in innovative activity.[65]

Announcements of electrical sound technology, particularly synchronous-sound cinema, frequently foregrounded this shift away from archaic methods of "cut and try" engineering. Promotional rhetoric asserted that supposedly unscientific and haphazard methods had been eliminated and that new acoustic devices resulted from the scientifically coordinated labor of numerous anonymous engineers. Aligning invention with a corporation, most prominently Western Electric / AT&T / Bell Laboratories, but also (for the Photophone or the Panatrope) General Electric, Westinghouse, and RCA, this rhetoric argued that the products of well-established, well-capitalized firms offered higher quality and fewer risks than those of less well-financed innovators. Period accounts of the development of synchronous-sound cinema mentioned Lee De Forest only as a solitary inventor, an important precursor who had ultimately failed to innovate sound films. Although variously celebrated as the "Father of Radio," De Forest by the late 1920s had become more a symbol of radio's origins than a representative of contemporary engineering practice.

Locating the site of acoustic innovation within large corporations rhetorically eclipsed not only the lone inventor but also the amateur home audio engineer who had been so central to the initial development and popularity of radio in the United States. While the popular and the radio press had previously valorized amateur radio enthusiasts through contests involving home-built radio receivers and the ability to tune in radio signals from distant cities and countries, announcements of electrical acoustics foregrounded the corporation in popular presentations of technological innovation. The innovation of electrical sound methods was different from previous instances of technological change, these announcements suggested, the product of technological refinement and mass production inaccessible to both the domestic audio enthusiast and the independent, entrepreneurial-minded inventor.

The shifting cultural salience of Thomas Edison during this period even more dramatically underscores a revision to American narratives of innovation. Throughout most of the first three decades of the twentieth century, Edison was central to popular constructions of science and invention. Within phonography, Thomas Edison's signature verified the achievement of fidelity through applied science. Appearing in almost every Edison advertisement as well as on each Edison phonograph and on every Edison recording, that signature, often accompanied by an image of Edison, asserted the prestige associated with his name and attested to Edison's much-publicized, alleged personal approval of each product bearing his visage and/or name. The signature as a signifier referenced a further signifier, namely "Edison," not the corporeal individual but the range of values associated with scientific progress, rigorous personal standards, and a larger mythology of industrial achievement central to cultural constructions of modernity and invention in the United States at the time. In, for example, "The Test," a 1915 two-page advertisement announcing Diamond Disc technology and, in fact, inaugurating the Tone Test demonstrations, Edison's figure looms in the foreground, larger than the "expert audience," larger than singer Alice Verlet, larger even than the phonograph itself.[66] Framed from the waist up, head bent forward in repose, Edison is represented in a position familiar within the culture: the great man listening to his epoch-making invention. (See fig. 2.7.) Images of Edison listening in a similar pose pervaded popular accounts of the phonograph from the outset, and reproductions of such images suitable for framing were available to consumers without fee from any Edison dealer.[67]

In feature magazine articles accompanying such images, "Edison" was repeatedly constructed through a series of common traits, frequently told apocryphal tales of his Huck Finn-like, mischievously entrepreneurial boyhood; of working alone into the night in his laboratory, a hearing-impaired man who had trained his acoustic abilities well beyond those of a normal consumer; of a man so committed to the quality of sound reproduction, to cultural uplift, that he personally approved every recording before it was mass produced; of a man who heard through his teeth, compensating for his hearing loss by biting into the phonograph in his study so as to register the minutest vibration. Such articles frequently included descriptions of the teeth marks on Edison's personal phonograph. "'I hear through my teeth,' said he [Edison], 'and through my skull. Ordinarily I merely place my head against the phonograph. But if there is some faint sound that I don't quite catch this way, I bite my teeth into the wood, and then I get it good and strong.'"[68] The teeth marks, like the signature on each Edison product, functioned as a physical trace of the commitment to quality promised to consumers who purchased an Edison product.[69]

FIGURE 2.7 Edison used both Tone Tests and publicized accounts of them to introduce new innovations in sound reproduction. Frequently Edison himself was featured as prominently as the technology. (Author's collection)

But while Thomas Edison was celebrated across the country during 1929 on the "golden jubilee" of his invention of the electric light, the nationally reported celebration was in fact a backward glance at an earlier and outmoded era of American ingenuity. Press accounts now labeled Edison's laboratory "primitive," and it was in fact re-created for Henry Ford's Dearborn museum.[70] Edison was still publicly valorized as "the scientific arch-benefactor of the Republic," or as "a symbol of America at its very best"; and at a testimonial dinner in Detroit on the night of the jubilee, President Hoover credited Edison with "being a leader in the modern method of invention by definitively organized laboratory research."[71] But announcements of electrical acoustics largely replaced the heroic celebration of the individual innovator with the anonymous research and development engineer firmly located within the modern corporation. While the reproduction of Edison's signature on a phonograph had in the past attested to its quality—in the words of Mr. Phonograph, "The name of my famous master is on my body and tells you that I am a genuine Edison phonograph"—now Western Electric's corporate logo declared a new era of invention and successful innovation.

National advertisements sponsored by Western Electric, and pervasive indications at local theaters and in local newspapers that a neighborhood theater was equipped with Western Electric equipment, cultivated the corporation's reputation as a leader in American innovation. Print advertisements instructed moviegoers to "Look for this Sign [the Western Electric Sound System]" as a guarantee of sound-reproduction quality.[72] Film sound was not simply another in a series of acoustic advances, but the product of particularly modern engineering techniques undertaken in the larger public interest. Heralding the cinema's conversion to sound with advertisements like "Sound Pictures . . . A Product of the Telephone," Western Electric invoked familiar promises of the greatest entertainers suddenly made widely available for everyone.[73] Importantly, though, the advertisement asserted a particular version of technological history: sound film originated from the corporate achievements of Bell Laboratories; the sound film was a logical, indeed almost inevitable, product of the telephone. Other advertisements from Western Electric also overtly linked the innovation of sound cinema to the substantial achievements of "the world's leading makers of telephones, switchboards, cable, trans-Atlantic telephone equipment, telephoto machines and public address systems."[74] Advertisements labeled the Bell Telephone Laboratories as a site where "many hundreds of trained scientists and their assistants are engaged in a continuous search for new materials and equipment for making America's telephone service more extensive, more efficient and more economical."[75] As Edward Craft asserted in "The Voice from the Screen," the "painstaking labors of many scientific workers" not only led to improved telephone service but also brought new acoustic technologies to the market.

This rhetorical construction of the corporation as the site of technological innovation should be understood, in part, as a product of an ongoing attempt to shape public perceptions of AT&T and its associated companies.[76] Pursuing a public relations policy intended to diffuse antimonopoly sentiment and related attempts to regulate the Bell System's increasing control over voice transmission in the United States, AT&T presented itself as a benevolent organization responsive to the public interest. Besides

the conscious attempt to embody the corporation through the female telephone operator (specifically in the female voice), the telephone company
also sought, in historian Stuart Ewen's words, to "portray the telephone in
general, and AT&T in particular, as the glue that holds a modern society
together."[77] The application of electrical methods to the phonograph and
the innovation of film sound provided opportunities to reassert the public service provided by the Bell System and its affiliates. Western Electric's
advertisements announcing emerging sound media provided rudimentary
lessons in mechanical, acoustic and electrical principles, but their primary
message emphasized that the latest acoustic innovations were the product of modern corporate-based research and development in general, and
AT&T in particular. An army of skilled engineers labored less at the service
of corporate profit than for a greater public good. In the words of one Western Electric advertisement, "science, art and business, working shoulder
to shoulder, have accomplished [sound pictures]."[78] Proclaiming the end
of "the rule of thumb" and cut-and-try engineering, and relocating popular perceptions about the site of technological innovation, AT&T sought to
elide monopolistic business practices and disguise for-profit initiatives as
service to a larger public good.

Utopian Visions

The initial four years of broadcasting history in the United States were
punctuated by a series of "event broadcasts" designed to demonstrate the
possibilities of electrical transmission of voice and music.[79] While studio-
originated broadcasts, whether of phonograph recordings or musical performances, talks of various sorts or even drama, made up the majority
of early broadcast content, the programming that generated the greatest
public interest and acclaim and served to most powerfully announce the
possibilities of the new medium remained descriptive accounts and broadcast content that served to "conquer distance" and bring geographically dispersed listeners into a virtual proximity with a live event. Most often, event
broadcasts also served to announce radio's potential to provide a national
mode of address through experimental networks that linked several broadcasting stations into a temporary chain. Broadcast accounts of championship boxing bouts, World Series games, and commemorative speeches by
political leaders, both caught the public's fancy and received substantial

promotion in the popular press and radio trade publications. Frequently accompanied by photographs and descriptions of large crowds assembled in public to listen to broadcasts projected through public address systems, press accounts of early event broadcasts not only publicized radio innovations but also promoted specific possibilities that radio seemed to promise. Sporting events figured prominently in such announcements, offering the drama of a highly anticipated and unfolding event, but political discourse also constituted a particularly powerful source of event broadcasts and their demonstration of radio's potential.

Despite countervailing discourses, participatory democracy and the notion of a political public sphere remained in the 1920s a powerful and pervasive fiction strategically invoked to serve a variety of ideological interests. Emerging media, especially radio, proved particularly susceptible to often-inflated claims about the capacity for communication systems to function as the instrumental fulfillment of utopian democratic urges. Often, event broadcasts became sites where innovators of new sound media yoked radio technology to an established rhetoric of political emancipation and consensual social desires. Advocates and promoters of radio and the other media of electrically amplified sounds sought to align technological change to the utopian promise of a(n) (ever-deferred) fulfillment of democratic ideals.

Radio promoters had for some time announced broadcasting with claims for its contribution to democratic processes. In a year-end survey of radio station attitudes toward political broadcasting during the 1922 election, *Wireless Age* found only one station that limited coverage to a single party. The journal concluded, "Broadcasting has brought into being an absolutely unbiased and impartial medium through which public expression may be given to subjects that have public interest."[80] National, networked broadcasts of the presidential nominating conventions of both political parties during the summer of 1924 provided a focus for subsequent assertions regarding radio broadcasting's social benefit.

Much of the nation was able to listen to the three-day Republican National Convention beginning on June 10, 1924, in Cleveland, Ohio. AT&T connected its two radio stations (WEAF, New York, and WCAP, Washington, D.C.) to sixteen stations between Boston and Kansas City using modified long-distance telephone lines. Denied access to AT&T's "longlines" plant, stations WJZ (Newark) and WGY (Schenectady) used telegraph lines leased for the occasion from Western Union to form a two-station chain for their broadcast coverage. With Graham McNamee announcing for the AT&T network and Major J. Andrew White presiding over the competing broadcast (both had gained some national renown from announcing previous event broadcasts), the audience heard a mix of convention speeches, music, and noise from the convention floor and explanatory commentary. With the exception of an ill-fated platform debate orchestrated by Robert LaFollette's supporters, the Republican National Convention functioned smoothly, providing little drama surrounding the inevitable nomination of Calvin Coolidge.

The Democratic convention that began in New York's Madison Square Garden on June 24 proved to be far more dramatic both as a political event and as a demonstration of broadcasting's potential. Over an unexpectedly long and contentious fifteen days of speeches and demonstrations, the Democratic Party eventually settled on compromise nominee John W. Davis on the 103rd ballot. By all accounts, radio coverage of the tumultuous event captured the public's attention. With virtually every roll call vote begun by the Alabama delegation's loud declaration, "Alabama, twenty-four votes for [native son Oscar] Underwood," networked, live radio broadcasts helped to transform Alabama's vote into a national catchphrase, destined to reappear in comic routines and vaudeville acts for a generation.

With twenty stations linked through AT&T's wired network (by some accounts, listeners as far away as Argentina tuned in via experimental shortwave broadcasts), and three stations connected through RCA's use of leased telegraph lines, radio coverage of the Democratic National Convention provided a powerful announcement, over two weeks in duration, of a future for networked radio broadcasting. Microphones both on the convention podium and distributed throughout the crowd allowed broadcast engineers to shift the audience's attention from speeches and seemingly endless roll call votes to the spontaneous (and sometimes planned) outbreaks of partisanship unfolding on the floor of Madison Square Garden. Throughout the pandemonium, McNamee and White, now well on their way to stardom in the new medium, provided explanatory and descriptive commentary. Unlike the Republican convention but very much like a World Series game or a championship boxing bout, the Democratic convention and radio's coverage of it provided geographically dispersed listeners with a virtual ringside seat at a live, unfolding event, the resolution of which was anticipated but delayed for over two weeks.

Press accounts often stressed the Democratic convention's qualities as an acoustic spectacle:

> The tumultuous clamor of leather-lunged convention fans rose in great waves of intonation that swept across the hall and broke in a mingled roar of stamping and shouting, and whistling and singing. The hilarious abandonment of a mob gone mad was truly infectious. Some one started a parade. Under the flag-draped roof of the Garden, five thousand people milled in the sweltering heat of a packed hall. As a single voice the yell rose from fifteen thousand raw-hide throats: "Hail! Hail! The gang's all here!" The convention racket reverberated long after the commotion had subsided.[81]

The convention's racket reverberated throughout U.S. culture in the form of often grandiose claims regarding new media's contribution to democracy.

While the innovation of public address systems had previously been credited with enlarging the audience a political speaker might address ("The Circle Grows Wider," in the words of a Western Electric advertisement), radio broadcasts were celebrated for an even greater affirmative effect on the inevitable forward march of democracy. Radio, advocates argued—particularly in the aftermath of the 1924 convention broadcasts—would lead to both a better-informed citizenry and a "cleaner," more honest, more transparent political process. Partisan extravagances of the press, such as political advocacy that masqueraded as news reporting, would disappear when citizens could hear a speaker and then measure their direct perceptions against the written coverage a speech received. Moreover, radio by its very nature would encourage a new mode of political oratory and promulgate a more rational and deliberative political process in which the genuine merits of an argument displaced emotional sway from the tools of political persuasion.

Promoters of radio marshaled an impressive array of political figures who echoed corporate spokesmen in hailing radio as the instrumental fulfillment of the Enlightenment public sphere. In a statement prepared for the radio journal *Wireless Age*, three-time Democratic Party presidential nominee William Jennings Bryan praised radio's ability to connect candidates with their constituents and even prophesied broadcasting's role in creating a new era of international understanding and world peace.[82] Because of radio's capacity to disseminate "an intimate knowledge of men and events," declared Michigan's Sen. Royal S. Copeland, "with greater intelligence than ever before the electorate will march to the polls next November and determine the fate of the country during the next four years."[83] Radio advocates claimed that the attention garnered by the convention broadcasts and, importantly, the drama of the prolonged Democratic Party selection process increased public interest in the political process. Because radio brought politics into the home, women were more interested in and better informed about politics. By shattering the gendered boundaries of "ward meetings, street corner harangues, or conventions," the summer broadcasts in particular and radio in general represented the most important step toward full political participation of women since gaining the franchise, or so claimed Christine Frederick on the eve of the 1924 election.[84] Sen. Edwin Ladd of South Dakota suggested that radio would cure the indifference and apathy of the average citizen and even exert a "leveling effect in the social order" as "farmer, laborer, and individual of limited means" would become as equally important politically as "the business and professional man and the multimillionaire."[85] The aftermath of the 1924 convention broadcasts thus became a site in which radio was rhetorically aligned to the Enlightenment public sphere. By educating and informing citizens and soliciting their active involvement in the political process regardless of gender, geography, or wealth, radio—these announcements claimed—promised an instrumental fulfillment of America's democratic aspirations.

But radio promised even greater contributions to democratic processes. Broadcasting's advocates often stressed the benefits that accrued when a political speaker addressed spatially absent audiences. Radio's annihilation of distance by bringing the voices of politicians into widely dispersed listeners' homes would not only create better informed and more active citizens, but it would also improve the quality of U.S. political rhetoric. The summer convention broadcasts illustrated that radio audiences would tolerate only a clear presentation of facts, claimed General Electric's Owen D. Young: "The old method of meaningless oratorical demonstrations, delivered from a platform, will not be of any consequence to a radio audience. This fact will undoubtedly improve the statesmanlike delivery of political speeches to the American people."[86] Radio promoters consistently emphasized that broadcasting would cultivate a more rational, deliberative political process. Commenting on the effect of the 1924 convention broadcasts, H. P. Davis of Westinghouse declared that "appeals to the emotions cannot be made by radio and there is no such thing as working up 'mob effects' on the radio listener seated in his own home. . . . The day of the stump speech, with the dramatic gesturing of the speaker and his frenzied words, the equally tempestuous parades of his cohorts; this whole appeal to the emotions is passing with the growing use of the radio telephone to broadcast over the entire United States. . . . The voters will grow more and more particular in making their choice."[87]

Political figures echoed these radio industry pronouncements. Assistant Secretary of War Dwight F. Davis and Secretary of Labor James J. Davis both saw the private space of the home as a site of dispassionate political deliberation. Radio provided a spatial buffer from the "enthusiastic mob" susceptible to "abuse and rant" at political meetings.[88] Similar sentiments were expressed by Secretary of the Interior Hubert Work: radio would provide greater intimacy between speaker and audience, and the comparative solitude of domestic listening would guarantee the triumph of reason over emotional sway.[89]

Event broadcasts in the early 1920s announced particular configurations of sound technology's potential and aligned emerging media with powerful preexisting utopian desires. While radio broadcasting—particularly *networked* radio—seemed to promise an instrumental fulfillment of participatory democracy, innovators framed other applications of electrical sound technology in terms of similar broadly consensual goals.

Many U.S. audiences initially encountered Warner Bros.' Vitaphone sound-on-disk system not through the recorded musical accompaniment and selective sound effects of the feature film *Don Juan*, but through a series of short, sync-sound entertainment films that preceded the Vitaphone features. The first of those short films, screened at the New York City Vitaphone premiere (and subsequently often initially screened at the various local Vitaphone "premieres" around the country), depicted Will Hays, head of the Motion Picture Producers and Distributors of America, directly addressing audiences. The Will Hays short subject celebrated and paid tribute to the Vitaphone technology. Like Edison's Kinetophone film "The Lecture," Hays's four-minute direct address presentation did more than simply demonstrate the successful synchronization between recorded voice and image of speaking body (although it did indeed function as such a demonstration). It also asserted a specific vision of the social value of this technological innovation. Hays's recorded speech sought to align the apparatus with a larger rhetoric of cultural uplift and a vision of technology contributing to the democratization of culture, a rhetoric that had previously celebrated radio and had accompanied the phonograph for decades.

Beginning with the assertion that the story of the screen was always more dramatic than the stories written *for* the screen, Hays foreshadowed, whether intentionally or not, the pervasiveness of behind-the-scenes accounts of the work done to accommodate technological change. Throughout the four-minute speech, Hays's rhetoric clearly conformed to the conventions of commemoration as he invoked the past and the future, locating the Vitaphone in a larger narrative of the cinema's importance to American life. That importance, Hays claimed, resided in the film's service to society as he cited the "immeasurable influence as a living, breathing thing" that the motion picture had "on the ideas and the ideals, the customs and the costumes, the hopes and the ambitions of countless men, women and children."

Having asserted the social contribution of movies in general, Hays turned to the specific benefits of the Vitaphone. Citing the invaluable role played by music in motion picture presentation, Hays claimed that the "motion picture is too a most potent factor in the development of a national appreciation of good music." Cinema's role in elevating the musical taste of the nation was now being expanded through the auspices of this technological change: "That service will now be extended as the Vitaphone shall carry symphony orchestrations to the town halls of the hamlets." Hays thus invoked a rhetorical trope familiar to announcements of acoustic technologies, namely their promise to collapse the distance (both geographic and in terms of access to culture) between the urban center and the dispersed hamlet. Hays next invoked another common rhetorical frame placed around acoustic technologies, their promise of transcending time. "It has been said," Hays intoned, "that the art of the vocalist and instrumentalist is ephemeral, that he creates for the moment. Now neither artist nor art will ever wholly die." After crediting the "long experimentation of the Western Electric Company and Bell Telephone Laboratories" and crediting as well the Warner brothers and promoter Walter Rich, Hays concluded his celebration of the Vitaphone by reinvoking the dominant theme of his brief remarks. "It is another great service," he enthused with arms outstretched, "and service is the supreme commitment of life."

The four-minute short subject demonstrated the ability to synchronize sound with image, and Hays's commemorative rhetoric located technological change within a larger historical narrative of the development of the motion picture: "Far indeed have we advanced from that few seconds of shadow of a serpentine dancer thirty years ago when the motion picture was born to this public demonstration of the Vitaphone." Projecting that narrative into a future of "greater and greater service," Hays portrayed technological change as nondisruptive; it posed not a radical departure from the historic or immediate past, but came as a logical development within the medium's evolution. In assigning social significance to the Vitaphone, Hays invoked two preexisting utopian promises previously assigned to acoustic devices: their ability to foster greater musical appreciation despite geographic location (the transcendence of space and the democratization of culture) and their ability to preserve for posterity the greatest performers and performances (the transcendence of time). Hays thus announced the Vitaphone in familiar and reassuring terms, framing it as the product of an ongoing commitment to shared consensual ideals ("service") and as the (further) fulfillment of utopian promises already assigned to acoustic devices.

Will Hays's voice was far from alone in echoing these preexisting assertions of the social value of acoustic technologies. During the 1926–1929 period in which electrical sync-sound film technology was successfully introduced, William Fox frequently proclaimed that his corporation's Movietone sound-on-film system would do far more than change the film industry and the experience of moviegoing. In frequent public pronouncements, Fox heralded a day in which, through the auspices of the sound film, the greatest educators and clergy, the most important political figures and philosophers, would speak to previously unimaginable audiences. In familiar hyperbolic rhetoric, Fox sought to frame sound cinema not only in terms of cultural uplift and edification but also as the technological fulfillment of the Enlightenment public sphere. Echoed widely in print coverage, Fox's announcements of the Movietone aligned electrical sound technology with visions of an informed citizenry, enlightened and participating in the democratic process.

"Science Finds, Industry Applies, Man Conforms"

Announcements of electrical acoustics built on previous methods of publicly presenting technological change to consumers, mixing a pedagogical mode of address with spectacular performances of new innovations. These announcements located the technological object, in addition to the sounds it produced, as the site of consumer attention and incorporated electrical sound media into a larger narrative of cultural change that collapsed history with progress. History could be seen as a series of evolutionary developments, ever greater accomplishments that offered the better, the faster, and the more convenient. Innovators sought not only to educate consumers about electrical media but also to align the new apparatus with specific, preexisting cultural values including cultural uplift, a democratization of high culture through the erasure of geographic distance and the transcendence of class distinctions. Presentations of technological change in a variety of formats invoked preexisting utopian visions of an informed and edified citizenry and the promise of an authentic public sphere that seemed perpetually on the verge of being technologically achieved. These announcements offered consumers reassurance that emerging technologies promised only benefits. They were, in effect, cost-free and what was lost were not old and familiar forms (the silent melodrama, local identity) but either the outmoded or the undesirable.

While often conforming to previously successful methods of introducing technological change, announcements of the new electrical media also brought subtle shifts in the ways in which technology was presented to the public. Foremost among these was the rhetorical work that relocated the site of invention and innovation to the research and development laboratories of large corporations. Such a rhetorical move echoed and in fact ideologically underwrote the increased importance of corporations in American economic, social, and technological life.

In a survey of corporate presentations at World's Fairs and Expositions, Roland Marchand tracks a shift in public presentations of the corporation from the end of the nineteenth century through the initial decades of the twentieth.[90] Noting not only an intensified participation in such exhibitions, Marchand identifies transformations in the style and the content of displays that indicate an increased concern with corporate image-making. Corporate self-representations blurred "distinctions between education and entertainment, between business and show business."[91] The circuit of amazement and explanation described by Tom Gunning shifted toward amazement and entertainment interpolated within an ideological agenda that legitimated an emergent corporate status quo. Many of the announcements of electrical acoustics detailed above participated in this shift. The earliest public performances of acoustic innovations (Lyman Howe, the Sullivan Brothers, Edison) and later corporate-sponsored public presentations (AT&T's 1915 transcontinental "Wonder of Wonders," "Finding His Voice") mixed entertainment with education about emerging technologies. Marchand finds that corporate exhibitions had by the early 1930s shifted: "A focus on transmitting of technical information gradually gave way to an emphasis on entertainment values, simplicity of message, and command over audience attention."[92] Yet Marchand notes that an educational impulse continued to inform corporate public presentations into the 1930s, and while often the messages were simplified (offering less a demonstration of an entire manufacturing process—a factory literally brought to the fair—than a more ideologically infused celebration of scientific progress and technological innovation per se), corporate exhibits continued to offer education along with entertainment and distraction. This shift toward entertainment had not yet been fully attained during the announcements of electrical sound technology. Instead, these announcements offered consumers both rudimentary lessons on applied scientific principles and, perhaps more importantly, a variety of ways to think about technological change. Those ways of thinking were dominated by utopian promises of technologically achieved consensual social goals and reassurance in the face of the potentially unfamiliar.

The methods of announcing technological change described above continued within the Bell System and its associated companies, perhaps most dramatically in 1933 when they introduced a binaural method of transmitting sound. Using a multichannel method of encoding, transmitting, and reproducing sounds, the system not only increased the frequency response and volume of existing public address and sound reproduction systems, but it also reproduced music in auditory perspective. In its first public demonstration at Constitution Hall in Washington, D.C., famed conductor Leopold Stokowski manipulated the system's reproducing controls while his Philadelphia Orchestra performed at the Academy of Music in Philadelphia. The public performance emphasized a stereo version of orchestral music transmitted across distances, but it also included an intermission in which a highly simplified explanation of the system's components accompanied elaborate displays of its capacities.

The intermission included five separate demonstrations of acoustic innovation (a more rigorous pedagogical explanation of applied electrical principles had been reserved for a meeting of the National Academy of Sciences two days earlier). The first transmitted to Washington from Philadelphia a fictional scene involving two workmen that offered the illusion of copresence resulting when conversation and sound effects were transmitted in acoustic perspective. A subsequent demonstration illustrated the creation of spatial effects as a soprano walking back and forth across a Philadelphia stage was reproduced in Washington. Much like some moments from Edison's Tone Test demonstrations fifteen years earlier, the lights in Constitution Hall were then extinguished and the Washington audience seemed to hear two trumpet players on either end of a stage until the stage was reilluminated to reveal only a single trumpet player in Constitution Hall accompanied by a second whose performance was transmitted from Philadelphia. Another demonstration illustrated the binaural system's frequency range through the use of electrical filters that progressively eliminated a single octave at a time and thereby emphasized the nine octaves reproduced with the system in comparison to the six octaves normally available through ordinary radio reproduction.[93] Familiar utopian assertions about the tremendous benefits such a system would have to musicians, composers, and audiences alike accompanied these demonstrations. Stokowski, in his remarks at Constitution Hall, suggested that the transmission of orchestral music through such a binaural system would not only lead to greater availability of quality of music throughout the land but also facilitate a greater variety in musical compositions and expansive opportunities for young composers, conductors, and musicians.[94]

This demonstration of the binaural transmission of music in the spring of 1933 continued the process of introducing technological innovations that I have described above even as it moved inexorably away from consumer pedagogy and toward more elaborate showmanship and wondrous technological display. The Bell System's exhibit at the Century of Progress Exposition in Chicago in 1933 similarly continued this mix, in the Bell System's own words, of a desire "to interest and to instruct."[95] In an elaborately designed display area, visitors to the Exposition experienced a variety of Bell System innovations. Besides another display of the possibilities of binaural transmission, audiences encountered demonstrations of innovations in telephone service, including a presentation detailing how a dial telephone worked. Visitors given the opportunity to make a souvenir long-distance telephone call could watch as the route of their call was illuminated on a large map of the United States. A demonstration of delayed speech featured an early application of magnetic recording, and a specially designed projecting oscilloscope presented both some of the essential characteristics of speech and music and the means through which the Bell System was able to transmit multiple messages simultaneously.[96]

The Bell System's Washington, D.C., public demonstration of binaural sound transmission and the corporation's exhibit at the Century of Progress Exposition in Chicago illustrate that many of the methods used to introduce electrical sound technology to the public in the late 1920s continued to shape public presentations of emerging technologies. Even while continuing to invoke a pedagogical mode of address, to align new technologies with utopian promises, and to celebrate progress, these public presentations importantly shifted away from the explanation of applied scientific principles and toward an ever-increasing opaque presentation of technological marvels.

Guidebooks for the 1933 Century of Progress Exposition included a slogan that concisely summarized the attitudes about American society and technology articulated by the preceding decade's announcements of electrical acoustics: "Science Finds, Industry Applies, Man Conforms."[97] But more than just a summation, a culmination of sorts, the slogan and the larger vision of technological change it condensed pointed to ongoing shifts in the meanings assigned to innovation that continue to resonate in our present day.

"Science Finds" Announcements of electrical acoustics, as well as preceding and subsequent technologies, foregrounded invention as an activity. While initial presentations of acoustic technology located the site of innovation at the workbench of the lone inventor, or more generally elevated the figure of Thomas Edison to heroic status, subsequent innovations and the rhetoric presenting them to the public shifted the site of innovation to the collective labor of anonymous teams of engineers working collaboratively in the well-financed research and development laboratories of major corporations. These announcements also presented a particular view of technological development and scientific progress. History was presented to consumers as a series of progressive innovations of which the latest technology was

the culmination of previous efforts. Eliding failures from such histories, history and progress were collapsed into a mutually reinforcing category. History was to be understood as a progressive evolutionary advance to a glorious present with technology poised to fulfill utopian promises, an advance that implied the prospect of an ever greater future.

"Industry Applies" This portion of the slogan echoes the use of technological innovation to embody the corporation as a central feature of modern American life. The engineer, the apparatus itself, and the corporate logo stood in for an often opaque series of economic relations. Images of large organizations functioning in the public interest and for a greater social good screened often monopolistic business practices. The corporation became a provider of American abundance, thereby foreclosing consideration of other economic arrangements, regulatory initiatives, or simple scrutiny.

"Man Conforms" While the notion that "Man Conforms" seems inimical to the utopian promises of liberation through technological change, these public announcements often functioned precisely to heal this apparent contradiction. Announcements of new technology sought to engineer consumers, preparing citizens for a new and enlightened consumption of goods and services. By establishing ways of knowing and appropriate ways of interacting with new technologies, these announcements provided an epistemological framework for understanding not only technology but also one's place in relationship to it. "Man Conforms" seems to suggest that the rhetorical strategies used to announce technological change ultimately worked, shaping inevitably and finally consumer encounters with and understandings of new media. But the relationship between consumers and an apparatus is not so easily engineered, and as recent work in the history of technology describes, consumers can play active and often unanticipated roles in technological change.[98]

Rather than transparent indications of consumer relationships to emerging media, we can trace in these announcements common beliefs about invention, innovation, technology, progress, and media that were specifically promulgated by corporate innovators. These announcements reveal some of the methods innovators of electrical acoustics used to incorporate emerging media into larger habits and practices, consensual values and narratives. The path from technology to media involved more than the economic and regulatory "closure mechanisms" described in chapter 1. For in addition to engineering a functioning technological apparatus and several applications with marketable identities, innovators also *sought* to engineer public perceptions of new media.

Part of the work of a subsequent stage of innovation involved discursive and material attempts to heal the potential gaps between the promises used to announce technological change and the actual practices that incorporated new sound media into smoothly functioning commodity relations. First, having made the machinery of sound transmission and reproduction an object of consumer interest (through consumer pedagogy), innovators sought to relocate the site of consumer attention away from applied electrical principles and to the sounds the machines produced. Second, under the guise of various formulations of "public service," major corporations also sought to elide the economic realities of increasing concentration of media ownership. Building on the utopian promises that announced technological change, corporate innovators sought to disguise for-profit relations as ultimately best serving the public interest by linking emerging sound media to a preexisting rhetoric of "musical democracy" and by equating listening with citizenship and democratic participation. I take each of these issues up in turn in the following two chapters.

Three

From Performing the Recorded
to Dissimulating the Machine

Economic forces interacted with often-inflated utopian promises of media's social impact to shape the identities of emerging acoustic media. But the values ascribed to them and their relationships to preexisting media forms and commodity relations (in short, their social identities) also both influenced and were themselves influenced by the actual physical design of sound technologies. A radio receiver as a material artifact, for example, embodied certain rhetorical claims about broadcasting's social function even as it made specific physical demands on the consumer who purchased it. Having engineered functioning devices that augmented sound in new ways, corporate innovators also turned to engineering consumers' relationships to those devices. Users of technology were not, however, merely passive respondents to electrical or behavioral engineering. Instead, patterns of consumer interaction with emerging sound technology actively contributed to its successful incorporation into American society.[1] Innovators of electrical acoustics responded to consumer behavior even as they tried to shape it. The "interface" between listeners and sound-reproducing technology was characterized by competing impulses: on the one hand, consumers became active performers of sound reproduction while, on the other hand, the successful incorporation of new technology into consumption patterns necessitated dissimulating the work of machines. Announcements of technological change had foregrounded the material qualities of emerging technology, calling consumer attention to applied electrical and mechanical principles. But part of the process of smoothly securing new media into commodity relations involved the effacement of the apparatus, and innovators of electrical acoustics sought to redirect consumer attention away from the machines themselves and to the sounds they reproduced.

This chapter begins with a survey of the process through which the *mechanical* phonograph was successfully redefined from both a business device (an aid to stenography, a tool for accurate transcription and record-keeping) and an instrument of public performance to a musical instrument offering domestic entertainment. Throughout the first two decades of the twentieth century, phonograph companies waged a successful campaign to make the phonograph a necessary component of the modern home. Preexisting technologies and practices exert an often powerful inertia on emerging media. That which is new is often conceptualized, imagined, and understood in relation to that which came before. Perhaps no more powerful example illustrates this process than the efforts to engineer an identity for the radio in American homes. When introducing broadcasting in the 1920s, innovators modeled radio on the phonograph (as a preexisting acoustic device), even as they also used it to stress that which made radio different. By examining in some detail a precursor to electrical acoustics, this chapter traces the links between the emerging medium of radio and the already-established phonograph.

I conclude with another case study from the innovation of electrical acoustics, in which the process of grafting emerging technology onto pre-existing spaces and practices was less immediately successful. While radio entered the American home following a trail blazed by the domestic phonograph, electrical acoustics' entrance into the projection booths of U.S. movie theaters followed a less clear and direct path. Initially at least, prevailing practices and ideas about film projection clashed with the physical design of sound technology. Into a gap between film industry goals and technological realities a series of discursive and practical struggles centering on the labor of the film projectionist sought to manage technological change. This chapter then examines three sites entered by emerging sound technologies and the manner in which struggles to establish an apparatus' identity are articulated through both the physical design of technological artifacts and consumers' performances in reproducing sounds. In all three cases, a desire to dissimulate the machine clashed with a countervailing tendency that viewed sound reproduction as a performance.

Constructing the Phonograph's Domestic Identity

Throughout 1922 an advertising campaign for Johnson's Polishing Wax published in *House Beautiful* depicted a woman of apparently moderate class status cleaning her wood floor. The advertisements, not surprisingly, stressed not only the ease of using Johnson's Wax and its fine results but also located an attention to floor care as part of the well-appointed home. In the background of each image, however, was a cabinet model phonograph. Anonymous as to its manufacturer, doubly silenced (it is clearly not in use nor the focus of attention), the phonograph, together with other details of decor, rendered typical the depicted domestic space. Simultaneously present (in these depictions of domesticity) yet absent (from the rhetorical focus of the advertisement and its claims), the phonograph had become a ubiquitous object in depictions of domesticity. These advertisements for Johnson's Polishing Wax attest to the phonograph industry's success not only in engineering sound-reproduction devices but also in shaping consumers' relationships to reproduced sound. The redefinition of the phonograph as a home entertainment device, undertaken before the advent of electrical acoustics, extended beyond design and performance innovations to also include constructing an identity for the apparatus within the home.

The phonograph industry's rhetorical construction of both the phonograph as a device and listening to the recorded as an activity were pursued through advertising discourse and public announcements of innovations, the physical design of the machines themselves, and highly publicized re-· visions to the recorded repertoire. Efforts to define the phonograph's domestic identity stressed multiple major themes: various versions of fidelity claims; the phonograph's identity as a prestige product, and its relationship to ongoing constructions of a broader musical culture in America; listening to the technologically reproduced as a form of musical performance; and the redesign of the apparatus itself so as to reinforce these other themes and to integrate the machine into home decor.[2]

Like all phonograph advertisements, Edison's stressed the fidelity of its machines and recordings to original acoustic performances.[3] With the innovation of Edison's Diamond Disc technology in 1915, those fidelity claims hinged on the inability to distinguish between the recorded and the live. "The Test," printed on two pages of *American Magazine* in December of that year, chronicles the making of "musical history" in Orange, New Jersey, when three hundred "phonograph experts" found themselves incapable of distinguishing between the sounds produced by Belgian singer Alice Verlet and the "re-creation" of her voice with the New Edison Phonograph.[4] (See fig. 2.7.) The text of the advertisement described a presentational procedure that would become common in public "Tone Test" demonstrations throughout the United States: the live performance of an Edison recording artist alternated with a recording produced by the same artist. The advertisement further characterized the experts' assessment of the Diamond Disc demonstration as provoking linguistic inadequacy: "'perfect' failed as a descriptive word." "It was not enough to call it 'human, life-like, natural,'" the advertisement continued. "This New Edison was *nature itself*. It *was* the artist in all but form" (emphasis in original). Like many other phonograph companies, and like the constructed nature of fidelity as a concept associated with reproduced sound, Edison's advertisements assigned a value to its products by collapsing the cultural distinction between machine and man. Similar to the fidelity claims of the Victor Talking Machine Company (described below), Edison's advertising linked two distinct categories of experience by bestowing on the phonograph that which attested to the uniqueness of humanity. The Edison Diamond Disc was "The Phonograph with a Soul." Phonographic devices offered not an encounter with a machine, but a kind of spiritual communion.

Edison's presentation of such fidelity claims also emphasized the wondrous achievements of applied science. Advertisements frequently referenced scientific principles, as promotional rhetoric sought to make product differentiation comprehensible through simplistic metaphors. Edison's vertical-cut system of recording allowed the diamond stylus to float over hill and dale without wear-producing friction—like "an automobile running over a hill and then into a valley." In contrast, Edison claimed that competing lateral-cut systems damaged recordings because the movement of the steel needle in its groove was like "a twisting river always wearing away its banks."[5] Edison's advertisements reproduced "questionnaires" in which the great inventor explained in the first person the nature of musical sound, the importance of overtones, and the ability of the Edison phonograph system to capture and reproduce each delicate overtone.[6] A pedagogical mode of address illustrated that consumer choice, with the help of a rudimentary review of the appropriate engineering principles, was in fact no choice at all.

Under the guise of scientific study, Edison's advertisements also encouraged consumer participation in the great man's philanthropic inquiries. "Will You Join Mr. Edison in an Experiment?" asked a 1921 advertising campaign.[7] The advertisement reprinted the Edison "Mood Change Chart," an instrument designed to measure and document the influence of great musical performances on auditors' mental states. The reproduced chart documented the test undertaken by William J. Burns, head of the Burns International Detective Agency, selected as the first subject "because he will be the least susceptible to emotion." Readers were asked to analyze their own psychological and physiological reactions to Edison recordings as well as the reactions of their friends. (Local Edison dealers could provide as many blank charts as needed.) Framing this "experiment" with selected quotations from Confucius, Martin Luther, Napoleon, and Ralph Waldo Emerson attesting to the therapeutic values of music, Edison's rhetoric invoked a need for music shared by all humanity, and then having evoked such a need with culturally validated voices, promised to fulfill that need through the recorded. Philosophers had asserted the value of music; now Edison, with the help of world-famed psychologists and everyday listeners, would "scientifically" measure that value. By June of 1921, Edison claimed that "the psychological research work, which we have been conducting for nearly two years, indicates that the well known and almost incalculable benefits of music can be derived, in full measure, from the proper use of this new instrument [the Edison phonograph]."[8]

Edison's "Mood Change Chart," like many of the rhetorical constructions of the phonograph as an instrument of entertainment for the home, located listening to the recorded within a larger cultural tradition valuing the spiritually uplifting qualities of great music. But here, such uplift was the subject of scientific inquiry, in which all could potentially participate and from which all were to benefit. Further, applied science served the public interest and took on philanthropic overtones through the auspices of the great man who "in order to keep his favorite invention within the reach of everyone, . . . sacrificed millions in profits he might have made."[9] Because of the remarkable and affordable fidelity of the Edison device, *every* home could benefit from the spiritual qualities of the greatest performers and performances.

While Edison's rhetoric often invoked philanthropy and applied science, Victor's publicity efforts more frequently cultivated a preexisting star system and sought to augment it with Victor-recorded performers and performances. Victor is often credited with the introduction of high-culture prestige in relation to phonograph recordings through its line of Red Seal recordings of opera and concert performers.[10] Although initially using British-produced matrices, Victor held its inaugural recording session for the Red Seal series on April 30, 1903, in a recording studio in Carnegie Hall.[11] Highly publicized releases of recordings by concert and opera performers, as well as the consequent price differentiation for these disks, were part of a larger redefinition of recording technology as an instrument of musical high culture. Victor would actively cultivate this cultural capital in succeeding years.[12] The company's 1904 exclusive contract with Enrico Caruso solidified its prestigious status. In April of that year, Victor used the entire back cover of the *Saturday Evening Post* to announce this exclusive agreement with the opera star.[13]

Phonograph historians have stressed the importance of Enrico Caruso to Victor's cultivation of a market identity for the domestic phonograph. Benjamin Aldridge suggests that Fred Gaisberg's 1902 recordings of Caruso for Victor's British subsidiary were perhaps the most important in the history of recorded music, not only because of the profit they generated but also because they influenced the Metropolitan Opera in New York City to recruit Caruso; they successfully undermined the prejudice with which big-name artists viewed the phonograph; Caruso substantially contributed to Victor's prestige, thereby aiding Victor in building its Red Seal catalog; and Caruso was able to stimulate the public's interest in serious music.[14] With such full-page magazine advertisements as "Both Are Caruso" (depicting the singer in costume and a large phonograph recording, 1915), "My Victor Records Shall Be My Biography" (attesting to the artist's legacy after his 1921 death), "Caruso Immortalized" (the Victrola "has bridged the oblivion into which [the singer] passed"), and "You Hear the Real Caruso," Victor's advertising campaigns prominently featured Caruso as an exclusive celebrity recording artist across several decades, even continuing after the singer's death.[15] But Caruso was far from the only celebrity figure heralded in Victor's advertisements as similar promotional efforts focused

FIGURE 3.1 Victor's advertisements emphasized the company's exclusive contracts with celebrities from the opera and concert stage. (Author's collection)

upon Sousa, Chaliapin, Rachmaninoff, Melba, and literally dozens of other widely recognized performers.[16] (See fig. 3.1.)

Often Victor's advertisements stressed its entire slate of concert and opera stars.[17] As early as 1903, Victor promised consumers in full-page advertisements "The Living Voices of International Celebrities." Cameo-like photographs of performers of international renown circled the text of the advertisement in which Adelina Patti and Sarah Bernhardt provided signed endorsements of the Victor technology.[18] Whether in composite photographs or artists' renderings, Victor's advertisements depicted the period's greatest entertainers as if they were assembled all in one place, frequently in the phonograph owner's home.[19] On the inside cover of the *Independent* of December 15, 1917, Santa Claus leads down an elaborate staircase in a wealthy home Caruso, Alda, De Luca, Farrar, Gluck, Kreisler, Martinelli, McCormack, Melba, Schumann-Heink, Victor Herbert, Harry Lauder, Sousa, and others. Their destination is a Victrola placed under the Christmas tree.[20] These advertisements emphasized Victor's exclusive contracts with the artists and, as if to reinforce that exclusivity, often warned consumers that the recordings could be "safely played" only on Victor machines.

Jenny Lind is only a memory,
but the voice of Melba can never die

Two voices of finest, purest gold.
One is gone forever.
The other lives for all time.
There was no Victrola to capture the fleeting beauties of Jenny Lind.
But Melba's voice will still be heard in centuries to come.
Today Melba herself thrills and entrances vast audiences throughout
the world. Happy singer and happy public, that her flawless, limpid notes
will flow forever in undiminished beauty from her Victor Records!
Practically every great singer and instrumentalist of this generation
makes records only for the Victor—thus perpetuating their art for all time.

Victors and Victrolas, $10 to $400. Victor dealers everywhere.

Victor Supremacy

FIGURE 3.2 Phonograph advertisements frequently mourned the loss
of voices that hadn't been recorded,
while celebrating the preservation
of contemporary voices for posterity.
(Author's collection)

Victor's advertising emphasis on Caruso and the other well-known performers with whom it had exclusive contracts clearly aligned the Victor
trademark with musical prestige and linked mechanical reproduction with
the accepted canon of musical art. But from its inception, discursive framing of phonography also mourned a loss—that which hadn't been recorded
(or that which hadn't been recorded with the latest device). Phonograph
companies frequently referred to lost voices of dead performers (both opera singers and great orators) while simultaneously invoking a promise of
future preservation.[21] Singer Jenny Lind appeared in several Victor advertisements mourning the transience of her voice: "All that remains of Jenny
Lind is her picture, her autograph, and memories dear to all who ever
heard her sing. Her greatest charm—her wondrously sweet and melodious
voice—is gone forever. How different had she lived in the present day!"[22]
(See fig. 3.2.) This promise of preservation, of technological embalming, in
the face of the passage of time and mortal transience returned again and
again accompanying subsequent acoustic innovations.[23]

But more than simply a technology that allowed voices to transcend
time, the phonograph was also rhetorically constructed as an apparatus

that eclipsed spatial limits. In a series of rhetorical moves that would be echoed subsequently with the introduction of electrical acoustics, phonographs were presented to the public as instruments that could collapse the distances between entertainment centers and geographically dispersed listeners. Phonograph technology could bring to the farthest hinterlands not merely traces, but the *actual* performances of the greatest singers and instrumentalists of the day.[24] This ability to transcend spatial boundaries promised a democratization of high culture, as the phonograph became another (mechanical) voice for edification and uplift on an ever-widening scale.

Like the Edison advertisements discussed above, Victor's assertions about the universal accessibility provided by recorded sound also stressed the fidelity of Victor products. In, for example, *Literary Digest* in 1918, Victor promised that, "Every home can have the best music—*on the Victrola.*"[25] Linking fidelity claims with exclusive artist contracts, the advertisement invoked the familiar assertion that the greatest artists selected Victor because it "satisfies their high artistic demands." Here, the fidelity of Victor recordings "parallels their actual performances on the opera and concert stages." Although acoustic fidelity is not a concept readily depicted visually, Victor's advertising campaigns often included illustrations of performers who magically appeared within the home. Such metaphoric depictions of fidelity underscored that mechanical reproduction collapsed space and made both contemporary and now-dead artists accessible to all. In "The Victrola Brings the Opera Right into Your Home," Victor promised consumers the greatest contemporary voices presenting the music of "Carmen."[26] A man and a woman dressed in tuxedo and gown sit in rapt attention, their gazes directed toward the Victrola, while floating above the scene like an apparition, ghostly onstage performers entertain an opera house audience. One version of the 1916 campaign "Will There Be a Victrola in Your Home This Christmas?" depicted well-dressed guests assembled in a wealthy home; located under the Christmas tree was the Victrola, and alongside the phonograph Caruso in his Pagliacci costume entertains the domestic scene.[27] Implicit fidelity claims suggested that the phonograph both transcended time and eclipsed space, bringing the greatest voices and performers into every adequately equipped home. Listening to the recorded with the Victrola offered not the audible traces of an absent acoustic event but copresence with the event itself.

Throughout its advertising campaigns, Victor linked mechanical reproduction to a wider, ongoing rhetoric of spiritual uplift through music. Promising "to cheer, refine, educate and uplift," Victor asserted that, "Listening to the Victrola fifteen minutes a day will alter and brighten your whole life."[28] The company actively cultivated a perceived connection between listening to the phonograph and education through a series of book-length publications. Such titles as *The Victor Book of the Opera* (1912), *The Victrola Book of the Opera* (by 1921, in its 6th edition), *The Victrola in Americanization* (1920), and *Music Appreciation for Children* (1923, 1930) constructed listening to the recorded as an edifying experience while simultaneously promoting Victor's product line.[29] Victor's catalog was an alphabetically organized reference work that included biographies of important composers and performers, brief descriptions and production histories of important operas, background information on specific types of music (for example, Gregorian chants), and a pronunciation guide. Over five hundred pages in length and itself the subject of magazine advertisements, the catalog provided an educational supplement to a list of recordings, their numbers and prices.[30]

Even more elaborate as a pedagogical accompaniment to Victor's line of recordings, *The Victor Book of the Opera* included the stories of seventy operas supplemented by descriptions of production histories, duplications of original programs, illustrations of stage sets and performers in costumes, and translations of lyrics. While clearly functioning as an elaborate promotion of the Victor line of recordings (each encyclopedic entry included descriptive and pricing information for the appropriate Victor disks), the lavishly illustrated publication adopted a pedagogical mode of address in keeping with the larger Victor effort to frame listening to the recorded as an edifying experience.[31] The publication and distribution of such reference books added to Victor's carefully cultivated prestige and located the phonograph within a larger, ongoing construction of musical culture in America. For example, a 1914 commentator in the *Musician* noted that the well-appointed middle-class music room must contain "reference books on music and other books of musical character [for they] are not alone decorative, but add a note of refinement and comfort."[32] With the formation of an educational department in 1911, and its subsequent outreach efforts, Victor sought to institutionalize outside the home the phonograph's successful association with musical training and broader cultural values of uplift.[33]

By presenting reproduced sound in terms of universal access to the greatest performers and by linking the phonograph to musical training, Victor invoked a preexisting and ongoing construction of domestic music in American culture. Called "musical democracy" by Craig Roell in his history of the piano, this related set of assertions stressed the morally uplifting qualities of music, the universal need for such uplift, the public service offered by commercial providers of music and musical instruments, and the universal accessibility of such uplift through both musical performance and consumption. Roell notes, "To widen sales and extend the moral and cultural benefits of music, to make music making and listening universally accessible—to establish a *musical democracy*—the industry adopted and helped develop business strategies that reinforced the emerging consumer culture. It achieved mass-marketing success through the sale of the player piano, revolutionary in that this machine allowed anyone to 'play' music skillfully without effort."[34] Although describing the piano industry in the United States, Roell might well be characterizing the phonograph industry in general and the advertising and actions of Victor in particular. The rhetorical framing of phonographic devices as musical instruments involved not so much the discursive *construction* of the machine as a discursive *reinforcement* both resonant with and responsive to other contemporary discourses about music, domesticity, and culture.[35] By invoking larger, ongoing constructions of musical culture, Victor's massive advertising campaigns and Edison's ubiquitous "Tone Tests" classified the ownership and reproduction of recorded sound as a musical activity consistent with bourgeois conceptions of taste and culture. Advertising and promotion labeled the phonograph a musical instrument of the highest order.[36]

These rhetorical efforts of the phonograph industry to construct a domestic identity for phonographic devices influenced, and was in turn further supported by, revisions to the designs of the machines themselves. In 1906, Victor dramatically changed the physical appearance of the home phonograph with the introduction of the Victrola. While prevailing phonographs featured an exposed amplifying horn and turntable, Eldridge Johnson's design for the Victrola folded the reproducing horn on itself and enclosed it—as well as the turntable, sound box, and tone arm—within a cabinet that contained the entire machine.[37] The phonograph became a piece of furniture. According to Victor's vice president Leon Douglass, writing in 1906, the Victrola grew out of a gendered perception of the marketplace, and the innovation's success took the company by surprise:

> I told Mr. Johnson that it was my opinion that ladies did not like the mechanical looking things in their parlors. Mr. Johnson improved on my cabinet and the result was the Victrola, an instrument fully enclosed in a cabinet which was an attractive piece of furniture. I ordered two hundred. Mr. Johnson was afraid we would not be able to sell so many and I was a little timid myself, as they cost so much that we would have to sell them at two hundred dollars each. We not only sold those but many millions more. We were obliged to use seven thousand men to make the cabinets alone.[38]

Besides the impulse to reengineer the phonograph to conform to perceptions of a gendered consumer's preference, the innovation of the folded-horn loudspeaker solved several perceived engineering problems as well as responded to the ongoing rhetorical construction of sound-reproducing devices as musical instruments for the home. Providing greater amplification, arguably improved fidelity, and rudimentary volume control (listeners could modulate volume by opening and closing the doors located in front of the amplifying horn), the folded-horn loudspeaker also made possible the encasement of an amplifying horn adequate in size within a cabinet that conformed to the preexisting design standards established by upper-class homes. A few years later, Victrola cabinets would be modified to suit virtually any preexisting home decoration scheme.[39]

Columbia soon followed Victor with the introduction in June 1907 of a cabinet-model, internal-horn Graphophone (eventually called the "Grafonola").[40] Conservatively designed to resemble a preexisting device, this initial cabinet Graphophone mimicked the physical appearance of an upright piano. The large, cumbersome Graphophone physically reiterated the phonograph's link to the prestige already associated with pianos, but few of the piano-shaped Graphophones sold. Columbia next produced an alternative "library-table style" more consistent with the Victrola's transformation of the phonograph into a piece of furniture. Despite the success of Victrolas and Grafonolas, Edison continued its commitment to a cylinder-based system and to machines with the traditional exposed horn until October 1913, when the company used a full-page advertisement in *Collier's Weekly* to introduce a cabinet machine that played only disks.[41]

By 1918 the phonograph industry had reengineered devices and shaped consumer perceptions so that the phonograph was increasingly accepted as both a musical instrument and a piece of furniture. That year the phonograph became the site of a debate about interior design when an article published in *Country Living* denigrated cabinet styles for clashing with contemporary decorative trends. The current designs of Victrolas and similar machines "did not meet the demand imposed by the modern apartment for space-saving double-duty equipment."[42] This inaugurated what came to be called "the flat-top craze," a desire for phonographs that would blend well with contemporary furniture styles and could function, when not reproducing sound, as a lowboy table. Despite the aesthetic judgments of tastemakers, Victor revised the Victrola not in accordance with the so-called flat-top craze but instead with the innovation of what were derisively referred to as "humpbacks," upright-shaped machines like the Victrola in which the lid was dramatically curved in order to *prevent* the machine from functioning as a table. Victor's employees retrospectively explained this choice as a design decision motivated by the company's view that its machines were musical instruments and not tables.[43] These later Victrolas, with their sharply curved tops, were particularly ill-suited to the "double-duty" of competing "flat-tops." However, by 1924, Victor finally marketed flat-top Victrolas encased in cabinets of varying designs.

While "flat-top" phonographs could conform to perceived middle-class desires for an apparatus that doubled as the ubiquitous hall table, other market segments with more elaborate interior design goals also found lacking the commercially available phonograph cabinets. According to some accounts, New York City interior decorators in the 1910s purchased phonographs for $50 to $70 and encased them in legitimate antique cabinetry or specially designed, faux-antiques, only then to resell the aesthetically modified machines for $1,000 to $2,000.[44] In 1923, Eleanor Hayden, a commentator on interior design, reported similar tales of cabinetry modifications. She noted, however, that whether antiques or copies, the decorative cabinets were designed without concern for acoustic principles and hence these hybrid devices "lost much in tone when they became beautiful pieces of furniture."[45] Whether such innovative, informal cabinet redesign did or did not take place, portions of the commercial phonograph industry in the 1920s addressed a perceived market for highly elaborate cabinetry consistent with the interior designs of the wealthy.

"When music delights the ear, why should a musical instrument offend the eye?" asked Hayden in *International Studio* in 1923. Hayden's article, "Phonographs as Art Furniture," attributed innovations in expensive phonograph cabinet designs to both a pervasiveness of musical appreciation and a growing concern among wealthy and middle-class women with home decoration. Rhetorically, Hayden's article collapsed distinctions between aesthetic appearance and the phonograph's acoustic performance. Claiming that an era was at an end in which cabinets and phonographs worked at cross-purposes (beautiful cabinets that were acoustically inadequate), Hayden suggested that the latest designs could become "a *harmonious* part of the furnishings of the room . . . rather than an *offensive note*," and that "a *mellow, soft toned* walnut chest . . . will *harmoniously fit* into one of the spacious, simple interiors made in the Sixteenth Century Italian manner" (emphasis added).[46] Having been accepted as a musical instrument, the phonograph required simple aesthetic modification to prevent its presence from striking a discordant note in the music rooms of the wealthiest consumers.

The phonographs described by Hayden were all produced by the Vocalion company and featured a wide array of elaborate designs intended for elite, wealthy consumers. Throughout the article, Hayden described both the phonographs themselves and the carefully constructed interior decor for which they were best suited. "So a beautiful, formal drawing room, an entrance hall of Louis XV or an elaborate Georgian effect might be created around the decorative Vocalion cabinet which is finished in Chinese blue-green lacquer with raised figures in silver leaf and placed on a base elaborately hand carved and adorned with glazed English silver leaf and with exquisitely hand chiseled key plates, corners and hinges of silver bronze."[47] The article itself, as well as the cabinet designs which it featured, attest to the degree to which the phonograph industry, through a variety of means, had successfully reengineered machines and influenced consumers' perceptions so that phonographic devices became prestigious products integral to the modern home. The process of selecting an appropriate phonograph cabinet could sublimate the creative act of music making into designing and implementing a mise-en-scène in which the phonograph performed.[48]

Preexisting class perceptions embedded within American consumer society influenced these debates over phonograph cabinetry and the innovations of multiple versions of essentially the same apparatus. By rhetorically aligning sound-reproduction devices with prevailing notions of prestige and status, the phonograph industry addressed multiple potential markets simultaneously even as it reinforced the perceived distinctions between those markets and their consumers. But while financial barriers no doubt made the more elaborate Vocalion models differently available along class lines, the expense of a Red Seal recording apparently did not. A 1910 Edison market research survey of phonograph industry sales anecdotally suggested that consumer purchasing did not precisely correlate with apparent income levels. "Red Seal and operatic records have a tremendous demand among the middle and laboring classes. Italians, Hebrews, French, Germans and Spanish being the biggest buyers. Many seemingly day laborers pay $3.00, $5.00 and even $7.00 for single records."[49] It is in this type of purchase that the success of rhetorically aligning the phonograph with notions of prestige can be measured. Publicly constructing sound reproduction as a prestige product addressed all social classes and influenced the purchasing decisions of consumers from a number of distinct socioeconomic groups.

Innovation of the Victrola and subsequent design developments eased the machine's transition into domestic space, physically resolving concerns that this modern device would disrupt the "harmony" of either the middle-class or wealthy home. Hayden's article points to the ways in which some social groups defined the "problem" of the phonograph's position in the home as requiring the dissimulation of the machine as a machine. Simultaneously eliding the technology itself and shifting the auditor's attention from the machine to that which the machine produced, cabinet-model phonographs rendered technology per se invisible and inaudible at the service of mimetic claims for fidelity to an original acoustic performance. This marked a shift in public encounters with the phonograph. Initial public presentations of the phonograph as a musical instrument (like Lyman Howe's and the Sullivan Brothers'), constructed as local cultural events, encouraged audiences not only to contemplate and to experience the performance of sound but also to marvel at the technological capacity of the device. Later coin-operated versions of the phonograph in arcades and phonograph parlors during the 1890s often included a strategically placed window through which the listener could watch the machinery in action. The phonograph consumer of the 1890s purchased not only sound reproduction but also a technological experience, an encounter with applied science as a kind of spectacle. Initial consumption of the phonograph hinged on both the performance of recordings *and* the apparatus itself. The product being consumed was both the sound and the machinery, the mechanical and acoustic principles that produced the sound.[50] But the innovation of the Victrola and subsequent incorporations of the phonograph into furniture participated in a larger ongoing process that disguised the phonographic apparatus itself, eliding the mechanical functioning of the machine and focusing consumer attention and interest instead on the sounds it produced.[51]

Attempts to erase the mechanical qualities of the phonograph included disciplining the behavior of phonograph salesmen. Emily Thompson reports that with the commercial release of Edison's Diamond Disc technology, Edison forbade dealers from referring to the apparatus as a "machine"; they were instructed instead to refer to it as a "musical instrument."[52] This practice of discursively shaping the public identity of the technology through naming extended to Edison's practice of calling disks not "recordings" but "re-creations." Edison dealers were also instructed to offer "recitals" featuring public performances of disk "re-creations" of Edison's finer artists. Tone Test programs invited the audience to visit not their local phonograph dealer but "our Music Room at any time to hear any selections you may choose."[53] Further, Edison's advertising copy continued this disavowal by sometimes including phrases such as "Mr. Edison's new invention [the Diamond Disc system] is not a talking machine."[54] In a series of national advertisements during the summer of 1921, Edison offered $10,000 in prizes in a contest to develop a new name for the phonograph. Suggesting that the term *phonograph* no longer did justice to his latest advances in fidelity, Edison declared, "I want a phrase, which will emphasize that our new instrument is not a mere machine, but that it is an instrumentality, by which the true beauties and the full benefits of music can be brought into every home."[55]

Although innovations like the Victrola and rhetoric like Edison's sought to dissimulate the apparatus, to elide the technology as a mechanical device, the operation of domestic phonographs reasserted the experience of technology *as technology*. No longer an object invoking wonder nor offering the experience of scientific spectacle, the domestic phonograph reconfigured as a musical instrument became an object out of which the astute home user, the appropriately disciplined, edified, and uplifted consumer, would coax the best sounds. Instead of interacting with another in a long line of technological marvels that characterized modernity, the phonograph listener was increasingly constructed and addressed as a musical performer. Anyone who has operated a phonograph from the period can attest to the fact that "playing" a Victrola, for example, was a labor-intensive activity. Recordings last between three and five minutes, so the listener/performer not only changed disks every few minutes but also made frequent selections from his or her collection of recordings. Further, the machine itself required a good bit of attention and monitoring as the spring-driven motor had to be rewound at least after every two performances and, to ensure maximum fidelity in reproduction, the needle changed after every two or three songs ("jewel" needles freed consumers from the need to frequently change their phonograph's steel needles). Different designs of needles produced different qualities of sound, so the Victrola performer could make aesthetic choices about desired acoustic effects. Should he or she select, for example, a Silvertone release, the Victrola's speed regulator had to be adjusted from 78 to 80 rpm for that specific recording and then readjusted for the next selection. Home audition was not the physically passive activity that radio listening would eventually become some fifteen years later as the design of the machine itself prevented the contemplative, stationary listening of the concert hall.[56]

Besides the design of the apparatus itself, period advertising and the nature of some of the recorded repertoire also constructed home listening as a kind of musical performance. Operating the machine became in some instances a creative, performative act. Edison and Victor framed phonograph entertainment as not only an accompaniment to dancing but also an opportunity for a pseudo-public performance of the recorded to one's assembled guests. As a musical instrument, the phonograph could provide a kind of "musicale," the edifying social event in which amateur musicians entertained a hostess's guests.[57]

In November of 1912, the *Edison Phonograph Monthly* (a newsletter to Edison dealers) announced that all subsequent cylinder releases would be accompanied by a four-page flyer providing information about the recording. The flyers provided details about "the lives of the great masters, their struggles and triumphs, pointing out the particularly interesting passages in a selection and explaining its significance and the thought which it is intended to express." The *EPM* stressed that dealers should use this new written accompaniment to each cylinder as a selling point: "The hostess in entertaining her friends can do more than merely put a Record on the machine with the remark 'Johnnie just loves this one'—she can relate interesting little anecdotes concerning many selections or their authors, increasing immeasurably the interest in the records."[58] Victor similarly designed its reference books, including *The Victor Book of the Opera*, to facilitate such spoken introductions to recordings.

The F. K. Babson catalog frequently provided descriptions of domestic social events in which the Edison phonograph occupied the center of attention:

> I have known some families in Chicago and New York whose regular Edison parties became the fashion. The hostess arranges a program of Edison music to suit a variety of tastes; there's a bit of refreshment, perhaps some dancing, and the whole crowd of citified people, accustomed to every kind of diversion, have enjoyed this so much that another Edison party is arranged for two weeks later at the home of a member of the party. And so on through the whole winter—a round of Edison receptions.[59]

Edison promotions depicted "The dinner complete with an Edison" and included a suggested list of recordings for entertaining. Advertising copy advised worried parents that the phonograph could help their daughters entertain young men. Providing an alternative to a "nickel show," the phonograph could aid those parents who "would probably prefer to have the girl stay at home." It could even compensate for a less-than-popular daughter: "Or, if no young men are in the habit of seeing your daughter frequently . . . give her the phonograph to entertain callers, and she will have plenty of occasion to remain under her own roof!"[60] More directly addressing rural consumers, the Babson catalog also described how an informed use of the phonograph could entertain the farmhands at supper or help to keep the young people down on the farm and thereby resist the lures of the distant city. A list of recordings identified nine Edison cylinders as suitable to "An Evening for the Young Folks."[61]

In a more popular vein, Babson's Edison promotions suggested an appropriate evening's entertainment in the form of a "Minstrel Show at

Home." Providing consumers with a selection of recordings and suggesting that Edison cylinders provided enough entertainment for a show that would last an hour and forty minutes in length, Babson reassured phonograph owners that these recordings provided both "Headliners" from the vaudeville circuit and clean, wholesome entertainment suitable for the entire family and neighborhood.[62]

While the physical design of mechanical phonographs demanded that listeners interact with the machine/musical instrument, promotional rhetoric provided advice and guidance about how consumers could present reproduced sound as a kind of domestic performance. Some innovations suggested that even the solitary phonograph listener became an active performer. Among the more extreme instances of the rhetorical framing of phonograph listening as musical performance was that which surrounded the innovation of the Graduola. This special feature added to the Aeolian-Vocalion line of phonographs amounted to a kind of product differentiation from competing, high-priced machines. The Aeolian Company, widely known as the producer of the most prestigious player-pianos in the United States (e.g., the Pianola), first released a phonograph line in the winter of 1914. Advertisements in the fall and the winter of 1915 heralded the Graduola as "that exclusive tone-controlling feature of the Aeolian-Vocalion. You will hear the music melt to your touch, then grow to strength again, graduated delicately by ever changing pressure of the hand—without the slightest muffling or dulling of its qualities!"[63]

The Graduola involved a relatively simple method of changing the output of the acoustic-reproducing horn by manipulating a shutter-like aperture at the base of the horn. A five-foot retractable cable enabled listeners/performers to sit in their armchairs and adjust the tone and volume produced by the machine. The Graduola responded to a perceived market desire for volume control (volume could be marginally adjusted on Victrolas and similar home machines by opening and closing the cabinet doors in front of the acoustic horn). But the feature offered something more than mere volume manipulation. Aeolian-Vocalion's advertisements rhetorically framed the device as providing the opportunity for *participation* in the construction of a home musical performance. The Graduola "enables one to take a personal part in their [records'] rendition." This emphasis on musical performance coincided with the larger tenor of Aeolian-Vocalion's advertisements in that, even more so than Victor, it constructed its product line as both musical instruments and furniture suitable for contemporary decor. Aeolian-Vocalion presented its machines as something beyond the normal phonograph—an elite product. Proclaiming the Aeolian-Vocalion as "virtually a new type of musical instrument" and as "the phonograph that makes YOU the artist," advertisements promised the consumer that with the Graduola, you "become your own conductor." In the words of the ideal consumer/performer described in the advertisements: "And so I found the new phonograph that gave to me at last the means to Voice the Latent Music-Instinct of My Soul."[64]

The Graduola advertisement, "In the Firelight Glow! An Evening Spent with My Aeolian-Vocalion," overtly constructed listening to the phonograph as a musical performance, albeit one with barely veiled onanistic overtones:

I heard every instrument swelling from the bosom of a wonderful harmony, and yet so separately beautiful that each separate virgin freshness was preserved in all its ravishment. . . . So I took the Graduola device in my hand—it is part of the Vocalion. I pressed it. Softly, beautifully—their strength proportioned to my most delicate pressure—those velvet tones ever so gently melted away. I swelled them forth again. I pressed them down to a very whisper of limpid beauty—for this remarkable device never muffles the Vocalion's unduplicatable tones—simply controlling them at their source.[65]

Throughout the first decades of the twentieth century, the phonograph industry redefined the device from an apparatus suited for business applications and public performances to a musical instrument for the home. Phonograph companies pursued a number of overlapping initiatives in redefining sound reproduction as a domestic activity. In promoting the fidelity of recorded sound by rhetorically seeking to elide the difference between the recorded and original performances, the phonograph industry sought to redirect auditors' attention away from the phonograph as a machine and toward the sounds it produced. Aligning the apparatus with existing definitions of domestic entertainment and elite musical culture (in effect mapping the phonograph onto preexisting discourses of musical democracy, uplift, and edification) made ownership desirable or even a necessity. Phonographs themselves were redesigned so as to disguise their components and mechanical functions. This dissimulation of the machine implicitly supported fidelity claims while also making phonographs consistent with preexisting (and changing) standards of interior design. By the mid-1920s, this rhetorical framing and physical redesign of the phonograph successfully redefined musical production in the home to include the *reproduction* of the recorded.

But at the same time, aspects of the phonograph's design—its requirement that consumers actively engage with the apparatus—combined with a different promotional rhetoric to draw attention back to the device itself. While the introduction of the Graduola most dramatically illustrates this tendency, Edison's insistence that salesmen refer to his device as a "musical instrument" for sale in local "music rooms" and period debates about the status of the phonograph as a musical instrument suggested that playing the recorded was an active musical performance. When Victor and Edison encouraged consumers to consider how they presented recordings to domestic audiences (and when they provided information and advice on how to accomplish this), they further emphasized the performative activity of operating a phonograph.

Immediately before the conversion to electrical methods, the design of the mechanical phonograph and consumer interactions with it existed in a conceptual and physical space characterized by two countervailing tendencies: on the one hand, discourses surrounding phonography and qualities of the machines themselves dissimulated the machinery of sound reproduction, its physical embodiment of mechanical and acoustic principles; but on the other hand, the materiality of the technology returned through consumer performances of the recorded repertoire. With the advent of electrical acoustics—specifically the introduction of domestic radio receivers—the radio industry inherited both of these tendencies. Radio's progress throughout the mid- to late 1920s involved multiple attempts to heal the apparent gap between efforts to dissimulate the machinery of sound reproduction and countervailing perspectives that viewed sound reproduction as an active performance.

Radio Enters the Home

Throughout late 1921 and into the mid-1920s, the radio industry identified essentially two distinct markets for radio technology: those seeking the component parts to construct their own wireless receivers and those seeking preassembled radio sets. The first group considered radio to be a scientific activity, a hobby of sorts through which they directly applied electrical principles. This definition of radio as an activity hailed radio enthusiasts (predominantly male) knowledgeable in the application of scientific principles and encouraged their engagement with technology as technology. The second consumer group experienced radio as a domestic entertainment device, analogous to, and in significant ways an improvement upon, the already established phonograph. This second identity for radio de-emphasized the technological qualities of radio receivers and stressed instead through this dissimulation the *service* that radio could provide. Rather than an opportunity to learn about and directly apply scientific principles, listening to the radio provided domestic entertainment.

Initially, RCA provided component parts to home engineers seeking to build (and then rebuild) their own receivers, and by mid-1921 the company distributed a series of catalogs and booklets promoting this activity.[66] Aided by a proliferation of magazines addressing amateur radio enthusiasts and newspaper coverage (including weekly radio pages or entire radio sections that often included detailed instructions for assembling receivers), RCA and other manufacturers stimulated a preexisting enthusiasm for home radio engineering that quickly outstripped the productive capacity of radio manufacturers. By 1922, RCA ran newspaper advertisements acknowledging its inability to produce enough vacuum tubes to satisfy consumer demands and promising that accelerating the productive capacity of factories would soon allow it to better serve the public.[67] Home engineers interacted with radio technology neither as an instrument of wonder nor as a domestic object providing a programming service, but instead as precisely a technological artifact the principles of which were well known and the subject of lively competition and debate. In addition to constructing functioning radio receivers, home engineers also often enjoyed early radio through so-called DX listening. A listener would carefully tune, retune, and adjust his home-built set so as to identify (or "log") the broadcasts of distant stations. Contests sponsored by radio periodicals involving aggregate logged distance, or furthest reception by home-built sets (sometimes with technological limitations such as the use of a single tube), actively promoted this alternative identity for radio.[68] But also in 1922, RCA attempted to expand this market for radio equipment and components to include the less technologically inclined and numerically far larger group of consumers who might purchase preassembled radio receivers. With a national advertising campaign inaugurated in October 1922 through advertisements in the *Saturday Evening Post*, RCA aggressively addressed this potentially larger group of consumers.

The difference between these two distinct relationships between consumers and radio was particularly vivid in the radio press. During the mid-1920s, a journal like *Radio Broadcast* uneasily synthesized editorial content that varied from detailed articles on the construction of radio sets and advice columns on antenna installation to barely disguised promotional pieces touting the emancipatory and uplifting possibilities of consuming the finest radio entertainment. The radio industry's shift in emphasis between these two identities was visually enacted on the covers of the journal *Wireless Age*. Initially depicting the technology of radio (vacuum tubes, enormous antennas, internal views of radio sets) and visually addressing home engineers, *Wireless Age* soon switched to cover illustrations exclusively depicting radio listening in a variety of domestic scenes.[69]

RCA's 1922 publication, *Radio Enters the Home*, illustrates the methods through which members of the wireless patent pool constituted consumer markets by simultaneously promoting multiple identities for radio. The RCA booklet openly acknowledged two distinct types of radio consumers. Prefatory remarks addressed "the amateur who, after studying [Morse] code, receives his first intelligible message in telegraphic code, thrills to a sense of scientific or mechanical proficiency," and noted that "technical portions of this book and catalogue are for the amateur." But RCA's promotional rhetoric primarily addressed "the family group, unconcerned and uninterested in code or technical study, [that] thrills in its turn to an entirely different emotion—an emotion responsive to the dramatic realization that it is now possible to spend an evening in your own home with great artists entertaining you 'in person'; with news of the day coming to you before it is printed in a newspaper; with statesmen and lecturers replacing each other in the program designed to please you."[70] For home engineers, the publication illustrated and explained the principles of wireless telegraphy and telephony and included a glossary of radio terms, charts depicting international Morse code and common wireless abbreviations, and even a blank form on which a DX listener might log the distant stations he heard. Promotional material depicted the wireless service provided by RCA through marine radio equipment, transoceanic telegraphy, and Radiogram communication with outbound ships. Even though the publication's title announced an emerging domestic broadcasting identity for radio, that identity coexisted with definitions of the medium as a point-to-point service provided by a large corporation and as a point-to-point mode of communication between properly trained and equipped ordinary citizens. RCA's description of the wireless applications of radio was, however, marginalized at the back of the publication.

Instead, RCA made prominent the emerging domestic identity for radio announced in the publication's title. The cover of *Radio Enters the Home* repeated prevailing visual strategies of phonograph advertisements by illustrating a domestic scene in which well-dressed adults congregated around an acoustic device integrated into the well-appointed home. While the publication clearly indicated that RCA had identified a domestic market for electrical acoustics, the products illustrated, described, and promoted in subsequent pages (a series of preassembled crystal detectors and vacuum tube regenerative receivers as well as various condensers, vacuum tubes, antenna equipment, for receiving and *transmitting* radio signals) nonetheless reveal an industry still struggling to accomplish the additional research and development necessary to service those consumers potentially daunted by electrical technology.

The Aeriola Grand, RCA's initial prestige-model domestic radio receiver, attempted to address the "family group" depicted on the publication's cover by directly mimicking prevailing models of phonographs. (See fig. 3.3.) While the name Aeriola (as well as RCA's subsequent Radiola models) clearly echoed the already widely successful Victrola (or, for that matter, Columbia's Grafonola phonograph), the physical design of these radio receivers virtually repeated successful models of domestic phonographs. When mounted atop an RCA-marketed cabinet, the domestic radio receiver conformed to the design standards for upright phonograph cabinets of the period. A hinged lid opened to reveal not the phonograph's turntable, sound box, and tone arm, but the Aeriola Grand's power switch, tuning mechanisms, and amplifying and detector tubes. Just like contemporary models of phonographs, reproduced sound emanated from a grille at the front and top of the apparatus, and beneath this grille two cabinet doors opened to reveal not the phonograph enthusiast's record collection, but the batteries that powered the radio receiver. In developing the physi-

RADIO CORPORATION OF AMERICA

THE AERIOLA GRAND

Simplicity of manipulation
and elegant appearance
make the Aeriola
Grand a popular
receiver

An instrument which may
well be accorded a place
in the most discrimi-
nately equipped
apartment

29

FIGURE 3.3 In designing the Aeriola Grand radio receiver, RCA mimicked the physical appearance of the already established domestic phonograph. (Author's collection)

cal appearance of the Aeriola Grand, RCA's designers conceptually gutted a cabinet-model phonograph, maintaining its external appearance, and installed the apparatus necessary to reception and reproduction of radio signals. If, as Victor's vice president Leon Douglass remembered, Eldridge Johnson's initial design for the Victrola arose out of a desire to create a phonograph that appealed to "the ladies," then RCA's Aeriola Grand design should be understood as not only an attempt to follow conservatively a previous acoustic device's successful transformation into a domestic artifact but also as a literalization of RCA's gendered conception of the domestic radio broadcasting marketplace.

RCA used more than cabinet design to link radio to the phonograph. It also addressed a mass market of consumers generally unfamiliar or unconcerned with applied electrical principles through rhetorical constructions of listening in the home. In describing the benefits of radio listening, RCA's *Radio Enters the Home* directly invoked much of the rhetoric used to establish the phonograph's domestic identity. Calling radio a "new, inspiring and powerful resource at the disposal of civilization," RCA credited broadcasting with "carrying into the homes of rich and poor alike a modern facility of pleasure and education which is binding the people together in a new and democratic brotherhood."[71] Like the phonograph before it, radio could exercise a democratizing influence that linked entertainment with edification. Throughout the publication, radio broadcasts were consistently referred to as "concerts," thereby invoking a rhetoric of cultivated uplift elsewhere characterized as "a richer and more complete home-life, with mental stimulus and pleasant relaxation, [which] has been made possible through broadcasting and its receiving corollary, the radio telephone receiver."[72] Like the Edison Company's insistence that its Tone Tests were "concerts," radio innovators painted listening to electrically reproduced sounds with the patina of high culture.

Stressing the radio's potential to support a "musical democracy" and the device's ability to eclipse the distance between entertainment centers and the geographic margins of consumption, domestic radio innovators also emphasized the device's ability to make the middle-class parlor a "safe" site for consumption of the best of urban culture. RCA's publication paired "mental stimulus" with "pleasant relaxation," and it thus redefined radio technology, in part, by locating the device within an ongoing set of traditional values. Edison's advertisements of 1918 in the F. K. Babson catalog had promised that the phonograph would successfully integrate modernity within conservative notions of domesticity. Children could be kept in the home and farmhands appropriately entertained through the phonograph's ability to bring the most wholesome cultural products of the modern city to everyone. Safely consumed within a domestic space that allowed parental oversight or insulation from perceived urban dangers, the radio, like the phonograph before it, could provide not simply the convenience of domestic entertainment, a convenient collapsing of spatial distance between cultural production and consumption, but also an acoustic experience more in keeping with preexisting constructions of domestic space as a safe haven from the modern world. Radio listening represented another step toward a long-sought-after synthesis of the greatest achievements of modernity and traditional domestic values.[73]

Initial attempts to market radio receivers as domestic entertainment devices went even further in evoking utopian promises. Broadcasting innovators also emphasized the technology's ability to accomplish that which the phonograph had tried and failed to achieve. Radio could deliver on the phonograph's unfulfilled research agenda of increasing consumer control of the sounds produced (both volume and tone control) and providing "more" sound (not only more volume but also an expanded frequency range, especially more bass tones).[74] Radio could also provide greater *immediacy* in bringing new entertainment trends to spatially dispersed consumers since broadcasting provided instantaneous access to the latest cultural productions. Phonograph companies had promised consumers rapid access to emerging entertainment trends by suggesting that recorded sound could bridge the distance between geographically dispersed listeners and entertainment centers. A 1922 advertisement for Edison recordings addressed a perceived consumer desire for rapid access to the latest in entertainment, promising that Edison was *"FIRST* with Broadway Hits."[75] The advertisement chronicled the debut of "April Showers" ("the fox-trot sensation of the season") at New York's Club de Vingt and, through the auspices of Edison Re-creations and airplane mail service from Long Island to Chicago, the song's introduction to "Chicago's smart set" only five days later at the Edgewater Beach Hotel. While the phonograph could bridge such distance, domestic radio promised to eclipse it through the immediacy of live broadcasting. By stressing radio's arguable improvements over the domestic phonograph, RCA and others foregrounded some of the sufficient differentiation necessary to inspire consumers to add another acoustic technology to their homes.

Both the naming and the initial design of RCA's Aeriola Grand illustrate that radio innovators looked to the phonograph when conceptualizing broadcasting as a domestic entertainment. Such conservatism also characterized rhetorical efforts to constitute a market of home listeners uninterested in radio's competing identity as a hobby or applied science. But these efforts did not immediately and unproblematically redefine radio's identity to include domestic applications. While the Aeriola Grand as an artifact outwardly invited its incorporation into the bourgeois home through a cabinet design that mimicked the Victrola, ultimately the device still required frequent and direct consumer engagement with the material trappings of applied science. The detector and amplifying tubes hidden beneath the lid of the Aeriola Grand required attention and periodic replacement. Listening to the early 1920s domestic radio receivers involved actively operating these devices, not only installing and replacing tubes and batteries but also frequently tuning and retuning the circuits to find and maintain a connection with a broadcast signal. While initial domestic radio receivers like the Aeriola Grand could be tuned with a single control, later innovations that provided greater sensitivity and access to distant stations required the simultaneous adjustment of multiple controls.[76] Further research and development was necessary before the emerging radio industry could successfully constitute and cultivate domestic listening as a significant market segment.

Innovations like the regenerative circuit, the neutrodyne circuit, and superheterodyne receiving circuits improved the volume, selectivity, fidelity, and consistency of domestic radio reception between 1922 and 1925. Increasing signal amplification rapidly freed listeners from the telephone headsets characteristic of several early inexpensive receivers marketed to domestic listeners. Developments in sound-reproduction technology replaced headsets with "gooseneck" horns that were themselves displaced by cone-type loudspeakers and, after 1925, by the electrodynamic loudspeaker. These innovations allowed listening to quickly become more fully a group activity, further displacing the home engineer with the family as the primary addressee of radio signals. The elimination of batteries as a power source through the introduction of receiving and amplifying circuits that could operate on direct and/or alternating current simplified the operation of radio sets and eliminated the potential for batteries to leak and damage consumers' other furnishings. New antenna designs and a related increase in receiver sensitivity allowed greater ease in the installation of radio sets as well as a less intrusive presence for radio in the home. Perhaps most important for universalizing broadcasting's appeal, the innovation of single-control tuning substantially decreased the amounts of patience, skill, and perceived technological know-how necessary to operate successfully a domestic radio receiver.[77]

These shifts in the design of radio receivers did not immediately become industry standards. By as late as November 1927, when *Scientific American* reported on the autumn styles announced at radio industry trade shows, reporter Orrin Dunlap, Jr., described a radio manufacturing and marketing industry in a state of flux.[78] While receivers that could operate on household current were displacing battery-operated sets, fully 70 percent of the new models on display at the Madison Square Garden radio show still relied on battery power. Single-control tuning, introduced the previous year, was becoming increasingly available on newer models, but had not become an industry standard. Cone loudspeakers were displacing acoustic horns ("They [horns] are vanishing as did the horn of the early phonographs"), yet fully 60 percent of available console-model radio receivers still contained built-in folded acoustic horns. Receiver designs increasingly concealed antennas, responding to perceived market desires for an apparatus that dissimulated its technological components. In Dunlap's words, "There seems to be a tendency this season to get away from the external loop [antenna]. Wherever loops are used they are usually advertised as 'concealed,' because it is said that housewives nine times out of ten vote against a receiver equipped with a visible loop."[79]

Dunlap's reference to commonsense wisdom about housewives illustrates that creating a domestic identity for radio broadcasting required more than simplifying radio receiver operation and improving performance. The radio, like the phonograph before it, had to be successfully integrated into prevailing standards of domestic design. While the Aeriola Grand physically embodies the extent to which early broadcasting innovators looked to the phonograph to design a device perceived as aesthetically suitable for domestic spaces, simply mimicking the appearance of phonographs did not ultimately prove to be a successful method of introducing this new acoustic technology into American homes. The "problem" posed by the aesthetic appearance of radio receivers, like that of the phonograph before it, became a site of contention in domestic design magazines. In 1924 one commentator declared about radio receivers, "but until very recently one had to look at a curious sort of box that was plainly an interesting mechanism, but equally plainly a thing of very little, if any, beauty, and certainly not a fine piece of cabinetwork."[80] Radio's success as a home entertainment device hinged on a dissimulation of the apparatus.

Radio innovators made some initial attempts like those surrounding both the phonograph and the telephone to disguise the apparatus in a variety of other objects. Radio speakers were disguised as globes, as books and bookends, and as other domestic objects. Ultimately though, the radio industry addressed these design concerns in two ways: first, through the miniaturization of radio receivers, and second, through the innovation of combined radio-phonograph devices. By making sound technology smaller, less obvious, less "intrusive" in the terms of the day, table-top radios could facilitate their incorporation within domestic spaces, and they had the added benefit of lower prices.[81] Atwater Kent's 1925 advertising campaign, which depicted radio receivers in the homes of prominent Americans, not only functioned as a celebrity endorsement for the radio manufacturer's receivers but also illustrated that newer, smaller radio sets could be tastefully incorporated into homes of varying domestic designs. Depicting Atwater Kent radios in the homes of, for example, Irwin Cobb, George Ade, and Booth Tarkington, the campaign suggested that modern acoustic technology could accord with spaces as diversely designed as a bungalow-style home or an Art Deco–influenced urban apartment.[82]

Long after radio's redefinition as primarily a domestic entertainment device, integrating the technology into the home continued to be perceived as a problem. In 1929 a different Atwater Kent campaign stressed a variety of radio cabinets to suit any decor.[83] As late as 1932, some commentators on domestic design still viewed the radio set as an intrusion into the well-designed home. Norma Kastl in *House Beautiful* framed the issue in decidedly gendered terms, attesting to the manner in which the initial gendered conceptions of radio's identities clung to the device, even amid its successful introduction into American homes:

> Men think of the radio in terms of reception, tubes, aerials, and the programmes that come in. But women, however else they may regard it, think of the radio as furniture—a piece of furniture for which a place must be found and which must be related to the other furniture and accessories in the room. And, alas, what a piece of furniture the average radio has been to fit into the room ensemble! The men who made those large, brown, awkward cabinets in which the majority of radio sets are housed had no thought of designing graceful furniture, but only of encasing the intricate mechanism within.[84]

Like accounts of consumers dissatisfied with the phonograph as an earlier generation of domestic device, Kastl invoked tales of women who innovated their own solutions to radio's aesthetic problem by incorporating the technology within existing antique furniture. Her article lauded large retailers (Macy's and Gimbel Bros.) for introducing more appropriately designed radio cabinetry. Radio receivers disguised within a variety of secretary-desks, lowboy tables, and end tables suggest that, by 1932, domestic designers were beginning to find ways to incorporate a properly dissimulated acoustic apparatus into any home or apartment.

While miniaturization and the introduction of table-top radios provided one method of incorporating acoustic innovation into domestic space, combining radios with phonographs also facilitated the conversion of American homes to electrical acoustics. Some manufacturers actually combined radio receivers with *mechanical* phonographs before the phonograph industry's adoption of electrical recording and reproduction. RCA offered consumers in 1922 an attachment that allowed radio signals to be acoustically amplified and reproduced through their existing phonographs. But in October of 1925, RCA announced that two of the dominant organizations in the phonograph industry, Brunswick and Victor, had selected the RCA Radiola for inclusion in their phonographs.[85] With the phonograph industry's adoption of electrical methods, the manufacture of combined radio-phonographs proliferated. Combination machines eliminated the redundancy of two different acoustic devices. Rather than a competition between two sound technologies for a place in America's parlors and living rooms, a single composite machine allowed consumers both the immediacy of radio and the illusion of choice and control offered by phonography. Period announcements of contractual arrangements between radio and phonograph manufacturers frequently addressed perceptions of technological redundancy. J. G. Paine, counsel for Victor, upon announcing the contract with RCA for radio receivers, stressed that "when one hears a remarkably fine performance through the phonograph he may repeat it as often as he pleases—the same artist, the same song, the same quality. One may hear the most excellent performances on the radio but he may never hear the same program again by the same artists under the same conditions. That is the part played by the talking machine. On the other hand a great speech, sermon or football game have their highest interest at the moment of their giving. That is the part radio plays."[86] One can read in Paine's announcement an attempt to police the identities of each application of electrical acoustics, to establish their differences, and to emphasize the unique qualities of each. Where the phonograph offered the listener an opportunity to control the selection and indeed repetition of a performance, radio offered the live unfolding of ongoing events. Eventually, the physical design of combined radio-phonograph machines echoed economic relationships between manufacturers, when in 1929 the most important American radio concern, RCA, purchased a controlling interest in Victor, then the premier U.S. phonograph organization.[87]

A redefinition of radio receivers as technological artifacts followed and reinforced radio's emerging identity as a medium of point-to-mass broadcasting. Both in terms of promotional rhetoric and the physical design of wireless receivers as artifacts, the radio conservatively mimicked the efforts of phonograph innovators who had introduced sound technology into domestic spaces. Invoking notions of "musical democracy" and the eclipse of spatial distance used to successfully redefine the phonograph's identity, RCA and other radio broadcasting innovators redefined radio as a technology, and radio listening as a behavior, to expand the potential market for electrical acoustics. This domestication of radio as a technology involved a significant rhetorical effort to reengineer applied electrical-acoustic devices in terms of prevailing notions of gender. The radio industry sought to shift popular conceptions of radio listening as an activity from that which was appropriate to a predominantly male-identified world of home engineering to, in period terms, a feminized domestic identity that dissimulated technological processes and emphasized the experience of radio as a *service*.[88] This redefinition changed consumers' interactions with wireless artifacts from the direct engagement with applied scientific principles to a more passive listening. As radio entered American homes, rhetorical engineering of consumer behavior interacted with the material reengineering of technological artifacts in an effort to shape a specific identity for radio broadcasting that itself was reinforced in the realms of broadcast programming content, economic agreements and, ultimately, public policy regulation.

While radio manufacturers, domestic designers, and consumers struggled to incorporate the radio into the American home, the U.S. film industry simultaneously struggled to successfully incorporate electrical sound technology into movie theater projection booths. Like the properly interpellated phonograph listener operating the Graduola or the home radio consumer struggling to tune in a distant broadcast, the movie theater projectionist was an active performer of sound reproduction. But the performance of reproduced sound in movie theaters was destined to follow the home engineer and DX listener in that performing projectionists were soon eclipsed by an emergent mechanization of film sound projection.

Movie Projectionists as Performers

Film studios and theater owners had clashed throughout the silent film era over the speed at which local exhibitors and projectionists presented films.[89] While some film producers and cinematographers sought to standardize camera speeds (an admittedly difficult task given the use of hand-cranked cameras), projectionists, often at the prompting of theater managers who sought to fit an overly long program of films into a predetermined schedule of starting times, operated film projectors at a much faster rate. Some period accounts suggest that while camera speeds were "standardized" around 60 feet per minute, projection speeds varied between 45 and 70 feet per minute.[90] This practice of "over-cranking" or "film racing" could have unintentionally comic results. According to one period account, "In the *Passion Play* [the projectionist] can make Peter act the part of a jumping jack and he can turn a horse race into a howling farce, by over-speeding and under-speeding. . . . Imagine the figure of the Savior carrying the cross at a gallop."[91] While it is doubtful that any projectionist who transformed Peter into a jumping jack or theater manager who insisted that Christ gallop with the cross to Calvary would have maintained his employment for long, disparities between the speeds at which films were exposed and were projected attest to an essential technological instability of the silent film text and to the absence of standardized exhibition practices.

For some time, revisionist film histories have begun to address this instability of the cinematic text, exploring a series of local exhibition and film accompaniment patterns.[92] Tim Anderson, for example, examines the manner in which the film industry identified as a problem sound accompaniment for nickelodeon-era silent films and sought to move *toward* the standardization of the performative aspects of film exhibition. Examining the trade journal *Exhibitors Herald-World*, Anderson tracks attempts to discipline musical and sound-effects practices in nickelodeons. Local exhibition practices threatened more than a film's status as a stable, nationally distributed commodity, however, and Anderson notes that "the popularity and prevalence of inappropriate, 'jackass' music among some audiences existed because it yielded other sets of semantic pleasures that were *not* sanctioned by the dominant narrative."[93] Nickelodeon music reformers addressed both a perceived threat to the admittedly precarious social standing of the cinema and concerns that some performance practices undermined the ability of film texts to circumscribe textual meaning and ways of viewing.

Hollywood's conversion to synchronous sound held the *promise* of eliminating locally divergent, often idiosyncratic film exhibition practices by standardizing the acoustic accompaniment of films and by "synchronizing" the rate of film projection with that of film exposure. However, in the short term at least, the conversion to synchronous sound obstructed the industry's drive toward standardization and exacerbated the possibility of local variation in motion picture presentation. The conversion to synchronous sound transformed the motion picture projectionist into a *performer* of sound, displacing live acoustic performance from the orchestra pit to the projection booth. For a brief period in the history of American cinema, the projectionist became an active and audible performer in motion picture theaters across the country.

> He approaches his task, not from the standpoint of a worker who is to receive a monetary consideration in the form of wages for a given number of hours of service, but rather from the standpoint of an artist, mechanically etching upon the silver screen a series of beautiful photographic images that are unfolding to his movie audience a visual impression of a beautiful story told with the aid of his mechanical pen. (William F. Canavan, November 1929)[94]

Such glowing descriptions of cinema artisans were typically reserved for the film director artfully orchestrating the combined talents of cast and crew, or less often for the cinematographer who, "painting with light," skillfully used the motion picture camera as his "mechanical pen." Surprisingly though, the film artist described above was in fact the movie projectionist, anonymously practicing his craft out of public sight in the projection booth. William F. Canavan, at the time president of the International Alliance of Theatrical and Stage Employees and Moving Picture Operators (IATSE), sought in November 1929 to reinforce an increased importance for projectionists within the U.S. film industry. In the midst of the industry-wide conversion to synchronous sound, Canavan glowingly emphasized the projectionist's daily contribution to the cinema. Canavan labeled the projectionist both an artist and a showman, claiming that "real showmanship is one of the most essential qualities for the real projectionist. He must be show-minded in all that the term implies, with a background of theatrical experience which will imbue him with that inherent theatrical spirit,—'The Show Must Go On,' no matter what may happen."[95] Not surprisingly, the president of the projectionist's union also emphasized the professionalism of the membership he represented. Canavan described an army of under-appreciated projectionists actively preparing themselves for the conversion to sound through "intensive training and study" despite inconsistent technical support from equipment manufacturers. IATSE's members were embracing technological change despite the inherent inadequacies of both laboratory-developed equipment and "far from perfect" installations of that equipment in movie theaters. In Canavan's words, "the projectionist is more of an idealist than a working man. He looks upon motion picture projection as a 'Specialized Art' and is ever striving to improve the quality of screen entertainment even though it entails a personal sacrifice."[96]

Canavan's depiction of the motion picture projectionist as a skilled artisan might be readily dismissed as an example of relatively transparent self-interest and self-promotion: the president of a powerful union celebrating that union's membership. Yet contemporary labeling of projectionists as skilled professional artisans was far more widespread than can be explained away as simply self-aggrandizing rhetoric. Instead, the period surrounding Hollywood's conversion to sound exhibited an industrywide concern with both the labor and the identity of the motion picture projectionist.

While sync-sound technology promised increased standardization in motion picture presentation, at least during the period in which the industry adopted the new sound technology, such standardization of reproduction could only be approximated rather than achieved, and even then approached only through careful supervision, management, and control of the labor of motion picture projectionists. In the gap between the film industry's desire for standardization and the realities of both sound-recording and sound-reproducing technology, the identity of the projectionist and the nature of his labor became a site of rhetorical struggle. Such a rhetorical struggle involved attempts both to shape the identity of the emerging technology and to discipline the industry employees operating it. Technological change within U.S. film, like any other technology-intensive industry, required not only shifts in "manufacturing" processes but also the negotiated adaptation of the workforce to new conditions of labor.[97] During the conversion of movie theaters to recorded film sound, not only did local musicians lose their employment, but also film projectionists suddenly took on increased responsibilities for the successful presentation of films. For a time, at least, motion picture projectionists became more than operators of machinery, and a discourse of professionalism and responsibility sought to encourage these historically elided laborers to inhabit a new workplace subjectivity.

Before the full-scale diffusion of sync-sound technology, motion picture projection practices were defined as a problem because of a perceived lack of performance standards, an apparent absence of an implicit code of conduct for projectionists and consequent fears of irresponsible behavior, and a resulting pattern of inadequate film presentations to the public. Frequently citing the lack of standardization in film presentation, representatives of film production companies and advocates of "better" film exhibition called for an industrywide concern with the labor of projectionists. Among the most active advocates for projection reform, F. H. Richardson, technical editor of the trade publication *Moving Picture World*, argued in a series of public presentations and trade press articles for careful attention to the often overlooked work of the projectionist.[98] Citing such improper practices as inadequate or careless maintenance of projectors, damage to release prints through both routine carelessness and excessive lubrication, inattention to the dimensions of the projected image and the effects of auditorium lighting upon the screen image, and the waste of electricity through inefficient operation, Richardson cautioned the industry about the need to enforce standards for public presentations of films. Calls for reform by Richardson and others invoked both economic and artistic damage done to the industry through inattention to the role of the projectionist in presenting films to the public. Period accounts frequently cited not only under-motivated and unskilled motion picture operators but also a resulting damage to films, to projection equipment and, most importantly, to the box office revenues of theaters.

Proposed solutions to the "problem" of projection hinged on fostering a sense of professionalism among these anonymous theater employees. In addition to providing consistent and well-informed oversight of their labor (through, for example, establishing a "Supervisor of Projection" within theater chains), industry voices suggested the formation of projectionist societies or educational committees by union locals. But the most frequently cited solution for projection reform involved cultivating a new identity for the projectionist. Before the conversion to sound, some argued loudly for a reconceptualization of the projectionist from a "mere" operator of machinery to an active participant in the production of entertainment. "If we wish to secure the best possible results, we must raise the standard of projection. To raise the standard of projection, we must raise the standing of the projectionist."[99] The quality of projection could be improved and the

presentation of films standardized, some claimed, by inculcating projectionists with professional attitudes about their labor.

For Richardson, the projection problem extended beyond individual
projectionists to the larger attitudes of the industry as a whole. Noting
"the almost total lack of any sort of publicly expressed appreciation of the
work of the projectionist" as well as the tendency to label them "operators,"
Richardson argued that the attitude throughout the industry was "acting
to stifle all pride in work and incentives to excel in it."[100] Richardson even
went so far as to advocate a publicity campaign to educate the public about
the importance of film projection—a campaign that would include movie
stars' testimonials to the important work being done by projectionists.[101]

The film industry's adoption of synchronous-sound technology exacerbated these preexisting concerns about projection and rendered more urgent the accompanying rhetorical appeals. The introduction of new technology into the projection booth caused such urgency precisely because
the goal of standardizing the presentation of motion pictures became even
more elusive as the tasks of projectionists underwent dramatic change.
With the new technology came new responsibilities for the motion picture
operator and, implicitly, new power within the industry.[102] Many of these
new procedures remained essentially inaudible and largely dissimulated,
unless something went awry. Besides routine maintenance and trouble-
shooting of the new acoustic equipment, projectionists checked and successfully cued soundtracks recorded on disks as well as prepared to im-
provise should the sound-on-disk system fall out of synchronization with
the projected images. For the sound-on-film system, operators exercised
increased care to keep the film itself clean as any accumulated dirt or oil
could clog the delicate reproducing slits necessary for converting the encoded soundtrack into electrical signals. Projecting sync-sound films sometimes involved installing a new aperture mask when showing films with
the Movietone sound-on-film system. The location of the soundtrack alongside the image required a mask to prevent the on-screen projection of the
photographically encoded soundtrack. The soundtrack's presence on the
celluloid consequently changed the aspect ratio of the film. In addition to
changing this aperture mask, projectionists also often changed lenses to
restore the on-screen image to its "proper" size.[103] Among the most difficult tasks initially involved successfully navigating changeovers from one
projector to the next during reel changes. Until cues for such changeovers

were standardized on release prints, projectionists developed a variety of individual strategies to guarantee smooth shifts from one reel to the next. These strategies extended from careful rehearsals to even punching holes in release prints.[104]

Period accounts of these new responsibilities often invoked the financial interest of theaters as well as the industry as a whole. Noting that poor or inconsistent projection detracted from the labor of all who worked on a film, calls for a new attitude of professionalism addressed not only the projectionists themselves but also, importantly, theater managers and owners. Warren Nolan, a publicist for United Artists Corporation, emphasized the urgency of improved projection standards in light of technological change: "The gentlemen in charge of projection of sound pictures,—to say nothing of those who record them,—have been constituted individual showmen by this new form of presentation. They have been given power, and they have been given responsibility. Timing and tone are in their hands, and artists on the screen and audiences in the seats are at their mercy."[105] The conversion to sound was held up as not only an opportunity to improve the knowledge and skill of the projectionist, and thereby institute improved projection practice, but also as an opportunity for the industry as a whole to reexamine the importance of improved standards of showmanship.

Ongoing economic consolidation and an increased emphasis on standardizing the product offered to consumers formed part of the industrial context within which Hollywood converted to sound. Technological change promised, on the surface, both to cut theater owners' operating expenses and to standardize the presentation of films. Writing in 1929, Harold Franklin, president of Fox West Coast Theatres, noted the musical benefits of sync-sound technology: "Music, as synchronized, is in closer unity with the situations pictured than was the case in former times. There is not, moreover, the distraction caused by the close proximity of musicians to the screen. The small towns, where inadequate orchestras used to render their ineffectual accompaniments to the silent pictures, have reaped the special benefits of musical synchronization. Music of the best calibre becomes available to every type of theatre."[106] Because of their potential to eliminate idiosyncratic live musical accompaniment and with it the employment of musicians in individual movie theaters, synchronous soundtracks promised both economic savings in the long run and greater standardization of exhibition practices.[107]

This desire for standardized film presentations directly clashed with some of the design features of the initial sync-sound-reproduction devices. The technology initially installed in projection booths presumed that the operator would be a *performer* of reproduced sound. This conception of sound reproduction as a performance originated in some of the components of film sound projection devices. Western Electric's initial Vitaphone system incorporated developments from public address devices designed within the Bell System and introduced and refined during the early 1920s.[108] Not only were individual components of the public address systems reapplied to motion picture sound projection, but so too were the assumptions surrounding the role of the public address sound engineer in performing electrically mediated sound. Western Electric's public address system provided the sound engineer with flexibility in operating the device since many public address circumstances required that the system be readjusted during performances. Especially in outdoor presentations, the public address engineer had to respond to changing conditions by varying the system's performance, such as increasing or decreasing the acoustic output of multiple loudspeaker horns. As a provision of system design, AT&T's engineers were dispersed throughout a large crowd for the purpose of monitoring sound levels and signaling the system's operator to make volume adjustments as needed during that presentation. Public address sound projection was thus conceptualized as a live technological performance with volume adjustments used to maintain a consistent, audible "broadcast."[109]

AT&T conceptualized sound reinforcement and acoustic projection as not fully automated processes but instead as the product of ongoing monitoring and adjustment, a live performance undertaken by appropriately trained engineers. In the Vitaphone sound-on-disk system, volume was controlled in the projection booth through a fader that attenuated the power in the circuit between the reproducing device and the main amplifying system.[110] In period descriptions of the Western Electric sound projector, manipulations of volume during the course of a sync-sound film presentation (called "riding gain") were described as a matter of course. These accounts cited four reasons for projectionists to "ride gain" during a show. Sound volume levels had to be modulated, first because of variations in sound energy requirements in the theater and, second, owing to variations in the size of the audience at any given time during a performance. While these first two reasons for riding gain emphasize the projectionist's response to changing acoustic conditions in the performance space (much like the public address apparatus designed and marketed by Western Electric), the other two reasons for volume variation during a film anticipated that the projectionist would contribute to the aesthetics of the motion picture. Projectionists were to ride gain in response to variations in the levels of recorded sound and, finally, it was "necessary to have some means of varying sound level in theaters because of . . . the desirability of level control during reproduction for the purpose of emphasis."[111] Volume adjustments during a performance required that the projectionist both compensate for deficiencies in the recorded soundtrack and introduce acoustic effects.

RCA's competing Photophone sound-on-film system also assumed that "the control of volume level is the chief concern of the projectionist during the operation of sound-reproducing equipment," and consequently the system provided a fader between the projector's pickup circuit and the first stage of amplification.[112] Period descriptions took volume manipulation for granted during the presentation of a synchronous-sound film as "few pictures are so recorded that they can be run through at a single fader setting."[113] Cue sheets, either provided by the film distribution company or developed autonomously by the projectionist after appropriate rehearsal, indicated when and how the necessary changes to volume should be performed during a presentation. The Projectionist Cue Sheets for Paramount's 1928 baseball film, *Warming Up*, instructed the projectionist to vary sound volume throughout six of the film's eight reels. Indicating a "normal" fader (volume) setting of "11," the cue sheet directed projectionists to adjust the "Fader up one point at Baseball Scenes [and] after Baseball Scenes back to 11." In addition, at several points during the film, projectionists were to adjust the "Fader down two points at singing—then back to 11 after singing."[114] This practice of scripting volume changes during the projection of films continued at least as late as the 1933–34 film season. In its exhibitor's pressbook for *Golddiggers of 1933*, Warner Bros. instructed exhibitors to "tell your operator—for best results move Fader up two points for the musical production numbers on 'Golddiggers of 1933' and also on the Trailer."

Throughout the early sync-sound era, projectionists labored to establish and then maintain appropriate volume levels during sound reproduction. In Richardson's words, "To [the projectionist] the 'fader' is literally the 'throttle.'"[115] That task was not as simple as it might at first appear. In describing the maintenance program provided to theaters by Western Electric's subsidiary Electrical Research Products, Inc. (ERPI), Coke Flannagan informed the Society of Motion Picture Engineers that, as of 1929, deficiencies in the operation of projector and sound systems posed a greater threat to performances than mechanical failures. According to the ERPI engineer, the weak link in Western Electric's sound-film system was the projectionist—specifically in terms of volume levels.

> Operation [of Western Electric projectors] and showmanship are so closely related as to be inseparable, the operation of the equipment presents a greater problem than is presented by apparatus failures. Although the system may function perfectly the effect in the theater can be unpleasant and unnatural because of improper sound level. In fact undistorted sound when too loud is at times more objectionable than distorted sound at proper level.[116]

Sound reproduced with too much volume exacerbated any acoustic flaws of a particular theater. Excessive reverberation became "further aggravating" when the system was operated at greater energy—sounds were sustained longer, reflecting from surface to surface before being absorbed to inaudibility. Excessive volume led to "unpleasant effects" and inaudible dialogue. ERPI's engineers cautioned projectionists to operate the Western Electric system at the minimum power level to insure "comfortable audition."

In October 1930, H. M. Wilcox, operating manager of ERPI, suggested that one of the greatest current obstacles to nearly perfect reproduction was a tendency to keep the volume too loud. Warning against establishing projection volume levels by starting out too loud and then lowering the fader to an appropriate setting, Wilcox suggested that volume levels be determined by starting the fader too low and then raising volume until an appropriate setting was reached. Establishing the "proper" volume thus ultimately became an imprecise matter resolved by the judgment of the theater manager, the projectionist, or both. (In fact, the Cue Sheet for *Warming Up* indicated that, "The fader markings shown herein were made at sound test in a fair sized Reviewing Room, where fader setting '11' proved satisfactory. This setting will possibly have to be changed to suit acoustics and other conditions at various theatres. This should be watched.") Wilcox further cautioned against adjusting volume to achieve adequate results for the worst seats in a house, such as in the rear of the balcony. The resulting excessive power would both increase surface noise and reproduce at a volume too loud for the majority of seats in a theater.[117] In addition to establishing appropriate volume levels, the sound-film projectionist had to modulate volume and the system's output energy in relation to the size of the audience at any given time, especially when audience size changed during a specific performance. Audience members functioned as acoustic absorbing material, and different fader settings had to be determined for empty, half, or full houses.[118]

Although technical and trade journals of the time indicate that manipulation of sound volume became a necessary component of the projectionist's tasks during the early sync-sound era, the rhetoric of professionalism and showmanship surrounding sound projection sought to limit "riding gain." Shifts in volume were to serve larger aesthetic purposes and not occur simply at the whim of a projectionist or house manager, no matter how well meaning. In November 1929, Haviland Wessells cautioned against every theater manager or projectionist attempting to become his own musical conductor. Wessells described an anonymous house manager rushing back and forth to the monitor button to lower the volume of the music whenever the "action of a picture pepped up and music became a bit louder," and then increasing the power "when the music dropped down with a slowing of action."[119] The same manager, claimed Wessells, "raises a whisper to a shout," effectively ruining planned diminution of volume during dramatic scenes. Such apocryphal cautionary tales suggest that, when possible, prevailing standards of showmanship dictated that house managers and projectionists should defer to the recording levels encoded on the soundtrack.

The active manipulation of sound volume described above was often intended to guarantee the audibility of recorded soundtracks. But volume control involved more than merely insuring comfortable audition and/or minimizing the potential deleterious effects of excessive reverberation. Despite the concerns expressed by Wessells, period accounts indicate that operation of sound apparatus frequently involved careful monitoring and the introduction of acoustic effects intended to enhance the aesthetics of recorded sounds. Both conditions of sound recording at the time and representational protocols then currently undergoing formulation constructed sound reproduction as potentially a compensatory activity. It was through the performance of reproduced sound that inadequacies in the recorded soundtrack could be disguised. In addition, both "realistic" effects and acoustic embellishments could be added by projectionists to the synchronous-sound film. "Serious attention is being directed to the importance of controlling sound level of music and speech to insure naturalness and *to augment the effect* of the action or dialogue upon the emotions of the audience" (my emphasis).[120] Such acoustic augmentation included not only volume effects but also frequency attenuation and compensation in projection for acoustic defects introduced during recording.

During the period surrounding Hollywood's conversion to sound, film industry trade publications featured advertisements for acoustic equipment like the Samson Qualpensator, a device offering not merely volume control but also the ability to accentuate or diminish different acoustic frequencies. Acknowledging differences in acoustic taste, advertising copy noted that, "Some like the bass notes emphasized; others prefer them softened. Still others prefer the treble, others both treble and bass notes, or even the middle register notes modified to their taste."[121] The Qualpensator could vary acoustic output "to please in any one of these ranges" and, further, "[it] will do much to compensate for the poor acoustical properties of a room."

RCA's Photophone system included in its normal amplification circuits a variable low-frequency attenuator, called a "compensator," that was designed to diminish low-frequency signals and thereby accentuate the higher-frequency sounds of any recording. Through listening tests at the time of installation in movie theaters, RCA's engineers established a normal setting for the "compensator" using a test recording, but thereafter, the "compensator" could be varied at any time by the operator. The device allowed projectionists to compensate for so-called "boomy" recordings (those in which lower frequencies predominated) and thus "with this device it is possible to increase the intelligibility of reproduced speech which for any reason is lacking in high frequencies."[122] Devices like the Samson Qualpensator or RCA's "compensator" allowed sound-film projectionists to accentuate or to diminish the range of audible frequencies reaching the film audience.

Besides adjusting the frequency response of sound-film projection systems either to suit the public and their supervisor's tastes or to improve the reproduction of so-called "boomy" recordings, the appropriately disciplined film projectionist also manipulated volume to compensate for acoustic problems encoded onto recorded soundtracks. "Only half the mixing is done in the studio . . . the projection room must do the rest," or so announced a January 1930 advertisement for Hardwick, Hindle volume faders published in motion picture industry trade publications.[123] In December 1929, J. Garrick Eisenberg, sound recording engineer at Tiffany Studios, explicitly outlined a series of instances in which the projectionist's showmanship responsibilities included careful acoustic performance to disguise a variety of deficiencies in recorded soundtracks. Eisenberg claimed that the projectionist was "responsible also to a large degree for the artistic rendition of the sound accompaniment."[124] Among the specific recording deficiencies cited by Eisenberg were abrupt differences in volume level from one reel to the next and problems in scale matching such that "when the close-up of an individual player is shown, quite often the sound level is not increased proportionately, with the result that the close-up illusion is destroyed."[125] He also described inadequate differentiation of recorded volume levels for special effects. In this latter case, Eisenberg noted that recorded levels for sound effects might be either too loud or too soft:

> Sometimes when some loud, terrifying noise is intended to accompany the action, the relatively small difference in recording levels allowed enfeebles the effect: one hears loud, raucous conversation, commands, shouts, then a supposedly huge explosion goes off with only a slightly louder "pop." The result is ludicrous. Equally bad are such effects as footsteps approaching along a gravel path which sound like a herd of pachyderms trampling down a forest of young bamboo, in comparison to the dialogue level.[126]

Again, the Projectionist's Cue Sheet for *Warming Up* reveals a trace of early sound-era projection practice. In Reel 8, the projectionist was instructed to increase volume in step with narrative developments. "At scene where Tolliver strikes out McRoe bring fader up one point at each strike—and hold till end of baseball scene—then back to Fader '11' for indoor scene." Here, the projectionist manipulated volume to accentuate the increasing tension of narrative events.

Under conditions like those described by Eisenberg and codified by the instructions for *Warming Up*, the projectionist literally became a performer of sound, carefully attending to the presentation of recorded sound so as to disguise recording errors and manipulating the volume of reproduction either for the sake of realism (achieving approximate acoustic matching for shifts in shot scale, diminishing unduly loud footsteps) or for the sake of dramatic effects (acoustically accentuating the abrupt sound of an explosion or increasing crowd noise at baseball game). Although often assisted in these representational tasks by cue sheets provided by film distributors, the projectionist also had to be guided by, in Eisenberg's words, "some sense of artistic appreciation."[127] An appropriately disciplined projectionist of the early sync-sound era necessarily combined electromechanical proficiency with aesthetic judgment. The professional identity projectionists were encouraged to inhabit combined the routine labor of attending to a machine with the creative performance of sound reproduction.

Period accounts by Eisenberg and others described the production difficulties surrounding sound recording and the irregular and inadequate sound levels that resulted. Such deficiencies in production practice were consistently framed as temporary aberrations caused by hurried production conditions in Hollywood or by the ongoing development of recording technique. While projectionists were thus charged with compensating for recording inadequacies, those inadequacies were nearly always carefully framed as temporary in nature with refinement and standardization of production and reproduction techniques the inevitable goal.[128] In Eisenberg's words, "Meanwhile the burden of carrying on under present conditions falls largely upon the projectionist, and if he is at all worthy of his salt, he will make every effort to cover up by skillful handing of the gain controls, the present limitations of the art."[129] In the short term it fell to the motion picture projectionist to perform sound reproduction and actively contribute to industry standards of showmanship. Such professionalism was indeed a short-term solution. Far more efficient in terms of the goal of standardization would be the eventual refinements to both recording technology and practice so that the performative labor required of projectionists in the early sync-sound era could instead be successfully encoded into the recorded soundtrack. Such a process would increase standardization of film presentations and also "de-skill" the projectionist workforce, providing leverage for theater owners seeking more restrictive and exploitative contracts.

Although standardization of film performances was a goal expressed within the film industry, variations in theater acoustics, the design of sound-projection equipment, the quality and volume of recorded soundtracks, and recording practices all made such a goal initially unattainable. In a 1930 editorial of the recently inaugurated trade journal *Projection Engineering*, Donald McNicol clearly described both the desire for standardization and obstacles to accomplishing that goal:

> If all theatres were alike in acoustics, if all projection equipment was in the same first rate condition, and if all projectionists were equally familiar with the artistic effects desired in a given picture, there would be little complaint in regard to lack of uniform results. That such variations do exist signifies that every exhibitor should take all reasonable precaution to compensate for irregularities or inaccuracies peculiar to the auditorium and his projection facilities.[130]

Into the gap between standardization as an industry goal and the realities of equipment, theater acoustics, and recorded soundtracks, professionalism arose as an alternative way of framing and disciplining the labor of the motion picture projectionist. This rhetoric of professionalism invoked standards of education and training as well as renewed pride in one's work, and it framed the labor of the projectionist as an artistic contribution to the industry, in some ways equal to the contributions of those engaged in film production. The projectionist was, for a time, categorized as a motion picture performer, instrumental in the creation of aesthetic effects.

A closer look at the technology of motion picture projection and the labor of projectionists in the early synchronous-sound era illustrate that the soundtracks for these films were not finally and ultimately determined by Hollywood's recording engineers, but were instead acoustically contingent on a number of factors, not the least of which was, in the words of the literature of the period, the level of professionalism and showmanship exhibited by long-forgotten projectionist-performers. Volume adjustments for the sake of audibility of dialogue, for example, compensated for sound technology's and sound recording engineers' inability to serve perceived narrative needs while volume adjustments for the sake of "creating effects" supplemented and indeed usurped the signifying potential of the synchronous-sound film by often emphasizing the spectacular qualities of recorded sounds.

Conclusion

Having announced new technology by addressing consumers as interested in the details of applied science and educating them about electrical acoustics, innovators of electrical sound technology then, through the design of home radios and various discursive means, sought to redirect consumer attention away from the apparatus itself and toward the sounds it produced. In a transitional period, domestic consumers—much like motion picture projectionists—became performers of reproduced sound, actively interacting with the technology of new media. While the Victrola and similar devices required of consumers a good deal of activity to play recorded sounds, late in the mechanical phonograph era (and while many homes were converting to radio), the Graduola overtly constructed listening to the phonograph as a performance indicative of musical taste and skill. Early consumers of radio—whether home engineers who constructed their own receivers or the well-disciplined listener who sought only to "tune in" an uplifting radio "concert"—actively engaged applied scientific principles embodied in an apparatus. Film projectionists also navigated change in their workplace by becoming performers of recorded sound.

The transition to electrical methods, then, involved a variety of sites in which individuals actively interacted with media technologies. The social meanings of sound devices emerged as the product of the overdetermined interactions between product design, human performance, and the rhetorical framing of media forms, including efforts to discipline these consumers of new media. Mechanical and electrical reproduction were not fully automated processes. But much like, first, the design of the Victrola and then the domestic radio receiver, innovators took as one of their tasks the dissimulation of the technology of sound reproduction and transmission. Relocating the site of user attention from the machinery of applied science to the sounds it produced, standardizing the labor of movie projectionists, disciplining active consumer performance, innovators of sound technologies sought to render seamless, automatic, and natural the representation of sounds. Technology, as the physical embodiment of applied scientific principles, receded into the background only to return again, albeit temporarily, with each subsequent media innovation.

The redefinition of the phonograph as a domestic entertainment device, radio's entrance into American homes, and discursive struggles to manage the labor of film projectionists offer three related perspectives on what we might refer to as the human interface—the space and practices in and through which individuals actively engage with technological artifacts. In each case, individuals became literally performers of reproduced or transmitted sounds. Each instance illustrates that the process of introducing a new technology into existing spaces, of grafting it onto existing social practices, required careful discursive management. Among the goals of that management was the dissimulation of machines and their labor. Whether through cabinet design intended to integrate invisibly a new acoustic device into the home, or professional protocols designed to insure invisibility and inaudibility of a projectionist's labor, the process of technological change involved the successful effacement of material traces of applied science. While the physical designs of sound technologies often required consumers to become performers of reproduced sounds, those performances were always carefully framed and discursively managed and also comparatively short-lived as media innovators successfully dissimulated acoustic machines.

Four

Making Sound Media Meaningful

Commerce, Culture, Politics

Through the announcements described in chapter 2, corporate innovators of new sound technology proposed certain identities for the apparatus
and its different applications. Those announcements did not ultimately
determine how emerging sound technology became meaningful. Instead,
electrical acoustics developed within a horizon of preexisting anxieties,
utopian goals, and overlapping categories of social experience. Seeming
to offer new solutions to perceived problems, emerging sound media
opened up new conceptual spaces in which material practices might be
imagined.

Attempts to understand the ways in which the new sound media were
made socially meaningful can be organized according to three intersect-
ing categories of social experience: commerce, culture, and politics. This
is not to suggest that culture is apolitical or that it is somehow immune
from the economic dictates of commerce, nor that the principles and practices of consumer society are in any way apolitical articulations. Instead,
with few exceptions, period rhetoric surrounding technological change and
the material efforts to deploy new acoustic devices sought to keep these
categories distinct, enforcing and perpetuating hegemonic notions that
these categories of social experience were in fact separate spheres of activity. Discursive constructs like "entertainment" served to enforce conceptual
boundaries between these spheres even as, for example, political discourse

moved ever closer to a commercial mode of address. Commonsense notions which held that the spheres of commerce, culture, and politics were separate provided a preexisting discursive grid onto which new media were mapped, a set of conceptual matrices through which the social functions of emerging media were understood and experienced. Running across and through these three spheres, neither wholly dependent on nor independent of them, neither wholly constituted by nor constitutive of them, a series of oppositions provided a further framework through which emerging media were understood. The differences between local and national, between rural and urban, and between the immigrant and the native-born provided a focus both to how emergent sound technologies were imagined and how they were materially grafted onto social relations.

A long-standing, overdetermined desire, a perceived need, for media technology to serve as an instrument of national unification provided the governing logic that drove the mapping of emergent sound media onto preexisting institutions, relations, and practices. To some extent, the entire history to that point of innovations in transportation and communication had promised a similar unification across distances. Radio's apparent ability to eclipse space—to conquer distance—was but an intensification of the earlier promises of mail service, telegraphy, telephony, even canals and rail service, or celebrations of the cinema's ability to serve as a "universal language." Each in its own turn and with its own specificity had appeared to transcend geography and to link dispersed individuals, groups, businesses, and markets into a web characterized by varying degrees of both scope and intimacy. As expressed in announcements of electrical acoustics, this governing logic promised to fulfill a series of consensual desires. Within the political sphere, emerging sound media might help to create the rhetorically longed for (albeit perpetually unfulfilled) enactment of a participatory public sphere. By bridging the gap between citizen and government, by educating the populace on important issues of the moment, the new sound media might facilitate a more inclusive, democratic, deliberative process. Further, sound media might function to "heal" a perceived fragmentary social body, by both promoting the assimilation of immigrant populations and bridging the gap between rural and urban populations. Finally, when informed by the governing logic of national unification, they might through uplift and edification promote the advancement of a shared national culture.

In her comparative analysis of broadcasting systems in the United States, Great Britain, and Germany during the Depression, Kate Lacey notes that "radio was seized upon as a tool that could bind the various constituents of the nation together," thereby addressing fears that the international economic crisis would further fragment precarious national and cultural identity.[1] She quotes Merlin Aylesworth, head of NBC, who in 1930 looked both back on radio's history and toward its future and described the unifying potential of radio: "to preserve our now vast population from disintegrating into classes . . . we must know and honor the same heroes, love the same songs, enjoy the same sports, and realize our common interest in our national problems." Lacey cites the faith, widely echoed during the Depression, that a shared culture could provide "the possibility of social cohesion to be achieved and sustained by sharing the same imaginary."[2] The intensity of such a desire for a shared imaginary ebbs and flows over time, often dependent on the degree to which difference is feared. Material economic and social conditions directly impact how nations and citizens perceive such social cohesions. Although no doubt greatly exacerbated by the stock market crash and ensuing crisis, both the longing for a social cohesion and the assertion that it could be achieved through a shared cultural imaginary promulgated by acoustic media had been central to discursive constructions of sound technology since the 1920s.

The use of American mass media as an instrument of national unification had a powerful precedent prior to the innovation of electrical acoustics. From 1917 to 1919 the Committee on Public Information, supervised by George Creel, carefully orchestrated the use of all media forms in support of the United States' aims during World War I. The Creel Committee made extensive use of motion pictures as part of a larger propaganda campaign that included all print media, posters, window displays, billboards, exhibitions, and various public performances, including "the staging of sham battles which had a tremendous appeal besides illustrating the modern methods of trench warfare."[3] Private media companies, including newspapers, magazine publishers, and the U.S. film industry collaborated closely with this government agency. (See fig. 4.1.)

The film industry's participation with the committee's campaign extended from both the national level of film production and distribution to the local level of film exhibition and even included the international distribution of U.S. entertainment films. Feature-length films produced by

I Am Public Opinion

All men fear me!

I declare that Uncle Sam shall not go to his knees to beg you to buy his bonds. That is no position for a fighting man. But if you have the money to buy, and do not buy, I will make this No Man's Land for you!

I will judge you not by an allegiance expressed in mere words.

I will judge you not by your mad cheers as our boys march away to whatever fate may have in store for them.

I will judge you not by the warmth of the tears you shed over the lists of the dead and the injured that come to us from time to time.

I will judge you not by your uncovered head and solemn mien as our maimed in battle return to our shores for loving care.

But, as wise as I am just, I will judge you by the material aid you give to the fighting men who are facing death that you may live and move and have your being in a world made safe.

I warn you—don't talk patriotism over here, unless your money is talking victory Over There.

I am Public Opinion!

As I judge, all men stand or fall!

Buy U. S. Gov't Bonds Fourth Liberty Loan

Contributed through Division of Advertising

United States Gov't Comm. on Public Information

This space contributed for the Winning of the War by

The Publishers of The Independent

FIGURE 4.1 Creel Committee advertisements invoked the power of public opinion to judge citizens' contributions to the war effort even as its multiple-media campaigns sought precisely to shape that opinion. (Author's collection)

the government, including *Pershing's Crusaders* (1918), *America's Answer* (1918), and *Under Four Flags* (1918), demonstrated America's war aims and interpreted "the meanings and purpose of democracy" to movie audiences throughout the United States and the world. The committee's share of income from domestic film screenings totaled close to $900,000, which was then used to underwrite free screenings of the films abroad. American film producers further collaborated with the government to ensure that a certain amount of propaganda film accompanied any international distribution of U.S. entertainment films and that only U.S. films portraying American institutions and life in a positive light were screened abroad. Film producers worked closely with the Creel Committee in producing short subjects written by committee employees, and the committee provided the four major newsreel producers with a quota of 500 feet of timely war film to be incorporated into weekly newsreel releases. Locally, individual film theater owners linked the screening of committee films with various fund-raising schemes for the war effort and even movie theater curtains became a promotional surface provided to the committee's Advertising Division.[4]

Perhaps most dramatically, the intermissions in movie programs became a momentarily politicized space in which the Creel Committee's

"Four-Minute Men" directly addressed movie audiences to extol democratic principles, the aims of the war effort, and an ethos of self-sacrifice. The "Four-Minute Men," approximately 75,000 volunteer speakers disbursed throughout the United States, individually prepared short talks based on "the policy, points of emphasis, lines of argument, and general information" contained in bulletins circulated by the Committee on Public Information. In summing up the committee's accomplishments, Creel and his staff estimated that the "Four-Minute Men" collectively delivered more than 750,000 speeches to a total audience of over 314 million.[5]

The efforts of the Creel Committee reaffirmed for both political and commercial interests the tremendous power of an orchestrated campaign of mass media to shape public perceptions, attitudes, and even actions. It underscored not only the political potential of film as a medium but also the way in which the film industry might function as part of a coordinated network of national and local interests. Through films and their distribution, advertisements and displays in theaters, and importantly the Four-Minute Men, the Creel Committee infused the space of movie theaters with an overtly politicized discourse, but one that could be rhetorically framed and perceived as a continuation of the service local theaters provided to their patrons. By blurring, indeed collapsing, the boundary between entertainment and politics, the Creel Committee provided a powerful historical precedent to both the advertising industry and to those who might seek to mobilize the mass media as political instruments. Finally, the Committee on Public Information—in both its practice and the ideological assumptions underwriting it—powerfully attested to the potential that mass media had to inculcate shared consensual values to a national audience consisting of widely disparate classes and groups.

The innovation of electrical methods of recording and disseminating sound through a variety of media forms provided a series of new tools, creating previously unimagined possibilities to enlarge the scope of political or commercial campaigns. While public address systems greatly enlarged the audience that a political speaker could address, AT&T's experimental public address demonstrations (notably the 1921 Armistice Day event and the internment of the unknown soldier) illustrated that audiences could be further expanded, both numerically and geographically, through the wired transmission of speeches to public address systems at distant locations. Radio, particularly when broadcasting stations were linked into a network, went even further in accomplishing such a national mode of address, and the innovation of sychronous-sound cinema promised that the direct local appeals of a "Four-Minute Man" might be augmented by a coordinated distribution of the recorded remarks of well-known speakers.

While the Creel Committee demonstrated the potential of a national media campaign to construct for citizens a shared national imaginary, it also underscored to advertisers and commercial media interests the power to shape consumer attitudes and behavior on a national scale. The belief that emerging media might ideally function as instruments of national unification served specific commercial interests. A national mode of address that hailed citizens, uniting them into a social body, might also hail a nation of consumers, uniting their consumption patterns and orchestrating their desires so as to provide greater economic efficiency and profitability. Media, when configured as instruments of national unification, might mimic the then-expanding retail model of chain stores, delivering not standardized consumer goods but ideas, information, and entertainment. And like chain stores in the retail sector, the deployment of sound media through national chains was consistent with the growth of large corporations, their increasing power and ubiquity in American life. The emergence of large media corporations, vertically integrated and extending across multiple media forms, in part created by the innovation of electro-acoustics, facilitated specific forms of a commercial mode of address that were particularly suited to consumer culture. Just as national sound media might promote the long-articulated promise of a "musical democracy," so too might they become instruments promoting a "consumer democracy," but one which privileged the standardization of entertainment commodities and the centralization of economic power in a small number of national media firms.

A particular logic of consumer society thus underwrote the deployment of sound technology. The institutionalization of network-affiliate relations in the radio industry solidified a specific economic model to the exclusion of other potential models, including nonprofit configurations. For the U.S. film industry, economic arrangements offering increased standardization at the levels of production, distribution, and exhibition as well as anticipated economies of scale in exhibition, provided the conceptual logic and economic rationale for the conversion to sound. Any short-term economic trauma caused by technological change would be overcome in the long run by the increased leverage and control that could be exerted by national producers/distributors over the conditions of film exhibition and by accompanying economic benefits of such standardization and control.

Both radio networks and movie theater chains navigated in a particular way the conceptual binary of local and national. Following the lead of chain stores before them, radio networks and movie theater chains sought to equate *quality* with *standardization*. Throughout the late 1920s, radio industry rhetoric extolling the virtues of chain broadcasts consistently denigrated local programming for its uneven, indeed amateurish qualities. In an identical rhetorical move, the U.S. film industry initially hailed the conversion to sound in terms of the standardization of musical accompaniment that came with prerecorded soundtracks. Recorded sound would render obsolete the wildly divergent, inconsistent, and substandard accompaniment performed by local musicians. Through the centralization of the production of sound, national standards of music would be elevated.[6]

For both the radio and the sound cinema, the issue was not one of local *versus* national but instead one of navigating the relationship between the two in a specific way. The local became an enactment or articulation of the national—not reducible to simply an iteration of the national but a coordinated expression linked into precisely a *network* of other such articulations. Like Creel's Four-Minute Men crafting their local talks based on the information and guidelines distributed nationally, each local instance articulated the national without being entirely reducible to it. At its most dense, in its most thorough and pervasive form, such a network of relationships looked something like the coordinated, multiple-media promotion of Amos 'n' Andy and RKO's release of *Check and Double Check*.

Electrical-acoustic technology participated in, exacerbated, and was itself shaped by a larger trend of consumer culture that mapped commodity production and consumption across the binary of local and national. Entertainment, specifically acoustic entertainment, became more fully entrenched within a particularly modern array of commodity forms and relations. The establishment of a spatial identity for radio and the institutionalization of radio networks, the commercial possibilities offered to both radio and the cinema by advertising, the relationship between mass media and both elite, high-culture forms and politics—in short, the nature of what was meant by "entertainment"—enacted in practice the discursive gap between technology's apparent utopian promise and commercial reality. Entertainment, as a meaningful category of social experience, became a site around which various forces struggled to shape emerging media forms.

Commerce

In August 1926 at the New York premiere of the Vitaphone, and at virtually every other subsequent Vitaphone premiere around the country, Will Hays, president of the trade group the Motion Picture Producers and Distributors of America (MPPDA), framed the technological innovation in terms of service. In a sync-sound, direct-address short subject, Hays predicted a future for sound films in which they would render "greater and still greater service as the chief amusement of the majority of all our people and the sole amusement of millions and millions, exercising an immeasurable influence as a living, breathing thing on the ideas and ideals, the customs and costumes, the hopes and the ambitions of countless men, women and children."[7] Picking up the chant that linked technological innovation to the fulfillment of "musical democracy," Hays asserted that "the motion picture is a most potent factor in the development of a national appreciation of good music. Now that service will be extended as the Vitaphone shall carry symphony orchestrations to the town halls of the hamlets." Concluding his remarks, Hays declaimed, "It [the Vitaphone] is another great service—and 'Service is the supreme commitment of life.' "

The true measure of the Vitaphone's innovation was the service it provided. Economies of scale and standardization became in the public's interest. The idea that the film industry provided consumers with a service was already well established before the conversion to synchronous sound. Trade publications directed in the first instance toward exhibitors stressed the local identity of theaters and the service that they provided to a community. For example, in December of 1926, *Exhibitors Herald-World* reminded its readers: "But the theatre owner is after all a public servant. It is his duty to serve his community according to its palate for entertainment, and his personal interest—good receipts—is satisfied in direct proportion as he offers an amusement menu that please his customers."[8] At both a local and a national level, the film industry, like all manufacturers and retailers of goods in the consumer economy, sought to cultivate a public perception of its efforts as the provision of a valuable service.

One assertion regarding the sound film's "immeasurable influence" that was easily overlooked in Will Hays's celebration of the Vitaphone played a crucial role in the process of grafting new media onto commodity relations. While sound film could shape the "ideas and ideals" and even "the hopes and the ambitions" of countless audience members, it would also shape their "customs and costumes." Sound media's potential to shape the purchasing patterns of U.S. consumers became a crucial site of contestation regarding the social roles played by, and economic relations governing, emergent acoustic media. Although Hays spoke specifically about the synchronous-sound film, his remarks resonated with ongoing developments surrounding radio broadcasting. Media could and would become increasingly central to the cultivation of a shared national imaginary, but the importance of commerce to that imaginary had yet to be determined. While the centrality of large corporate interests in the innovation of electrical acoustics, the increasing concentration of media ownership amid technological innovation and change, and the ethos of federal regulation that followed and reinforced corporate innovation rather than challenging it, seemed to guarantee that media innovation would follow the logic of big capital, the *manner* in which media would participate in and perpetuate commercial culture was not a foregone conclusion.

Once electrical acoustics was conceptualized as a point-to-mass system that distributed sound—in short, broadcasting—struggles over identity did not subside but instead intensified. The overriding issues of this identity struggle involved from whom and from where such transmitted sound would originate and to whom it would be addressed. Three competing identities conceptualized broadcasting spatially, and the diffusion of radio throughout the 1920s and into the early 1930s became a process of negotiation and shifting emphasis among these different spatial conceptions of the broadcaster-audience relationship. Simultaneously viewed as a local service, as the conqueror of distance, and as a national mode of address, radio broadcasting's shifting identities affected radio programming, regulation, industry economics, further research and development, and audience listening habits.

Radio broadcasting's identity as a local service hinged on the initial massive and geographically dispersed proliferation of radio transmitters. By the end of 1922 the Department of Commerce had licensed over 500 stations for broadcasting. Initial broadcasters included an eclectic array of commercial, nonprofit, and educational institutions as well as individuals. The continuing and largely superfluous debate over the identity of the "first official" broadcast station attests to the diverse identities and local nature of the origins of radio broadcasting. While Westinghouse engineer Frank Conrad is often cited as the first regular broadcaster (whose success in his garage-originated broadcasts demonstrated to his employer that regularly scheduled broadcasts could stimulate sales of radio equipment), other contenders include a Detroit newspaper, Lee De Forest's 1910 New York broadcast of the Metropolitan Opera, the University of Wisconsin, or even Reginald Fessenden's Christmas Eve 1906 transmission of music over an adapted wireless telegraph device.[9] Large electronics firms as well as local electronics retailers and department stores, newspapers, colleges and universities, churches, local municipalities, and even wealthy individuals all began broadcasting early in the 1920s. Initial regulatory assumptions governing U.S. radio policy acknowledged the local, spatially dispersed identity for broadcasting. The "ether" belonged to no one, or rather to everyone, and initial regulatory principles required the Department of Commerce to license for broadcast all those who applied. The local identity of radio guaranteed a mix of programming every bit as eclectic as the operators of broadcasting transmitters. Amateur musical performers shared the airwaves with phonograph recordings, locally originated radio talks, and even remote broadcasts from local entertainment venues or after-dinner oratory. Radio defined as a local service also addressed diverse ethnic groups in their own language.

In addition to its identity as a local service, radio broadcasting also was conceptualized and experienced as the conquest of distance. When defined as primarily the transmission of sounds across great distances, radio involved a contrasting series of listening behaviors, programming content, technological innovations, and consumers' relationships to technology. This identity for radio broadcasting addressed so-called DX listening, a consumer practice and relationship to technology in which the radio receiver as an artifact took precedence over the programming being transmitted. Radio enthusiasts competed with geographic distance, with technical limitations of radio components, and with each other in assembling a radio set capable of capturing signals of distant origination. DX listening focused on exploring the geographical limits of available technology, seeking to expand those limits in an attempt to achieve ever greater distance in reception.

Listening became a display of technological proficiency by home engineers who designed, built, and perpetually rebuilt home radio receivers in an effort to log the signals of distant broadcasters. "Silent Nights," the voluntary suspension of broadcasting by local stations at prearranged times so as to minimize interference and thereby enable local DXers unfettered access to the signals of distant broadcasters, institutionalized radio's identity as the conquest of distance. Some municipalities continued these "Silent Nights" at least into late 1925 as a ritualized survival of an earlier construction of the consumer's relationship to electrical-acoustic technology. DX enthusiasts actively lobbied radio stations through the radio press for established standards of announcing that would frequently and clearly identify both a station's call letters and its location of transmission.[10]

During the initial years of broadcasting, the popular and the radio press frequently contained accounts of the role played by radio in expeditions to remote locales such as the Arctic or the Amazon, thereby rhetorically foregrounding radio's scientific conquest of distance. These articles described the ability of radio to maintain contact with explorers and even instructed home listeners on how they might best capture the transmissions back to civilization made by members of heroic expeditions. Radio defined as the conquest of distance also informed promotional efforts to assert the social value of broadcasting. Written accounts of radio progress throughout the 1920s frequently stressed the ability of broadcasting defined as a domestic entertainment device to ease the isolation of farmers. Radio's conquest of distance provided information as well, and again the farmer was frequently cited as enjoying the benefits of instantaneous price reports from distant markets.[11]

Radio broadcasting conceptualized as a mode of national address provided a third, competing identity for the emerging medium. Early in the 1920s, as various commentators grappled with the question of what radio broadcasting could become, many identified the technology's potential to address simultaneously large, geographically dispersed audiences. Related to broadcasting defined as the conquest of distance, radio's identity as an instrument of national address imagined the medium as a technological augmentation to citizenship, enabling the simultaneous presentation of information to the dispersed populace. Premised in part on a view that broadcasting might be modeled on public address, this view of the technology (in a powerful rhetorical move the effects of which continue to resonate today) collapsed listening with participation. During deliberations about a major public policy issue, the president and other leaders could take their case "directly" to the people, ideally *all* the people. Radio, then, could create a more transparent and direct manifestation of democracy by facilitating a politician's ability to address directly the citizenry.[12]

In addition to information, news, and political rhetoric of national significance (or political rhetoric *made significant* simply by reaching large, dispersed audiences), radio programming accompanying this identity could also provide citizens with simultaneous "participation" in leisure and entertainment. AT&T's initial demonstrations of networked broadcasts included not only events like the 1924 National Defense Test Day but also prize fights. Broadcasting events of various kinds, from political conventions to World Series games and boxing matches, demonstrated the potential for networked radio transmissions to promulgate a national mode of address.[13]

Throughout the mid-1920s radio's public identity shifted from a wondrous scientific marvel to a more common, albeit not yet commonplace, modern source of domestic entertainment. As radio was increasingly and more pervasively incorporated into daily life, its identity as "the conquest of distance" became rhetorically subsumed within broadcasting's spatial identity as a "national mode of address." But the notion that radio offered listeners "the conquest of distance" continued to inform a series of types of broadcasting events, and advocates of network radio rhetorically framed broadcasts of national sporting events, national political events (especially presidential nominating conventions), and even live, networked remote broadcasts (Hollywood premieres and musical performances) as the acoustic eclipse of space. The perception that broadcasting had technologically collapsed the distance between a listener and an event became less central rhetorically than the accompanying notion that such broadcasts simultaneously addressed and could thereby (re)constitute a geographically dispersed national public. While innovators still celebrated radio broadcasting for its ability to bridge the space between an individual auditor and a distant event, the link radio technologically forged *between* dispersed auditors became even more central to popular presentations of radio's identity. Radio, claimed prominent promotional rhetoric, could offer a collective, national experience.[14]

A national mode of address, however, also offered advertisers a national audience of consumers. First through the institution of indirect sponsorship (in which programming was linked only to the name of the sponsor as a public benefactor), and then through direct advertising that overtly promoted specific products, advertisers appropriated radio's national mode of address and deployed it at the service of an expanding consumer society.[15] A medium that rhetorically promised to unite a nation of citizens inexorably moved toward addressing a nation of consumers; a national mode of address rhetorically aligned with political democracy could easily serve so-called "consumer democracy."[16]

Federal regulations governing radio reinforced the commercial developments that privileged radio's identity as primarily a national mode of address. The formation of the Federal Radio Commission and its initial frequency allocations functioned conservatively to marginalize radio's identity as a local service and to reinforce the increasing prominence of national broadcasting. When the FRC reorganized the broadcast spectrum in 1927, it allocated the overwhelming majority of highly coveted clear-channel frequencies to affiliates of radio networks or broadcasters operating high-powered transmitters. The remaining approximately 600 broadcast stations shared frequencies with nearby stations, thereby reducing their hours of operation, potential audience size, and commercial viability. A shared frequency limited a broadcaster's available funds for programming and for updating equipment, insuring that such stations would operate at a comparative disadvantage into the future. Ultimately, the FRC's frequency allocations rhetorically and practically aligned the public's "interest, convenience and necessity" with the transmission of commercially produced entertainment originating from national entertainment centers and transmitted to geographically dispersed homes. Public service became aligned with commercially produced, nationally known entertainment. Radio historian Jeff Land notes:

> The great financial windfall that advertising brought to the major commercial stations led to the consistent technical upgrading of radio studios. This in turn increased the production quality of the music and drama programs the chains could offer their affiliates. It is in this dialectic between enhanced production quality and variety, underwritten by burgeoning commercial sponsorship, that the ideology of popular entertainment as public service finds its grounding.[17]

The principles underwriting the rhetoric of technical necessity and public service embodied in FRC policy drew heavily, albeit selectively, on the competing spatial metaphors commonplace in the early 1920s that initially underwrote conceptualizations of broadcasting's potential. High-powered broadcast transmitters and radio networks facilitated the conquest of distance and guaranteed a national mode of address. FRC rhetoric frequently invoked the remote citizen, spatially isolated in a rural area, and the necessity to ensure that such listeners received the same type and quality of radio programming as did those living in metropolitan areas. Rather than advocating and embodying the principles of local origination privileged in earlier, competing models of radio (i.e., AT&T's 1923 internal broadcasting plan which viewed radio as primarily a local service), federal radio regulations from 1927 on supported and reinforced radio conceptualized as a medium of national unification achieved through a national mode of address.

By the end of the 1920s, the competing spatial metaphors that provided potential identities for radio in the early 1920s had been successfully arranged into a hierarchy, and broadcasting was widely accepted as an instrument offering an acoustic national unification. The development of a political economy of radio dominated by advertiser-supported programming and network-affiliate distribution successfully combined two of the original spatial metaphors used to make sense of broadcasting in the early 1920s. The system of network-affiliate relations offered both a national mode of address and, in residual form, affiliates as providers of a local service. Development of broadcasting scheduling conventions codified the relationship between these two competing identities such that they became overlapping and apparently complimentary instead of conflicting. Network broadcasts claimed to offer dispersed listeners the "best" of national entertainment and culture while retaining the local specificity of community or regionally originated programming framed by a local mode of address. This hierarchical, mixed mode of address also, of course, served the interests of an ever-burgeoning consumer economy in which both manufacturers of nationally known brand names and operators of locally based retail outlets could address consumers each in their turn throughout a broadcast day.

The resolution of radio's three spatial identities according to a specific hierarchy was driven by the overdetermined desire to deploy sound technology as a unifying apparatus—a desire to produce both a shared national imaginary and national networks of material practices (i.e., consumer goods and their circulation). Radio broadcasting, when premised on a model of dispersed local providers linked into a network by wired transmission, seemed to promise audiences a guarantee of quality programming—the fulfillment of the "musical democracy" promised in announcements of electrical sound technology—while simultaneously holding down the costs faced by program providers. A local NBC affiliate, like the downtown Woolworth's or Rexall, mediated between a national provider (of programs) and geographically dispersed consumers. Such a model for broadcasters (or retailers) seemed tailor-made as a communications medium that might link the manufacturers of goods with consumers.

The consolidation of ownership within the film industry also appeared to offer such a marketing model and a medium that could connect producers and consumers of all types of goods. National providers of cinema were linked to consumers through a chain of widely dispersed local theaters, the best of which were owned by the producers/distributors themselves. It is not surprising then that, taking its cue from developments in radio, the U.S. film industry explored in 1930–31 the possibility of commercial sponsorship of films. The movie theater, like the ether and the home radio receiver, might become a medium through which national manufacturers of goods addressed widely dispersed consumers.

In the aftermath of technological change, increasing media consolidation, and, importantly, the stock market crash of 1929, several of the vertically integrated U.S. film companies experimented with the production, distribution, and exhibition of commercially sponsored films. While a system of advertiser sponsorship gained increased hegemony in the radio industry, cinema explored the possibility of selling audiences' attention to advertisers by incorporating advertising and even sponsored films into the entertainment program offered at local theaters. Short subjects, designed to entertain as well as to promote nationally distributed products—much like concurrent advertising on radio—contained varying degrees of overt marketing. Called "cinemadvertising" by *Time* and "advertalkies" by Marsh K. Powers, president of the Powers-House advertising agency, this new type of short subject addressed audiences as consumers but also conceptualized moviegoers as a kind of product: a captive audience whose attention might be sold to advertising agencies and manufacturers of nationally distributed goods.[18] Sponsored films offered vertically integrated film companies a further supplement to their income (the production of short subjects could be underwritten by advertisers, and they could be charged for the exhibition of the films), and these films offered the manufacturers of nationally produced goods a new medium through which to address consumers. Unlike radio, which could offer advertisers no guaranteed count of the audience addressed, movie theaters could provide film sponsors an accurate assessment of the audience for commercial messages.[19]

By July of 1930, Paramount already had sponsored films in production and Warner Bros. had organized a separate subsidiary, Warner Bros. Industrial Pictures, Inc. (built around its acquisition of the Stanley Film Advertising Company), with plans to be in full production by the fall of that year.[20] By distributing the sponsored films to the theaters they owned or controlled, Paramount (with approximately 1,500 theaters) and Warner Bros. (with about a thousand) could offer advertisers a "guaranteed circulation" of their commercial messages. In an advertisement in *Fortune* magazine, Warner Bros. promised potential sponsors an "audience of 5 million people in the finest moving picture houses."[21] According to Warner's plan, advertisers would pay $5 per thousand admissions, and they could opt for a national release of their sponsored film or specify a particular territory (or even individual theaters) so as to facilitate and coordinate tie-ups with local retailers of their products.[22] Paramount's initial sponsored film release, "A Jolt for General Germ," an animated promotion of the disinfectant Lysol and its germ-killing powers, was shown in some 1,200 Paramount theaters and remained in distribution until it had reached a total box office of 5 million. For this guaranteed exposure to film audiences, Lehn & Fink paid Paramount $25,000 (a rate of $5K per million viewers).[23]

By March of 1931, a number of national manufacturers and advertising agencies had signed agreements with Paramount and Warner Bros. for the production and exhibition of sponsored films. The Texas Company collaborated with Paramount on cartoon comedies while Liggett and Meyers promoted their brand of Chesterfield cigarettes with a series of shorts called "Movie Memories," which recycled footage from the first decades of cinema. The advertising agency Lehn & Fink contracted with Paramount on behalf of their clients Hinds Honey & Almond Dream and Pebeco Tooth Paste. Paramount also had contracts for sponsored films with Westinghouse while Warner Bros. had signed up the Lambert Pharmaceutical Company for entertainment films promoting Listerine. Sponsored films were in production for the India Tea Company and retail chain A&P (their initial release, "On the Slopes of the Andes," promoted grocery sales via a travelogue about coffee cultivation).[24] By early April of 1931, *Business Week* reported that eight national advertisers had signed up for the production and exhibition of sponsored films and that a total audience of at least 28 million people had already seen the films and "reacted favorably."[25] According to *Business Week*, interest in the sponsored film experiment was spreading, with RKO considering sponsored films and even informal talks among the major studios about enlarging the distribution of the films by screening them in each other's theaters.[26]

In some locales, advertising in local movie theaters had been a long-standing practice. Thousands of theaters across the country had for years supplemented their box office income by projecting illustrated slides promoting local businesses. But such screen advertising remained representationally distinct from presentations of films, and local exhibitors retained control of both the practice and all of the profits that it produced. Before Paramount's and Warner Bros.' nationally orchestrated sponsored film experiment, overtly commercial films had intermittently appeared in some U.S. theaters, but lacking access to the national distribution apparatus and theater chains of the vertically integrated studios, producers of commercial films had to settle for the irregular distribution they could arrange with independent theaters owners or small local and regional chains. Automobile manufacturers Studebaker and Graham-Paige, for example, were able in 1929 to get their general sales films exhibited in local theaters.[27] Such intermittent and scattered use of sponsored sound films was a far cry from the systematic approach proposed by two of the major studios in 1930.

In imagining sponsored films, Paramount, Warner Bros., and advertisers could look to the success of the short subject "The Nation's Market Place," a commercial film produced for the New York Stock Exchange, that had already garnered substantial theatrical exhibition (3,500 showings to an estimated audience of almost 2.2 million movie theatergoers).[28] Several major studios, particularly Fox, had previously released some propaganda films in their theaters in the years following the industry's collaboration with the Creel Committee. In 1929, William Fox, in a desperate attempt to maintain control of his company, took his case directly to film audiences in a series of synchronous-sound short subjects imploring audiences to purchase his company's stock. While Fox's appeal to moviegoers ultimately failed (he lost control of his company), the film industry as a whole was more successful, through a series of short subjects, in attacking the institution of daylight savings time in California. Fox also released and nationally distributed in its theaters a series of film shorts featuring "Mr. Courage and Mr. Fear," the point of which was to suggest, in the words of one account, "that the present country-wide financial collapse was not due to overproduction, the tariff, or European conditions. No, the collapse was merely a matter of mind! With a little courage, according to Fox's 'Mr. Courage,' we all could start inflating the prosperity gas bag once again by simply buying more goods."[29]

The incorporation of such overt propaganda films into movie theater programs suggested to the studios that the public might be ambivalent about, if not actually accepting of, the presence of sponsored films in film programs. Radio had also demonstrated that advertising could be successfully incorporated into entertainment, or rather, that consensual notions of entertainment might be redefined so as to include commercial sponsorship. By centralizing production and distribution of commercial short subjects, the studios could offer advertisers a systematic improvement over both the previous haphazard exhibition of sponsored films and radio's inability to provide an accurate assessment of audience size. National theater chains provided the ability to address audiences not as a local community (as illustrated slides had done) but as a national body of consumers.[30]

Period commentators and the studios themselves clearly viewed sponsored films in relation to radio advertising. *Business Week*, in an article titled "Talkies Adopt Radio Methods in New Sponsored Programs," suggested that just as in radio programs, an announcement of a film's sponsor might appear only at the beginning and ending of the film. The film industry journal *Projection Engineering* characterized such an indirect approach as "courtesy" films, "analogous to radio programs presented by business firms which are without reference to the company during the course of the entertainment."[31] In the trade journal *Advertising & Selling*, Peter Andrews used the language of radio to suggest that distribution of sponsored films through the major Hollywood producing and distributing companies allowed for "the entire country [to be] reached through a *national hookup* of theaters" (emphasis added), and he further cautioned that too much advertising could cause movie audiences to experience "a turn-the-dial reaction" so that "a suppressed 'shut off the station' desire may blossom forth later in inclinations to shun the theater in which the advertising was seen."[32] For both advertisers and much of the film industry, the solution to "turn-the-dial reactions" resided in adopting the indirect approach of some radio advertising:

> The sponsored movies skillfully get around the human desire to receive entertainment instead of advertising by following the radio broadcasting method of sugar-coating advertising with thick layers of amusement. Except for the title, "Sponsored by the makers of Chesterfields," it would be almost impossible to know that "Such Popularity" was an advertising film. The continuity has nothing to do with cigarette advertising. The advertising is more subtle; the film is supposed to promote the sale of cigarettes by building up good will in the same unobtrusive manner that radio advertising is supposed to do—and does![33]

While radio advertising and sponsorship established an initial commercial identity for an emergent medium, the introduction of sponsored films to American cinema dramatically reimagined this medium's existing identity. Promoters of sponsored films advocated indirect advertising so as to minimize the change that sponsored films meant for audiences.

Critics of sponsored films echoed much of the antiadvertising rhetoric that surrounded radio, while defenders of the experiment countered with equally familiar claims about "quality" entertainment. Carl Laemmle, president of Universal studios, appealed to theater owners "not to prostitute their screens to such pictures" and also urged producers/distributors to halt the practice. Laemmle warned that it was "a serious mistake to figure that because radio broadcasts contain advertising it is all right for the movies to do it," further warning that "in the long run it will degrade the movies and earn an ill-will which will drive millions from attending the movies."[34] In May of 1931, Nicholas Schenk informed theater managers within Loew's/MGM's chain that the company was actively fighting against the appearance of sponsored films. "We are definitely opposed to commercializing the screen," Schenk wrote. "An advertisement on the screen forces itself upon the spectator. He cannot escape it . . . yet he has paid his admittance price of entertainment alone." Schenk reassured managers in the Loew's chain that MGM was actively persuading other industry leaders and theater operators to resist the incursion of sponsored films. "We are continuing these efforts to free the screen from anything but entertainment. We are hopeful that the entire industry will take a stand against screen advertising and not attempt to force our theatres into that field."[35]

W. D. Canaday, vice president of Lehn & Fink, one of the first advertising agencies to aggressively pursue sponsored films, countered that these films' "advertising purpose has been completely subordinated to the requirements for a strictly entertaining feature" and besides, Canaday continued, movie audiences were already accustomed to ten minutes of "poorly presented or uninteresting" coming attractions, themselves blatant advertisements.[36] Canaday and other ad agency representatives sought to frame sponsored films as an entertaining supplement to any film program, arguing with a certain familiar tautology that if the amount of advertising in the films caused audience rebellion, ad agencies would simply make the films even *more* entertaining and less overtly commercial.

In publicly discussing sponsored films, advertisers often went to some lengths to identify certain self-imposed *limits* on both the types of sponsored films to be produced and their incorporation into larger media campaigns. Lehn & Fink cautioned local retailers to avoid overt references to "A Jolt for General Germ" in point-of-sale or shop window displays lest consumers recognize the commercial goals of the film and feel resentment. Instead local Lysol retailers were instructed to prominently display the product promoted in the film a week before, and a week after, the sponsored film's run at their local theater. The short subject's "entertainment"—conforming to the conventions of the animated film and produced, in fact, by the Fleischer studio—would create brand awareness that the prominent visible display of the product could exploit without undermining the film's integration into an entertainment medium.[37]

This indirect approach seems to have found favor in the motion picture trade press. Sponsored films were reviewed like any other short subject although the reviews indicate that the comparative absence of overt advertising became a crucial criterion of evaluation. In its review of Paramount's "Hurry, Doctor!" the *Motion Picture Herald* noted: "Although this is a commercial short, it is so handled that theatre patrons won't resent its appearance on the program. It shows Texaco Oil, which it advertises, coming to the aid of a 'sick' car, which shivers and cuts up in amusing fashion. The product is only mentioned three times, but unobtrusively enough so as not to disturb the entertainment value of the subject."[38] Less than a month earlier, a second Paramount release, "Suited to a T," received a harsher judgment: "Here is a good example of altogether too much advertising in a commercial short. There is little originality in the cartoon, and everything is subordinate to the India Tea company advertising, even to the accompanying dialogue. There is no question from the opening shot just what is the purpose and idea of the animated. It may pass in certain situations where the patrons are not too particular about looking at straight advertising in their short fare at the theatre."[39] The motion picture trade press, studios like Paramount and Warner Bros., advertising agencies and manufacturers collaborated to introduce sponsored films into American cinemas by conforming to the initial pattern of radio advertising and limiting the degree to which the films themselves flaunted their rhetorical purpose.

At more candid moments, however, the advertising industry voiced less commitment to the integrity of film entertainment and acknowledged that, much as they had done and were doing with radio, advertisers would test the limits of audience tolerance of an overt commercial address. In *Advertising & Selling*, Peter Andrews wrote: "Thus, advertising is inserted cautiously, at first only at the beginning and the end of the picture, and in some cases only at the end. If audience attitude is thought approving, more advertising is kneaded into the picture until a saturation point of advertising insertion is noted."[40] Some anecdotal evidence, widely circulated in the period, suggested that film audiences were generally receptive to sponsored films. In 1931, Darwin Teilhet reported that an advertising agency polling audiences after they had seen a sponsored film found only 18 of 191 audience members had an unfavorable reaction to the inclusion of an advertising film in their evening's entertainment.[41] If commercial sponsorship could prove acceptable for radio and film entertainment, some advertisers saw virtually no end to how pervasive such sponsorship might become. In an article that uncannily describes the contemporary pervasiveness of brands, Marsh K. Powers wrote,

If, as now seems possible, the continued popularity of sponsored entertainment, whether on the screen or on the air, is contingent upon repressing the sponsor's message to the briefest of poster copy, and if, under this limitation, it still proves itself a sales stimulant, then a sales-and-advertising principle has been established that must inevitably be applied in the widest variety of ways. The "commercial sponsor" will flourish like a green bay tree and the high rewards in sales volume will accrue to those sponsors who best foretell the temper of the public mind and, as a result, undertake those services for which the public will be most deeply grateful.[42]

Powers predicted a future in which commercial sponsorship might underwrite major league baseball (the Canada Drys vs. the Lucky Strikes), book publishing (with print ads on book jackets, flyleaves, and center-spreads), and even Broadway plays (Ethel Barrymore's new play "sponsored by The Austin Aircraft Corporation, builders of the famous Midget Aerocoupe, orchestra seats $1.00").[43]

While such pervasiveness of advertising, commercial sponsorship, and brands did not immediately flourish, in at least one notable instance an advertisement became itself a commodity. In January of 1931, music publisher Leo Feist, Inc., aggressively linked the publication and promotion of a new song to Chesterfield cigarettes. Called "They Satisfy" (the slogan for the Chesterfield brand), the fox trot was written by Gus Kahn and Carmen Lombardo. Appearing in Feist's advertisements alongside songs from the Goldwyn films *Whoopee* (featuring Eddie Cantor) and *One Heavenly Night*, "They Satisfy" was also promoted nationally through advertising for both the song and the cigarettes in seventeen newspapers and the magazines *American Weekly* and *Saturday Evening Post*.[44] While theme songs had for some time provided cross-promotion opportunities for film, music publishers, and phonograph companies, the Feist song aggressively expanded the principle to link entertainment commodities with other products within a commercial public sphere.

Despite Powers' prescience in describing contemporary branding at the turn of our century, existing historical accounts correctly identify cinema's adoption of direct advertising as a short-lived practice. Period accounts of the eclipse of sponsored films did not, however, attribute it to audience rebellion or even to the grumbling of independent theater owners or producers like Carl Laemmle. Instead, the print media (apparently seeing another threat to their advertising revenues) were credited with ending the film industry's sponsored film experiment by threatening to limit coverage of Hollywood and thereby curtailing a long-standing source of free publicity for the industry.[45] But the speed with which this practice was eclipsed in no way diminishes its importance as a symptom of struggles to determine media identity and the convergence of distinct media forms in the aftermath of technological change.

While on the surface sponsored films threatened to undermine the rhetorical assertions of service forwarded by the film industry and they appeared to challenge prevailing notions of entertainment, the experiment actually conformed to a logic of media "networks" and their relationship to commerce that was fast becoming the status quo within the radio industry. Broadcasters' success in defining the medium in a particular way—as advertising-supported, as a system of national program providers linked to listeners through a chain of local stations—validated the potential for cinema to adopt a new, supplementary identity and economic model. Because of the increasing concentration of ownership in the film industry (both prior to and exacerbated by the conversion to sound), major vertically integrated film companies could imagine themselves according to a chainstore model (similar to that of radio). Consolidation offered not only economies of scale but also a more fully integrated position in consumer culture. Existing links between cinema and the producers of consumer goods could be made more systematic and more profitable through media consolidation and a gradual redefinition of the nature of the "service" cinema provided. Radio was successfully undermining conceptual boundaries between entertainment and commerce through a particular mediation between the national and the local. Cinema too would contribute to this process in the 1930s and beyond. For the short term, however, that process did not include the incorporation of direct advertising into the film program.[46]

Cinema's relationship to advertisers of consumer products and its role in a larger consumer culture shifted slightly with the end of the sponsored film experiment. Systematic efforts to display consumer goods within the mise-en-scène of feature films linked manufacturers, advertising agencies, and film producers in temporary partnerships that connected film and filmgoing with other commodities and consumer purchases through both national and local tie-ins. Although not without their critics, these practices drew slightly less attention than sponsored films. Having failed to apply the direct advertising of commercial sponsorship for radio, film producers settled for a cinema-specific version of radio's indirect advertising.

Reports in 1933 indicated that several studios actively solicited recognizable, nationally distributed products from manufacturers for use as props in the production of feature films.[47] One account described a preproduction process at Columbia Pictures in which scripts were analyzed and a list compiled of every opportunity to feature in the film's mise-en-scène a recognizable product. Columbia then solicited the participation of advertising agencies in obtaining the products.[48] In September 1934, S. Charles Einfeld, head of advertising for Warner Bros., acknowledged in the journal *Sales Management* the creation of a special department within the studio to solicit the participation of national advertisers in tie-ups with its feature films.[49] The studio's systematic approach to cross-promotion sought to regularize the practice, to rationalize film's relationship to other consumer products through particular articulations of a relationship between the national and the local.

Warner Bros.' collaboration with Quaker Oats to promote the Joe E. Brown comedy *Six Day Bike Rider* exemplifies the process that linked films to consumer goods in coordinated national and local campaigns. Before the film's release, Quaker Oats ran a national newspaper campaign organizing "Joe E. Brown Bike Clubs" for children in cities with a local Warner theater. The cereal manufacturer then gave away bicycles to children who submitted two box tops and a short essay on "Why my mother makes me eat Quaker Oats," guaranteeing that at least one child in each bike club was a winner. Allowing local cooperation between theater managers and grocers, the bike clubs as well as the provision of point-of-sale advertising materials (including life-size color cutouts of Joe E. Brown) systematized local versions of a simultaneous national print advertising campaign that linked the film with Quaker Oats.[50]

Hollywood's intensified and more systematic approach to product tie-ins formalized a spatial conception of consumer society much like that of radio's networked relationships between national broadcasters/advertisers and local transmitters/consumers. Not surprisingly, the film industry, via its spokesman Will Hays, framed these commercial relations as providing a valuable public service to both American consumers and American industry. In an article published in *Good Housekeeping*, Hays argued for the positive influence of motion pictures on American life. Suggesting a link between the successful struggle for women's suffrage and film depictions of "the abilities, talents, and importance of woman," Hays told his predominantly female readers that motion pictures often played a role in shaping public attitudes and perceptions.[51] Cinema's contribution to the growth of U.S. political culture was, Hays suggested, surpassed by a more practical contribution to the daily lived experience of women. The motion picture contributed to "freeing woman from sheer drudgery" by depicting and popularizing a series of new, efficient, labor-saving devices for the home. "Fifteen years or so ago the motion picture began showing kitchens where power and its devices were in use—always, of course, as incidental to some story. In pictures orderly and effective home-keeping equipment for sweeping, scrubbing, washing, stirring, mixing, sewing made their appearance, and gradually it became plain that these things freed woman from enervating toil" (45). The movies, then, had provided a valuable public service by promoting domestic consumption.

Hays's rhetorical purpose in the *Good Housekeeping* article clearly involved an attempt to address common criticisms of the film industry at the time. The article took on such volatile contemporary subjects as gangster films, the movies' role in the lives of children, lapses in taste evident in some Hollywood advertising, and even the industry's vertical integration and concentration of ownership. But Hays's defense of the film industry emphasized the sound film's contribution to the development of a shared national culture that was articulated, perpetuated, and nurtured by the consumption of consumer goods. Hays suggested that the cinema's contribution to a shared national imaginary—expressed and experienced through consumption—was most dramatically illustrated by changes in the lives and habits of rural Americans. "One need not go far back in memory to recall when the rural dweller was easily and amusedly distinguishable because of crudity in dress and manner. Today there are few 'rubes' because there are so few areas not reached by the cinema. The result has been a definite socialization of the hinterland, through gradually absorbed knowledge and opportunity to draw comparisons" (130). Hays's presentation of the cinema's service to the nation echoed and reinforced advertising agency perspectives on film's power to shape consumer purchases. "The motor car took the farm people to the picture show, and automatically there followed appreciation of tools and civilization's progress. Safety razors were accompanied to the farm by the electric light, bottled gas for cooking, mechanical refrigeration, radio, custom-made clothing, and other features of a well-ordered life. In the language of the advertising expert, the motion picture created considerable consumer demand" (130). The indirect consumer address of product placement in films fueled the "social revolution" that swept through small towns and farms. Rural modernization, embodied in the consumption of a range of nationally produced goods, directly resulted from film's ability to shape consumer attitudes. "Printed or spoken sales argument may be resisted," Hays argued, "but specific illustration, unobtrusively, continually presented, imperceptibly produces a longing that finds expression in purchase" (130).

Emerging media in the 1920s and early 1930s took on meaning in relation to both a larger consumer culture and a long-standing perception that media's proper function might be to unify the nation. "So the motion picture, powerfully aided by radio, has helped tremendously to rebuild rural America during the last twenty years," Hays declared. "In its new vocal form it is beginning to affect our speech variations, tending to produce something closer to a national unity of language" (131). Emergent sound media helped to constitute a "shared national imaginary," largely and increasingly dominated by commercial interests. Through radio sponsorship, first sponsored films and then an intensification of product placement and tie-ins, broadcasting and film industries addressed audiences not only as consumers of entertainment media but also as citizens of a larger consumer culture.

This commercial mode of address was not, however, without its critics. Bruce Lenthall traces two critical responses to precisely this development in terms of radio. Critics focusing on "mass culture" decried the shared culture promulgated by radio (and, I'd argue, other now-related media forms) for its uniformity, its stasis, and its debased quality. A different group of critics attacked the political economy of emergent media that made possible such dreams of a shared culture: concentration of media ownership, oligopolistic media corporations, for-profit mass media, and media governed by the market to the exclusion of other possible values. These critics astutely attacked the degree to which the radio loudspeaker had become a mouthpiece for the interests of capital.[52] (See fig. 4.2.) Advertisers and commercial entertainment industries invoked the rhetoric of "service" to inoculate their audiences against both of these critical perspectives. It provided a rhetorical shield against a critique of the concentration of media ownership, and it also sought to reassure audiences that entertainment media conglomerates would not, indeed could not, participate in a cultural debasement since they functioned only to serve the consuming public.

Successfully masking for-profit activities as serving the public interest, this rhetoric sought to portray all broadcast programming as "educational"

FIGURE 4.2 When critics of early radio broadcasting portrayed the medium as a purveyor of political nonsense and debased cultural forms, they often agreed with radio's celebrants that the medium might have a profound effect on the shared political and cultural life of the nation. (Author's collection)

COMPETITION
—Fitzpatrick in the New York *World*.

and posited the free market as a guarantor of democratic values. Like Will Hays introducing the Vitaphone in 1926, CBS president William Paley—in 1934 testimony before the Federal Communications Commission that was also widely circulated in pamphlet form—stressed the service corporate media provided to audiences and to the culture as a whole. Through a series of adroit rhetorical moves, Paley argued that the *only* way to fulfill radio's potential as a tool of education lay down the path of commercial, for-profit programming. "Because radio is a sound business enterprise," Paley claimed, "it is able to make, and actually does make, such a continuously effective contribution toward our nation's cultural development."[53] In addressing the educational potential of radio, Paley posited an "American sense" of education in opposition to an elitist, European, old-world "aristocratic educational system." From this practical American perspective, "all broadcasts which tend to develop in our nation a unity of national sense and feeling may be considered to have an important educational value, whatever their subject."[54] Through this opposition, Paley and commercial radio could collapse the distinction between entertainment and edification, indeed claiming that "the criterion of success in such educational programs [is] a presentation so dramatic that the listener could distinguish it from pure entertainment only with difficulty."[55] Thus, CBS's rhetoric of service to the nation short-circuited a discussion that pitted culture versus commerce, insisting instead that the "sound business enterprise" of commercial radio would foster a democratic culture in which the marketplace determined what may or may not be of "important educational value." The market, according to Paley, could by definition *only* function at the service of the greater cultural good. Indeed, from Paley's perspective the commercial marketplace was the only "democratic" means of guaranteeing that radio would fulfill its regulatory mandate and serve the public's "interest, convenience and necessity."

The economic model of "chains" as a networked articulation linking the local with the national (whether embodied in broadcasting or film theaters) provided the structural possibility of a shared acoustic imaginary. Listening could become participation in a shared culture. But the process of mapping sound technology onto U.S. society did not stop here. Emerging media conglomerates, as part of their rhetoric of service, also grappled with competing visions of a less overtly commercialized culture. The utopian promises used to announce technological innovation still clung to sound media despite those media's susceptibility to a predominantly commercial mode of address. Acoustic media forms might deliver on some of the grandiose promises of cultural uplift and participatory democracy. While emerging media were being successfully mapped onto and were themselves exacerbating a commercial public sphere, simultaneous struggles sought to determine sound technology's relationship to cultural and political public spheres.

Culture

In her study of culture as an idea during the first half of the twentieth century, Susan Hegeman argues that, during the 1930s, ideas about and experiences of "culture" were often articulated in spatial terms. Hegeman attributes this, in part, to "the complex relationship between modernism as a practice and the experience of modernity itself. This experience," Hegeman continues, "rather than reflecting a seamless parade of jazz, cars, and steel, is, rather, marked by a perception of uneven development, and even friction, between those sites of modernization's greatest impact and the places that it touched less completely."[56] Although Hegeman describes a slightly later period, this perception of uneven development underwrote the grafting of emerging sound media onto social relations. Rhetorical constructions of radio broadcasting as, first, the conquest of distance and, later, as an instrument of national unification, as well as the film industry's suggestion that motion pictures had brought the benefits of modernization and consumer culture to the hinterlands, claimed that sound media might redress the uneven development characteristic of modernity. In a spatial conception of the nation that consisted of centers and a vast periphery, emerging instruments of mass media could function powerfully to bridge that gap, linking in thought, habit, practice, and culture a vast nation that had been unevenly experiencing dramatic change. While several of the New Deal's programs would begin to address at the level of infrastructure the symptoms of uneven development (the TVA and rural electrification among others), emerging sound media from the start took up where the phonograph had left off and promised a national unification at the level of access to culture.

Hegeman notes that some voices of what she calls the urban elite openly expressed disdain for the hinterlands (Mencken on the "booboisie" and his reference to the South and Midwest as the "Sahara of the Bozarts" or Harold Ross's declaration that the *New Yorker* was "not for the old lady in Dubuque").[57] But other voices within the larger culture industry actively sought to adapt their products to a perceived notion of the culturally different, often idealized, Midwesterner. Motion picture trade publications asked, "but will it play in Peroia?" and film historian Henry Jenkins has described how, with the conversion to sound and an attempt to adapt to a national audience, Eddie Cantor underwent a "desemitization."[58] The culture industry (including advertisers) used the new media of sound to actively adopt the role of mediators, bridging a perceived cultural gap that separated urban centers from the margins. When differences in taste and in culture are understood spatially, when the gap between "high" and "low," or even between "high" and "middlebrow" is understood in relation to the physical distance between consumers and centers of cultural production, then mass media that promise to eclipse space take on particular saliency as instruments of unification. A popular song, massively and more or less instantaneously available over the radio, on phonograph recordings, and in a film, drastically raises the stakes for those concerned with the deleterious effects of "low" cultural forms even as these media also provide a potential instrument of cultural uplift and edification. The new media conglomerates of the late 1920s and early 1930s navigated a precarious path between maximizing profits through the dissemination of standardized entertainment products and rhetorically positioning themselves as responsible corporate citizens who provided a service to the nation by promulgating elevated cultural standards.

Like the mechanical phonograph before it, the introduction of electrical sound technology seemed to offer the instrumental fulfillment of a longed-for "musical democracy." Assertions regarding the cultural significance of new sound media reconfigured the spatial binary *national:local* as *urban: rural*, and new media seemed to promise to bring the best of the city and its elite cultural forms to the hinterlands. As has been noted, the farmer became a crucial rhetorical figure in the initial debates about radio that claimed the medium's identity resided in its capacity to conquer distance. In a discursive construction that directly echoed earlier efforts to redefine the domestic phonograph as a musical instrument, many radio advocates highlighted broadcasting's ability to end rural isolation. Long nights on the farm—particularly in the winter—might soon be accompanied by the uplifting sounds of urban symphonies and operas, or by edifying talks featuring the nation's great speakers. Radio networks like NBC and CBS consolidated their power by seeming to provide the best in national programming to geographically isolated populations.[59] Not surprisingly, the U.S. film industry followed the rhetorical trail blazed by advocates of both the phonograph and the radio, arguing that the conversion to synchronous sound provided previously unimagined possibilities for the broad dissemination of culture. The centralization of sound production and the technological standardization of musical accompaniment practices would not only put an end once and for all to local, substandard musical performances in movie theaters but also allow even the most rural of filmgoers consistent access to the best orchestral and vocal performances in the land.

But if sound technology enabled the broad dissemination of high culture from urban centers to rural margins, it also threatened to spread broadly the perceived decadent and debased qualities of *some* urban culture. Emergent sound media and their increasing importance in American life exacerbated preexisting anxieties about the degradation of some popular culture forms. While the United States' musical elites initially embraced radio as an instrument of cultural uplift, seeing in the medium a capacity to foster the growth of musical appreciation for the classics, they soon attacked radio for its promotion of base and degrading musical culture.[60] Radio became a primary site around which was waged a battle over the cultural merits of different types of music, but this battle resonated across a variety of sound media. For every argument stressing the utopian possibilities of sound media, counterarguments seemed to warn against dire dystopian developments. Radio or the sound film might widely disseminate the best of a cultural heritage, or even improve the average American's use of language, but acoustic media also "threatened" to introduce and even valorize movie slang and thereby corrupt the national tongue.

Frank Biocca notes that during the early period of radio broadcasts, spokesmen for musical elites attributed to radio broadcasts of classical music a "triumph over jazz"[61] But while classical music received more air time than popular music up through the beginning of regular network broadcasts in the 1926–27 season, by 1930 " 'light' and 'variety' music programs dominated the airwaves with a ratio of almost 2 to 1 in air time over classical music."[62] This development refueled dire pronouncements regarding the deleterious effects of jazz, echoing many of the antijazz sentiments from the 1920s. On the pages of newspapers and journals, advocates of classical music and elite high culture railed against the acoustic corruption of U.S. culture. The expansion of radio and the sound film lent a greater urgency to their warnings. Emergent acoustic media thus became a site on which was waged an old battle regarding sound, cultural legitimacy, and the relationships between commerce and culture.[63]

Lawrence Levine has argued that jazz and Culture—as labels—took on meaning in an antithetical relation to each other. The emergence of jazz, Levine writes, "as a distinct music in the larger culture paralleled the emergence of a hierarchical concept of Culture with its many neat but never precisely defined adjectival boxes and categories."[64] Perhaps no sound practice from the early electrical era generated more scorn than so-called "jazzing the classics." It became a rhetorical touchstone for dire warnings regarding sound technology's threat to cultural standards and good taste. Debates about jazz and its relationship to culture coincided with the earliest developments of radio and the sound film.[65] Two events, reported in a radio publication, illustrate some of the preexisting anxieties surrounding jazz and culture that were exacerbated by emerging sound media. In May 1925, *Radio Broadcast* columnist Jennie Irene Mix published a brief note on the lawsuit filed by a Spokane music teacher seeking $10,000 in damages from the leader of a local jazz orchestra.[66] The suit alleged that the practice of so-called "jazzing the classics" constituted a crisis in that "the public has received a perverted idea of classical music, insofar that children may no longer desire a musical education." Characterizing "jazzing the classics" as "the greatest outrage perpetrated by jazz orchestras," Mix concluded, "Would that Mr. Woodward might win ten times ten thousand dollars!" In the same column, Mix revisited an earlier report that, on Christmas Eve, several readers of *Radio Broadcast* had heard "Silent Night" "jazzed" over the airwaves of station WTAM in Cleveland.[67] Admitting that she had not heard this "sacrilege," Mix reported with pleasure that apparently WTAM was not the culprit and reprinted the carefully worded, emphatic denial by station manager S. E. Baldwin. While *Radio Broadcast* effectively absolved WTAM of guilt in the matter, the fact that some station, somewhere, had on Christmas Eve indeed broadcast a jazz version of the hymn was left unquestioned. "Probably a number of people heard more than one station at that hour," Mix concluded. "[This is] the leading fault of radio at present." The Spokane lawsuit and the broadcast of a jazz rendition of a Christmas hymn illustrate that emerging sound media not only promised edification and uplift but also potentially exacerbated what was perceived as an ongoing debasement of cultural values.

Perhaps the antithesis of so-called "jazzing the classics" was the transmission via emerging sound media of the operatic canon. Warner Bros.' initial Vitaphone short subjects included a significant number of recordings by opera performers. In fact, the premiere Vitaphone program accompanying *Don Juan* in August 1926 included shorts by Metropolitan Opera stars Giovanni Martinelli (a selection from *I Pagliacci*) and Marion Talley (a selection from *Rigoletto*).[68] Lending prestige to the technology itself, to the studio, and to any theater that showed them, these opera short subjects and other recordings of culturally valorized performers built on and reinforced consensual notions of high culture. Like the importance of Caruso and grand opera recordings to the redefinition of the phonograph as a domestic musical instrument, these Vitaphone short subjects drew upon the preexisting prestige associated with both opera as a performance genre and its existing star system. Many of the performers in the Vitaphone shorts prior to the summer of 1927 would have been recognizable to U.S. audiences, if not from their phonograph recordings then from their appearances in Victor's prominent national advertising campaigns. Besides Martinelli and Talley, Anna Case, Frances Alda, Ernestine Schumann-Heink, Benjamino

Gigli, and Giuseppe De Luca joined myriad concert performers, including
Mischa Elman and Efrem Zimbalist, in performing for the Vitaphone.

Warner Bros. also followed the phonograph's lead by including in its Vitaphone opera films explanatory information that narrativized musical performances. Title cards provided "libretto" information in the same way that
Victor's publication of *The Victrola Book of the Opera* or printed information
on the backs of phonograph recordings sought during the 'teens and early
twenties to structure the experience of listeners as an edifying entertainment. By late 1929 and early 1930, however, such libretto information or
other framing material began to disappear from Vitaphone's recorded presentations. Musical performances were left to stand alone, decontextual-
ized, as it were, from the narrative and performance text in which they
originated.

Relatively quickly, Warner Bros. developed a system of labeling all its
Vitaphone performance recordings, categorizing them according to genre
designations (Comedy, Drama, Male Vocal, Opera, "Flash," etc.). Such cat-
egorizations also became a way for exhibitors to select from Warner's growing catalog of short subjects the types of entertainment suitable to a given
program or venue.[69] Coincident with the inauguration of this system of labeling, the number of releases designated as "opera" recordings rapidly declined, and soon even Vitaphone exclusive artist Giovanni Martinelli's performances were designated as "Male Vocal." This shift involved more than
simply generic labeling. Martinelli's Vitaphone performances after April
1930 less frequently reproduced the operatic canon and instead shifted toward concert performances, often sung in English. For Vitaphone release
no. 1162 (recorded February 1931), Giovanni Martinelli in "The Ship's
Concert," the catalog copy announced: "The famous opera star emerges
for the first time from the field of grand opera to that of lighter music
and he sings in English."[70] Subsequent releases—such as Vitaphone no.
1174 (semiclassical ballads, in English, recorded December 1930); no. 1213
("popular Italian songs," recorded February 1931); no. 1226, "The Trouba-
dour" ("popular semi-classical Spanish songs" accompanied by dancers,
recorded May 1931); no. 1245, "Giovanni Martinelli in Gypsy Caravan" (set
in a Gypsy camp with songs in English, Russian, and Hungarian, recorded
April 1931)—appear designed to broaden the popular appeal of Martinelli
beyond the narrowly defined audience of opera lovers. Catalog copy for the
latter addressed perceptions of the limited appeal of high culture, remind-

ing exhibitors that, "One does not have to be a lover of classical music to be able to appreciate this beautifully produced short."[71] Although still deemed popular enough by Warner Bros. to warrant appearances in short subjects, Martinelli and his Vitaphone repertoire had shifted. By release no. 1245 in May 1931, much of the high-culture tone characteristic of the short subjects accompanying *Don Juan* had disappeared from the Vitaphone repertoire.

Warner Bros.' presentation of opera to U.S. film audiences was a comparatively short-lived phenomenon. Charles Wolfe suggests that when Vitaphone relocated its recording facilities from New York to Los Angeles in the summer of 1927, opera stars refused to travel to California to record, but even when East Coast production resumed, traditionally identified "high culture" remained relatively absent from Vitaphone's releases.[72] The initial prevalence and then relatively rapid disappearance of the operatic canon from Vitaphone's releases attest to a degree of indecision as to the nature of sound film entertainment. Vitaphone took, in part, its sense of the recorded repertoire from the phonograph industry and, like the previous technology, used operatic performances to add prestige to the sound film while also implicitly forwarding fidelity claims and temporarily embodying in the recorded commodity some of the rhetorical promises that had been used to announce technological change. Opera short subjects could enact the rhetorical promises of "musical democracy" by making nationally available the greatest vocal artists, regardless of community size and location. The presence of opera in a local film program not only addressed opera lovers in a particular community but also, by applying a patina of high culture to this emergent acoustic media form, implicitly asserted to the community at large the social benefits of technological change. In recording opera performances, Warner Bros. followed a path already paved by the phonograph—particularly the Victrola—asserting that sound reproduction offered substantial social benefits. Ultimately, Warner Bros.' Vitaphone opera shorts might best be understood in a manner analogous to Victor's admission to the phonograph trade in 1905 that while profits resided in popular recordings, "there is good advertising in Grand Opera."[73] Wolfe concludes that rather than collapsing the boundary between urban and rural as the rhetoric of "musical democracy" promised, Vitaphone's opera shorts might well have reinforced it. "Opera shorts thus may have been targeted for urban markets, where an enthusiastic audience for these vocal performances encompassed ethnic areas as well as the cultural 'elite.' From this perspective, the divide between opera and 'popular' music appears less a function of class stratification than of urban-regional boundaries."[74] Warner Bros.' comparatively short-lived production of opera shorts might then also be understood as a response to perceived uneven "cultural development"—a response somewhat at odds with the commercial goal of standardized entertainment products.

A closer look at the developing Vitaphone repertoire and the release strategies of Warner Bros.' short subjects suggests that after a period of what might be understood as experimentation drawing on the preexisting success of the phonograph, Warner Bros. quickly moved to standardize the production of short subjects and to emphasize more popular rather than elite entertainment forms. The earliest Vitaphone shorts were characterized by a tremendous variety in the recorded repertoire and by a significant amount of titles presenting opera, concert, and symphonic performances. The Vitaphone release catalog's chronological listing of short subjects suggests that the marked decrease if not outright eclipse of opera films coincided with a move toward greater standardization of Vitaphone's production through the release of a specially marked *series* of short subjects. While the recorded repertoire still included tremendous variety (animal acts, playlets, musical performances, drag acts), that variety mimicked the vaudeville bill rather than the concert hall. Vitaphone's recorded repertoire was first managed by the institution of categorization to aid theater owners in constructing a balanced program and then further regularized through the inauguration of series like the 12-part "Adventures in Africa" travelogue, Bobby Jones in "How I Play Golf" (also twelve parts with cameos by Warner Bros. stars), CBS announcer Ted Husing in "Sportslants," syndicated newspaper cartoonist Robert L. Ripley in "Believe It or Not," and, of course, the Looney Tunes animated series.[75]

Within the commercial cultural sphere, jazz and other popular forms would largely triumph over symphonic music and opera, as was the case with Vitaphone's short subjects, but U.S. acoustic culture also often suggested a possible synthesis of what were elsewhere seen as inevitably conflicting musical forms. As radio and the sound film became incorporated within a broader political economy of media, they seemed to promise a resolution to cultural clashes. Such a synthesis, initially enacted in variety radio programming or the short-lived cycle of Hollywood revue films that mimicked radio broadcasts, serially combined high culture and more popular performance styles. This combination of elite and popular forms would be eventually codified and regulated during a broadcast schedule largely through the distinction between sponsored and sustaining radio programs. Sponsors predominantly offered consumers the popular while radio networks picked up the banner of culture through high-profile sustaining programs designed to do as much for network radio's reputation as it did for the cultural enlightenment of radio listeners.[76]

While Jennie Irene Mix's *Radio Broadcast* column of 1925 concisely indicated the cultural crisis surrounding and exacerbated by emerging acoustic media, it also included the solution, or rather, a *particular* solution that ultimately served the interests of forces seeking to consolidate the U.S. radio industry. In the same column that reported both the Spokane music teacher's lawsuit and the "jazzing" of "Silent Night," Mix outlined an argument for a particular political economy of radio broadcasting that emphasized national radio networks, the centralization of program production, and advertiser support (sponsorship that would eventually grow into full-blown advertising). "We are confident," Mix wrote, "that hundreds of thousands who have until now always referred to such people ["lovers of good music"] as 'highbrows' or 'poseurs,' are going to go over to these very ranks when they find through experience that love of good music is no more a pose than is the preference of living in a neighborhood where the surroundings are beautiful to the eye rather than in one where ash and garbage cans predominate."[77] The solution to the perceived cultural crisis might reside in the very media that seemed to be exacerbating it. Mix further reported the results of a study by AT&T broadcasting manager John A. Holman, who claimed with statistical certainty that public tastes were improving because of radio, particularly the national broadcasts originating in AT&T station WEAF (New York). Holman claimed that in January of 1923, approximately 75 percent of radio fans favored jazz but that by January of 1925 that percentage had fallen to a mere 5 percent.[78] A more economically rationalized (commercial) use of radio as a public resource promised to serve the cause of uplift and edification, effectively resolving anxieties about cultural degradation.

In a May 1925 survey of broadcast programming trends, James C. Young overtly linked "quality" programming to commercial sponsorship.[79] Such a perspective, repeated frequently—indeed pervasively—in the mid- to late-1920s, took on the status of "conventional wisdom." Radio, indeed all sound-reproducing media, could promote a more culturally educated populace. In the pages of radio journals, when programming did come under critical scrutiny, the medium's potential for raising national standards of taste remained largely unquestioned. The issue, as radio developed in the United States, became how best to fulfill such potential. In an August 1927 column proclaiming "Radio Is Doing a Good Job in Music," John Wallace cautiously assessed broadcasting's progress: "We do not delude ourself that radio has been able to contribute a highly sophisticated taste in music to very many people. But (what is vastly more important as an opening wedge), it has given a taste for music to millions of people who previously hardly knew it existed."[80] Asserting that "so far only the surface of music's possibilities has been scratched," Wallace concluded by noting the recent appointment of nationally renowned conductor Walter Damrosch as "Musical Counsel" for radio network NBC.

Broadcast programming like, first, the Atwater Kent show and the Victor hour and, later, Damrosch's "Music Appreciation Hour" or the Leopold Stokowski concerts, fulfilled a cultural function, introducing some listeners to classical music and broadly reinforcing the normative taste of high culture. By 1926, the music journal *Etude* presented broadcasts on stations WIP (Philadelphia) and WLS (Chicago).[81] But in the context of grafting new media onto existing social relations, these programs also fulfilled a powerful rhetorical function, implicitly arguing for commercial sponsorship of programming broadcast nationally by embodying the ever-promised utopian potential of sound media.

When the realities of 1920s radio broadcasting content failed to live up to this discursively constructed "potential," supporters of a particular political economy of broadcasting pointed to emerging radio networks as a panacea for culturally valorized musical performances. The national distribution of broadcast content enabled by radio networks guaranteed the uplift of public taste, or so the argument went, since national distribution allowed the greatest performers to address a near-national audience. Much like the film industry's claim that recorded soundtracks would end aberrant and amateurish live musical accompaniment in local theaters, network radio broadcasts could displace from the ether the degraded and the banal. National broadcasts could educate the public, elevate their taste, inspire young musicians, and create new concert- and opera-goers, silencing the cacophony of lower-powered stations broadcasting amateur recitals, or worse yet, jazz.

The clash between jazz and Culture was also enacted and tenuously resolved through the career and music of bandleader Paul Whiteman. Whiteman rose to national prominence at precisely the moment in which the new acoustic media were being grafted onto social relations. Like other entertainment figures—notably crooner Bing Crosby, comedians Amos 'n' Andy and Eddie Cantor, and singer Al Jolson—Whiteman's presence extended across multiple media forms and his pervasiveness and popularity were due in no small part to the intermedia commodity made possible by economic relationships within and between emerging media conglomerates. A lightning rod, of sorts, Whiteman attracted disdain from both sides of the jazz versus Culture debate. His importance, his visibility, and his acoustic pervasiveness in the midst of electrical sound technology's emergence both illustrate the salience of this conflict and provide an example of a different strategy to engineer a synthesis between high and low culture.

Rather than perpetuating jazz's real or perceived acoustic assault on the musical canon or the larger Eurocentric values which it embodied and promulgated, Whiteman sought to construct a synthesis of jazz instrumentation and technique and that musical canon, essentially a synthesis of European and American traditions. On February 12, 1924, Paul Whiteman's attempted synthesis of jazz and Culture reached a culmination of sorts when he presented a "symphonic jazz" concert at New York's prestigious Aeolian Hall. Offering a "history" of jazz that concluded with the premiere of George Gershwin's composition "Rhapsody in Blue," the concert also announced the emergence of symphonic jazz, an infusion of jazz into the Western musical canon.[82] In attendance at Aeolian Hall, leading figures of elite culture (Damrosch, Kreisler, Stokowski, McCormack, Rachmaninoff, Heifetz) mingled with noted American writers including Gilbert Seldes, Deems Taylor, Carl Van Vechten, and Heywood Broun.[83] In Whiteman's music and particularly in the discourses that surrounded it, the "dangerous," indeed "narcotizing," effects of jazz were tamed and diluted. Syncopation (now without the sin) served national interests by promoting a shared acoustic imaginary.

Whiteman in his 1926 autobiography *Jazz* repeated many of the hyperbolic critiques of jazz's dangers and, equally hyperbolic in his defense of the music, claimed it as an antidote to the damage done to humanity by the standardization of a machine age. "Spiritually," Whiteman claimed, "jazz is saving America from calamity."[84] Admitting to the critics' claims regarding the intoxicating potentials of jazz, Whiteman attributed to that intoxication an invigorating effect: "I firmly believe jazz has the power to take people out of a dull world, to shake them up, to give them an intoxication of rhythm and movement which makes them happy, makes them better, makes them live more intensely."[85] But if jazz somehow arose from and responded to the changing world of modernity, it necessarily contained within it a readily perceived threat to more traditional values and experiences. Jazz, particularly when pervasively disseminated by the new sound media, could embody the uneven geographic experience of modernity. In his analysis of Whiteman, Gerald Early suggests that "symphonic jazz" intervened in the preexisting cultural clash that pitted "city versus small town, agrarian values versus urban trends."

> Symphonic jazz was not simply a bow to the highbrow culture that condemned jazz as unartistic noise but to the frightened small-towner, philistine and otherwise, who was alarmed by a secular music that seemingly had no restraints and no aim but seemed to possess enormous emotional and commercial power. Symphonic jazz was an attempt to make jazz less the cultural assault against white, middle-class, Christian, small-town taste than it appeared to be.[86]

As such, Whiteman's music seemed tailor-made for an emerging acoustic imaginary that, for commercial as well as cultural purposes, sought to bridge the apparent gap between rural and urban.

By removing the sin from syncopation, by purging the bodily from jazz, Paul Whiteman's music simultaneously provided an antidote to the debased practice of "jazzing the classics" while synthesizing the "best" of the New World (jazz) with the Old World (the classical repertoire, symphonic instrumentation). Whiteman's music, the 1930 revue film *The King of Jazz* in which he starred, and virtually every promotional article surrounding the bandleader reasserted and valorized a synthesis between the otherwise opposed jazz and Culture. Whiteman's symphonic jazz further sought to bridge the geographically imagined cultural gap that separated the urban modernity of cultural production from the vast hinterlands populated by more conservative consumers.[87]

In 1935, Disney's Silly Symphony cartoon "Music Land" reenacted the clash in the wider acoustic culture between symphonic music and jazz.[88]

The animated short pitted denizens of the Land of Symphony against those of the Isle of Jazz in a revision, of sorts, to Shakespeare's *Romeo and Juliet*. The film focused on the tale of two young lovers from each island who are separated by both geography (the Sea of Discord) and apparently incompatible musical cultures. The mise-en-scène of each island imaginatively exaggerated commonsense cultural connotations of each musical style. The Land of Symphony, with its castle of organ pipes instead of crenelated towers, was in every way refined and measured. The Queen (a cello) dozed on her throne while harps waltzed with stringed instruments for her entertainment. The Isle of Jazz, with a flashing neon sign and a saxophone skyline, was a far more raucous and indeed lascivious setting. Banjos, brass instruments, and hula-clad ukuleles danced and cavorted for the pleasure of the King (a corpulent alto saxophone with pencil-thin moustache). The King participated in the proceedings, his joy registered with booms and crashes from his elaborate drum-kit throne.

The Prince of the Isle of Jazz (a tenor sax) sneaks off to visit his beloved the princess (a violin, of course) in the Land of Symphony, but he is caught by her mother the Queen and imprisoned in a metronome. A musical war results, in which jazz volleys fired by the Isle of Jazz's trombone, tuba, trumpet, and clarinet cannons are answered with the full force of the Land of Symphony's pipe-organ artillery, playing the *1812 Overture*. An off-target volley frees the Prince from his prison, and trying in vain to restore the peace, he rushes to join the endangered Princess, now adrift in the Sea of Discord. When both of their boats capsize, the reunited lovers appear lost until the rival King and Queen arrive on the scene for a last-minute rescue. King and Queen make eye contact in a moment of musical flirtation. To further solidify the musical rapprochement, the wedding between the Prince and the Princess is in fact a double ceremony also uniting the rival leaders—all of which is performed on a newly erected "Bridge of Harmony" that now spans the Sea of Discord and links the two lands.

"Music Land" draws its comic logic from references to larger discursive and practical struggles in U.S. acoustic culture. It visually and acoustically invokes commonsense critiques of each musical form: symphonic music is staid, overly measured, effete, cerebral, and pompous; jazz is unrestrained, lascivious, loud, brassy, and bold. That cultural resonance is particularly explicit in the cartoon figure of the King, for the Disney animators designed the King of the Isle of Jazz to resemble Paul Whiteman. "Music

Land" suggests an imaginary resolution—a synthesis—between the otherwise antithetical understandings of jazz and Culture, but the dialectic enacted in the Disney film remained unresolved elsewhere. The relative cultural value of jazz continued to be an object of debate for the next several decades. Whiteman's symphonic jazz and some of the "sweet jazz" bands it spawned became objects of disdain for the urban cognescenti and even for geographically dispersed enthusiasts of "hot jazz." The synthesis Whiteman sought would be most lastingly enacted through the institutionalization of sustaining programming on the radio networks. Presented to the public by radio networks as a crucial component of their service to consumers, sustaining programs represented a material resolution (a containment, in fact) of otherwise apparently conflicting cultural categories. But sustaining programs provided a resolution that ultimately served the emerging political economy of U.S. broadcasting.

Proponents of jazz often celebrated it as a uniquely American form of musical expression. Defenses of jazz took on a distinctly nationalistic tenor.[89] In a 1924 article illustrated with photographs of Whiteman's band leaving for Europe, conductor Hugo Riesenfeld wrote in the *Outlook*: "As a player of jazz no one can take the place of an American. His sense of rhythm is remarkable—stronger than that of any other nationality—and he can express that rhythm with a peculiar tang and buoyancy that cannot be duplicated. . . . The rapid, pulsating life in this country is largely responsible for the peculiar quality of its musical expression. The American has taken what was originally a Negro rhythm and grafted on to it the natural tempo of his own existence. He has produced something that is distinctively his own."[90] Composers like Leopold Stokowski (conductor of the Philadelphia Orchestra) and critics like Gilbert Seldes celebrated jazz as an American form of expression. Paul Whiteman, like other defenders of jazz, praised it as a uniquely American essence, thereby constituting an acoustic nationalism—or sound as a particular articulation of nation. In his autobiography, Whiteman claimed that jazz expressed in musical form an American essence; it expressed the soul of America; it contributed to a "racy, idiomatic, flexible American language all our own, suited to expressing the American character"; jazz was "the great American noise"; and, ultimately, jazz represented America's first cultural contribution to the world.[91]

The rhetorical construction that portrayed Paul Whiteman's music as a uniquely *American* synthesis of jazz and classical music reveals the implication of emergent acoustic media within discourses of nationalism and American identity. The shared imaginary offered by sound media could be uniquely American; it could unify what was perceived to be a dangerously fragmented social body. Such issues surrounding acoustic culture's relationship to national identity further implicated emergent sound media in the rhetorical sediment if not the actual reality of a participatory political public sphere.

Politics

Having announced new electrical sound technology as an instrument that might fulfill the perpetually deferred promise of a participatory public sphere, corporate innovators and emerging media conglomerates sought to map sound media onto existing political practices. These efforts largely stressed two related concerns: the amplification of political rhetoric—its increased scope and reach—and the preparation of social subjects to be citizens, that is, their construction as appropriate audiences for this amplified political rhetoric and even as political actors in a public sphere. Preexisting cultural anxieties about immigration and assimilation figured prominently in discussions of the political uses of emerging sound media, as the nation was often portrayed as a precarious combination of disparate groups.

For George Creel, reflecting on the accomplishments of the Committee on Public Information, immigrant Americans had posed a particular challenge to the committee's goal of national unification and ultimately proved to be one of the campaign's greatest success.

> Even as the Committee on Public Information claims success in its fight for public opinion in other countries, so does it advance its pride in the unity and enthusiasm that marked America's war effort. The foreign born, feared with respect to their ignorances and prejudices, were brought into closer touch with our national life than ever before, while absolute openness and honesty in the matter of official news, the eloquence of speakers, the story of motion pictures and posters, all combined to banish the ignorances and indifferences that threatened the full loyalty of the native born.[92]

Creel's final report emphasized the committee's success in promoting the assimilation of the "foreign born" and credited the aggressive use of print media with enlisting immigrant support for the broader war effort. In the aftermath of the Creel Committee, electrical acoustics and its different media applications appeared to offer new and expanded methods of hailing the immigrant and interpellating him or her into the national social body.

Although the Creel Committee claimed success in incorporating the "foreign born" into the U.S. war effort, anxieties about immigrants persisted in the war's aftermath. Following the precedent of Creel, some saw in emerging media, specifically radio broadcasting, the potential to constitute a national public and thereby solve one of the abiding "social problems" of the day. In 1923, J. M. McKibbin viewed radio as a tool of assimilation, hyperbolically framing the issue in terms of crisis and contagion:

> To-day this nation of ours is slowly but surely being conquered. . . . Millions of foreigners were received into this country, with little or no thought being given to their assimilation. What is the result? To-day we see it everywhere, in Little Italy, Chinatown, Slovakia, etc., each retaining, as far as possible, the customs, language, and traditions of its mother country. Each a parasite living upon the national resources and under the protection offered by America, yet giving little or nothing in return. It is this process of nationality isolation within one country which is ruinous.[93]

For McKibbin, the solution to this "crisis" resided not within the power of a single individual leader, no matter how charismatic or powerful ("Perhaps no man could mould these one hundred and twenty millions of people into a harmonious whole, bound together by a strong national consciousness"), but crisis could be averted and the contagion cured through the auspices of technological intervention. McKibbin suggested, "But in place of a superhuman individual, the genius of the last decade has provided a force—*and that force is radio.*"

McKibbin, and other commentators like him, believed broadcasting configured as a national mode of address could bind together a nation of diverse individuals by inspiring a "national consciousness" characterized by duty, pride, and shared commitment:

> If properly employed, radio will cause the indifferent or antagonistic American unconsciously to become familiar with our government, its people, and their ideals. . . . Radio is too large a force to deal with the

many petty social and political differences of village and town—it deals with matters of state and nation, with matters of international importance. The "listeners-in" are lifted from that common plane of trivial interests and unconsciously made to realize that they are (collectively speaking) responsible for the guidance of this nation; that its troubles are their troubles, and that its achievements are due, in part, at least, to the conscientious fulfillment of their duty.[94]

McKibbin's rhetoric, albeit a good deal more hyperbolic than that of many period commentators, illustrates the manner in which electrical-acoustic technology and the act of listening were grafted onto and understood in terms of powerful preexisting sentiments regarding the relationship between citizen and nation.

Assimilation through listening was thought to function unconsciously, inevitably, and even, perhaps, coercively. In 1922, Robert Olyphant, president of the Sons of the Revolution, wrote: "The voice speaking the English language and the voice propagating the American ideals and literature will go into the homes of these foreign born citizens. They will be listening to the wireless voice and benefiting from it, whether they want to or not."[95] Olyphant early on suggested that radio listening could remove language barriers but also cultivate a shared sense of history and, importantly, a shared internalization of "American" values.

Commercial uses of emerging sound media, particularly when employing a national mode of address, claimed to constitute acoustically a listening subject (although actually inhabiting that subject position is a different issue) by incorporating an auditor within a larger, national, posited audience. Listening promised to eclipse not only geography but also class, education, ethnicity, and the myriad other components of identity and difference that must be elided for participation in an idealized "public" of equals. Under such circumstances, the issue of assimilation, of incorporation into the body public, became particularly trenchant. The voices that emerged from loudspeakers could interpellate the listener, despite his or her specificity, within the national body, or so the argument went.[96]

In his study of local broadcasting in Chicago during the 1920s and early 1930s, Derek Vaillant describes the constitution of a "sound of whiteness" which united ethnic Chicagoans as "racialized whites" despite their ethnic differences. Radio's "sound of whiteness" resulted from the systematic exclusion or marginalization of African-American voices and culture from radio programming, combined with the broadcast of both blackface minstrelsy and African-American jazz and blues transformed into popular syncopated dance music produced by and for whites.[97] Local radio in Chicago, much like Paul Whiteman's symphonic jazz, provided a cross-cultural unification as "the disembodied cultural presence of African-American jazz on radio and its popular elaboration by white performers supplied another means of unifying ethnic audiences under the cultural sign and sound of whiteness."[98]

Emergent sound media in the late 1920s and 1930s were widely hailed as particularly potent instruments that might promote the assimilation of U.S. immigrant populations. Although desires for and proclamations about the national unifying potential of new sound media circulated widely, shaping both perception and practice, the same media also opened up possibilities to unite more localized and distinct populations.[99] While local Chicago radio could collapse perceived boundaries between white ethnic groups, "hailing" them as precisely white, Vaillant details a different, simultaneous cultural function of listening. Some local broadcasters in Chicago actively cultivated a sense of local and ethnic community, both reproducing aurally some of the spatialized ethnicity of the city *and* fostering a shared sense of ethnic identity between spatially dispersed members of groups, thereby reconstituting a virtual, acoustic community. "In the 1920s and early 1930s," Vaillant writes, "local broadcasters counterbalanc[ed] the assimilating pressures of commercial advertising, Americanization campaigns, and the homogenous fare of 'chain' and early network broadcasts."[100] Radio station WIBO, for example, fostered a Swedish identity both in Chicago's Andersonville neighborhood and for listeners as far away as Iowa, Minnesota, and Wisconsin.

Other local stations sought to address Chicago's ethnic populations through specific, regularly scheduled programs. In the summer of 1929, for example, labor-owned WCFL offered Chicago's listeners a weekly "German Radio Hour" on Sunday evenings and a Wednesday night "German Street Band" broadcast.[101] In the following year, the station's schedule supplemented the German programming with Sunday afternoon broadcasts that included "Brunswick's Polish Hour," "Brunswick's Lithuanian Radio Hour," and "Brunswick's Lithuanian Music Hour." Weeknight programming included an hour-long "Spanish Musical Program" and a "Jewish

Musical Hour."[102] In his history of WCFL, Nathan Godfried argues that these broadcasts not only reflected the diversity of Chicago's many ethnic neighborhoods but also the Chicago Federation of Labor's commitment to linking ethnic concerns with class-consciousness and progressive organizing, in opposition to more assimilationist tendencies within the AFL. Godfried's account describes these programs (as well as similar weekly broadcasts addressed to Italian and Irish audiences) as consisting of "music, dancing, singing, occasional dramatic skits, conversation and public announcements of ethnic festivals and community meetings," often sponsored by and produced in collaboration with local ethnic groups and local newspapers printed in listeners' original languages.[103]

Although an anomaly in that it was the only station owned and operated by organized labor, WCFL's programming is in some ways exemplary of larger developments in the late 1920s and early 1930s.[104] The station's programming schedule sought to address serially multiple perceived constituencies, that is, multiple "publics." Individual programs actively cultivated specific identities whether based on class (a regular half-hour broadcast on Thursday nights by Painters' Union, Local 194), ethnicity, or geography (a regular 15-minute morning "Farm Talk" by a representative of the Farmers Union). Simultaneously, however, WCFL also contracted with NBC's Blue Network to provide national programming that was both event-based (championship boxing matches or the inaugural ceremonies for President Herbert Hoover) and weekly (the Sunday night broadcast of Roxy and his Gang).[105]

Radio was not the only medium that reinforced Chicagoans' ethnic identities. In the winter of 1931, the Studebaker Theater in Chicago screened the German-language UFA-produced operetta *Zwei Herzen Im 3/4 Takt* (*Two Hearts in 3/4 Time*). The film proved popular enough that the theater held it over for a total of five weeks.[106] German-language films played in many urban areas during the early sound era. That same winter, the Little Theater in Baltimore screened the German film *Three Loves* for two weeks.[107] Baltimore's Little Theater and other venues that screened non-English-language films (Spanish-language movie theaters in Tampa, for example) should be understood as alternative practices (that is, not fully oppositional). Like Chicago's ethnic-oriented radio stations, non-English-language films functioned on the margins of an acoustic imaginary that increasingly erased ethnic differences as part of its national mode of address.

The emerging corporate-dominated status quo of mass media was neither monolithic nor monolingual, but ethnic broadcasts and phonograph recordings or international films could be contained within a larger political economy of media that, in the first instance, promoted a national mode of address and hence viewed ethnic-identified practices and texts as addressing "market segments."

Radio would remain arguably polyvocal. Ethnic, educational, or otherwise "specialized" programming continued to be broadcast, but it was increasingly marginalized by commercial developments and the radio regulations that followed them. In the cultural and commercial spheres, as well as in the political, citizenship increasingly hinged on the erasure of difference. Rather than celebrating or constituting difference, the emergent acoustic imaginary of U.S. media sought to elide many types of difference, hailing idealized consumer-citizens. While these ideal consuming-citizens might be gendered, might be either Democrat or Republican (they were generally *not* socialists), might be either a child or an adult, they largely lacked racial or ethnic identity or geographic specificity.[108]

Emerging sound media were understood as instruments that might more fully assimilate immigrant populations into a national culture, indeed into consumer culture. But these media would also change some of the practices of electoral politics. During the mechanical-phonograph era, politicians running for national office sometimes used phonograph recordings to take their case to the people.[109] The introduction of electrical acoustics provided new media for U.S. politicians to expand their audiences and influence. With the redefinition of radio as broadcasting and its proliferation in the 1920s, political phonograph recordings became less common. Radio broadcasting could promise both immediacy (the live broadcast) and the potential for a near-national mode of address. Eventually, the sound film provided, in synch, the sounds and images of a politician. Under these circumstances, phonograph recordings as a political medium seemed comparatively less efficient for politicians and less compelling to listeners. Although after the conversion to electrical methods, the recorded repertoire of the major phonograph companies included very few political recordings, portable sound trucks—including those that also presented motion pictures—were still incorporated into American campaigning.[110] *Motion Picture Herald* reported in the fall of 1930 that four trucks specially equipped with RCA's Photophone sound-on-film technology were touring rural New

York State in support of the Democratic ticket, presenting matinee indoor performances and evening performances outside.[111]

Although radio journals advocated and often emphasized the role radio might play in increasing citizen participation in the political process, often the same journals took a dim view of radio stations operated by municipal governments. For many radio advocates who emphasized the role the medium could play in enabling the dissemination of political discourse, the implicit model for radio's role in politics largely relied on retaining the ownership of the means of broadcasting within the hands of *private* and not public concerns. For example, the programming of New York City's municipal station WNYC was frequently labeled as highly partisan political propaganda.[112] If government ownership and control of the means of communication ran the risk of falling prey to blatant political partisanship, advocates of the emerging commercial political economy of radio suggested that the free market might—as it did with cultural programming—guarantee that radio as a powerful political tool would be used for the greater good. Nurturing that "greater good" also involved protecting radio from the influence of so-called "special interests."

As the FRC deliberated and sought to reallocate the broadcast spectrum, the mainstream publication *Radio Broadcast* took an editorial stand that advocated the elimination of some stations. Not surprisingly, the journal largely supported networks, their affiliates, and the larger regional independent stations. "There is no excuse for the existence of a station which serves only a special and limited interest—to the exclusion of general educational and entertainment services to which the broadcasting band should be devoted." Invoking the familiar argument of spectrum scarcity, *Radio Broadcast* advocated that stations be eliminated that represented "perfectly legitimate special interests, but used the entire time of a private broadcasting channel for the benefit of only a fraction of the audience and for only a small part of the total hours of the day." These special interests included "Labor groups, sectarian religious appeals, socialism, Mormanism, atheism, vegetarianism, and spiritualism."[113] That organized labor ranked at the same level as vegetarianism might seem particularly striking today, but as advocates of the commercial development of radio began to gain hegemony, as radio began to be identified with a free-market determination of the public interest, convenience, and necessity, the "voice of labor" seemed particularly discordant with the emerging acoustic imaginary so firmly grounded in consumer society. Celebrations of radio's contribution to a political public sphere stressed expansion and inclusion, but in the actual "marketplace" of ideas, not all beliefs were of equal "value."

While radio's potential ability to provide "demagogues" with a broad audience certainly prompted significant concern and editorial posturing, especially in the early thirties, the commercial broadcasting industry was well positioned both structurally and rhetorically to exercise control over access to the ether.[114] Under the guise of their emerging definition of "service," broadcasters could strategically exclude either the voices of labor or voices from the right. An address by Socialist Party leader Norman Thomas, scheduled to be broadcast over WEAF, was cancelled in 1926 after Thomas's arrest for speaking on behalf of striking Passaic (New Jersey) textile workers.[115] With profit rather than the free expression of ideas as the abiding criteria for program content and a regulatory framework in which the "public interest, convenience and necessity" was increasingly defined as serving the majority of the audience rather than "special interests," networks and large independent commercial stations could largely control which speakers and ideas contributed to the emerging acoustic imaginary. Controversial speakers, when profitable—either commercially or as political capital—could attest to the "openness" of the U.S. broadcasting system. Their exclusion would illustrate a socially responsible use of a public resource rather than political censorship. In the rhetorical battleground over radio censorship, commercial forces could largely stretch the definition of service to meet their needs.

Reporting in *Collier's* in the fall of 1927, William G. Shepherd suggested that the economics of the radio industry guaranteed that all political views would be aired (provided a spokesman could conform to NBC and Merlin Aylesworth's criteria of a "responsible, reputable, and representative person").[116] Shepherd argued that because NBC was owned not by the public but by radio set manufacturers General Electric, Westinghouse, and RCA, and because all substantial radio content, even the controversial, sold radio sets, NBC would fairly air political broadcasts from all responsible perspectives. This economic argument—essentially that the structure of the marketplace guaranteed fair and equal access to the ether for all—echoed the radio industry's self-congratulatory pronouncements that the marketplace guaranteed that all programming would be educational.

Ultimately and not surprisingly, the utopian promises of new media's contribution to a participatory public sphere went largely unfulfilled. Three overlapping developments essentially undermined the often grandiose announcements of sound media's "emancipatory" potential: rather than participation, listening became consumption; economic developments surrounding *commercial* media essentially excluded many voices and political positions; politics and political rhetoric adapted to the commercial orientation of sound media.

Electrical acoustics provided new opportunities for the dissemination of political discourse, but while announcements of new technology promised enhanced public participation, that participation in practice was largely limited to listening, a form of consumption. When Frank H. Cromwell assumed his duties as the newly elected mayor of Kansas City, his experiences exemplified the uneasy coexistence of commerce, entertainment, and politics that characterized emerging sound media. Cromwell, described as a man who had "specialized in the butter and egg business," experienced his initial days in the mayor's office with profound confusion as he struggled to understand the duties of different branches of city government. After some study, Cromwell decided to use radio to educate his constituents. According to one account: "Here was a real problem—and one of universal application. The merchant, to exist, reasoned the mayor, must sell his wares, and to sell his wares he must advertise. The city, with a great stock of wares to sell, also should advertise. The voter—the ultimate consumer at the city store—must know what is on the counters and shelves. He must be informed of the 'service' offered by his city."[117] After a series of "civic radio nights" (talks by representatives from different branches of city government describing the services they provided), Cromwell even had sessions of the municipal court broadcast over a local station.

As the account suggests, Cromwell's approach collapsed citizenship with consumption, effectively blurring the boundary between politics and commerce. The city became like a department store, the voter a customer, and even municipal court proceedings might be experienced like a dramatic broadcast. The equation of citizenship with consumption particularly suited a deployment of radio in which a national mode of address constituted listening subjects by incorporating spatially dispersed auditors within a larger, nationally posited audience. Cromwell enacted this model earlier and on a more local level than did the nationally broadcast political conventions, network-sponsored programming and, eventually, FDR's "fireside chats" of subsequent years, but in many ways his efforts exemplify the blurring of distinctions between politics and entertainment that would increasingly characterize U.S. media.

In his study of interwar radio, Douglas Craig tracks a shift in perceptions of radio's effect on U.S. politics. Much like Frank Biocca's conclusion that "music elites" were initially enthusiastic about the possibilities of radio, commentators in the 1920s optimistically predicted that the new medium would revolutionize politics, enhancing public participation in the political process. Craig notes that by the late 1930s more realistic assessments suggested that broadcasting had not demonstrably increased political involvement, but had significantly increased campaign expenditures during the 1920s.[118] Craig and the period commentators he quotes suggest that radio's greatest influence involved the nature of political oratory. Perceptions that the medium addressed domestic audiences with a kind of "intimacy" exerted a pressure on political speakers to adopt more measured, less hyperbolic rhetorical styles. Craig locates radio's influence on American campaigning as part of a larger shift dating at least back to the McKinley-Bryan campaign of 1896 in which advertising approaches took on increasing importance. "This became the twentieth-century norm," Craig asserts," transforming voters into consumers and policies and politicians into competing commodities."[119]

With the development of radio's, particularly *network* radio's, commercial identity, U.S. politics would increasingly have to adapt to the medium and its commercial format. Political leaders had already been compelled to adapt to the *technological* requirements of new media. The period was filled with apocryphal tales such as the necessity of surrounding Al Smith with chairs to prevent him from straying too far from the broadcast microphone.[120] But by the presidential election of 1932, the conventions of commercial radio as an emergent medium clashed with some of the qualities of U.S. political practice. For radio to function as a disseminator of political rhetoric, for political leaders and aspirants to make use of its capacity as an instrument of national address, the presentation of politics, its rhetoric and spectacle, its performance and substance, would change so as to conform to radio.

As described in chapter 2, the live radio broadcasts of the presidential nominating conventions in 1924 had underscored some of the political possibilities of radio broadcasting, and these broadcasts, particularly that of the 2-week-long Democratic convention in Madison Square Garden, had functioned to "announce" radio as a medium offering a national mode of address. By 1927, plans were under way for broadcasting the next nominating conventions in a manner more consistent with the developing conventions of radio entertainment. Rather than the continuous, live broadcast of convention proceedings punctuated by contextualizing remarks made by an on-site announcer, the convention broadcasts of 1928 would have singers and entertainers standing by in radio studios so that "when convention affairs are dragging or are of such a nature that they might not hold the interest of listeners it will be possible for convention leaders to shift to the studios and furnish music instead of discord or draggy proceedings to the listeners."[121]

The clash between politics and radio as a commercial entertainment industry became particularly vivid in October of 1932, when President Herbert Hoover stepped before the microphones of NBC to address the nation. In what was essentially a campaign address, Hoover exceeded his scheduled time allotment and continued to speak for an additional half hour, thereby extemporaneously preempting the regularly scheduled Ed Wynn comedy program. Political commentators made much of the potential damage to Hoover's campaign. The *Nation*, for example, noted that "votes lost to Hoover multiplied too fast for computation. Ten o'clock: the candidate solemnly labored point number seven; too late to hope for even a fragment of Ed Wynn. What did N.B.C. mean by this outrage? Whose hour was it anyhow? Ten million husbands and wives retired to bed in a mood of bitter rebellion; no votes left for Hoover."[122] The entertainment industry trade publication *Billboard* reported on the event in an article titled "Political Talks a Pain to Radio." Noting the difficulty of limiting political speakers to their scheduled time period, *Billboard* observed that "it is not so bad where the time runs over into a sustaining program period, but when a commercial is scheduled it is a different matter."[123] For the radio industry and for entertainment-minded listeners, a political orator's greatest sin was to impinge upon regularly scheduled entertainment programming. In all, Hoover spoke for one hour and forty-five minutes that night, effectively canceling Ed Wynn's Texaco "Fire Chief" show and running fifteen minutes into the "Lucky Strike Show," thereby costing NBC an estimated $12,000 in advertising revenue and, according to the *Nation*, costing Hoover significant political capital.

Some components of the entertainment industry openly acknowledged that the apparent clash between commercial and political programming was more systematic than the disruptions posed by long-winded candidates. In October 1932, ASCAP notified broadcasting licensees that they could deduct any fees collected for political broadcasts from the net amounts used to determine the station's 3 percent music licensing fee. One report of the policy change noted that, "ASCAP believes that stations do not care for the money received from either party, since such broadcasts do much to disrupt the regular commercial programs and discourage rather than increase listeners."[124] Entertainment as a socially constructed category was in part based on its ability to dissimulate any connection to the political, and for political discourse to find a place in commercial mass media, it had increasingly to mimic the form of the entertainment commodity.

In 1927, radio manufacturer Frank Reichman suggested that American political parties were lagging behind big business in making the most of the advertising potentials of radio. Noting that Eveready, Maxwell, and Ipana all sponsored "Hours" over the radio, Reichman suggested that the political parties might consider sponsoring a "Republican Hour" and a "Democratic Hour." He suggested that the Democratic Party sponsor a series of performances of Paul Whiteman since "they have established themselves in the minds of the public as a political organization with advance ideas" and because they had "a jazz mayor in New York and a syncopated governor in the Empire State." For the Republicans, Reichman proposed "an old-time orchestra such as is now touring vaudeville . . . [to] keep in the minds of the public the rustic simplicity of the Coolidge regime." Reichman didn't neglect the socialists, proposing a program of "modernistic music such as is sponsored by the 'League of Authors and Composers.' "[125] Reichman's suggestion is not quite so far-fetched as one might imagine. By the end of the next decade, the left was making active use of musical performers to foster the goals of the popular front and, increasingly, mainstream political parties cultivated strategic alliances with entertainers and shaped their political messages with the language of mass media and advertisers.[126]

While radio broadcasters and politicians struggled to determine what role the medium might play in American political life, the motion picture industry largely solved the "problem" of politics by purging it from the sound film. Although in announcing the innovation of the Movietone sound-on-film system, William Fox had promised that sound films would deliver to the nation and preserve for posterity the voices and visages of the greatest political leaders, as the film industry converted to sound it cautiously navigated the boundary between entertainment and politics.

While Warner Bros. had used its Vitaphone short subjects, particularly "high Culture" performances, to introduce sync-sound film technology, Fox relied heavily on newsreels and short speeches of international leaders to introduce its competing Movietone technology. Many of Fox's earliest newsreel subjects seem selected so as to demonstrate the range of sounds that could be reproduced through the Movietone process, but other early Movietone releases did reference, if not actually conform to, the vision of a technologically facilitated public sphere. In the late 1920s, Fox Film released a series of recorded speeches by world leaders, international celebrities, and U.S. politicians. European heads of state and members of royal families joined Americans like President Herbert Hoover, General Pershing, Chief Justice Taft, Thomas Edison, Calvin Coolidge, and Alfred E. Smith to appear in Movietone short subjects directly addressing American film audiences. But such speeches often largely consisted of banal pleasantries and references to the "miraculous" Movietone process. Ultimately, the short subjects audibly lent prestige to Fox's Movietone sound-on-film technology while simultaneously reconfirming an ongoing process in which the U.S. newsreel served commercial entertainment rather than political or informational ends. Much like Warner Bros.' Vitaphone opera short subjects, Fox's presentation of direct-address speeches by political figures and other celebrities merely referenced utopian claims used to announce technological change. Fox's Movietone speeches, like other Hollywood sound newsreels that would follow in their wake, constituted a small step in the larger ongoing process that moved American political participation further toward the consumption of a spectacle.

By 1931 the film industry as a whole largely viewed the newsreel as one component of a larger entertainment package and not as a contribution (whether real or imagined) to U.S. political life. Summing up the current newsreel situation, industry analyst Terry Ramsaye wrote, "While Fox has strongly played world coverage, Paramount Sound News has tended toward the development of spectacularly staged special events. Pathé News, under a new policy and some degree of alliance with the impending development of the Trans-Lux coin-in-the-slot newsreel theatre, appears to tend heavily toward the quest of celebrities and screen interviews."[127] Sound newsreels essentially contained little that was newsworthy. Emphasizing celebrity and spectacle with occasional brief segments depicting national or international events, Hollywood's early sound newsreels tended to limit rather than to inspire political discussion.

In 1931, Fox instructed its theater managers to censor potentially controversial material from newsreels shown in their theaters. Reminding managers that, "It is our business to show entertainment—and nothing else but entertainment—on our screens," Fox cautioned managers to go to any lengths to avoid "demonstrations or irritations" among audiences.[128] The company was quite specific about the types of news material to be avoided and direct in its instructions to managers: "You are herewith very pointedly instructed to delete all subjects of a controversial nature on prohibition, pro or con; all subjects which can be construed as Bolshevist propaganda; all political speeches which take sides on matters of public interest; shots showing breadlines; and economic discussions on which the country or your particular patronage is divided." Reprinted in a national trade publication directed to theater owners and managers, the lesson of Fox's caution would not have been lost on the industry as a whole and can be taken as a statement of broader film industry attitudes toward political content. "We don't say that such matters are not interesting," the corporation declared, "but the danger is that they are liable to become too interesting." Explicit instructions to exclude "all political speeches which take sides on matters of public interest" offers a dramatically different perspective on the potential of the sound film than that expressed by William Fox some four years earlier when he announced the innovation of Movietone with grandiose promises of an authentic public sphere enabled by technological change.

The Fox Corporation's directive that theater managers avoid controversy in newsreels reflects a broader attitude within the film industry that sought to avoid political controversy. In October of 1932, for example, *Billboard* announced that Columbia Pictures was withholding its release of the film *Night Mayor* around New York until after the election and that due to disappointment with the box office performance of *The Phantom President*, Paramount planned to avoid political satire in the future.[129] Just as the U.S. film industry "learned" to avoid the kind of overt commercialization of the sponsored film experiment, subsuming through product placement the promotion of commodity culture and national products into the mise-en-scène of its films, so too did the industry carefully navigate the semipermeable boundary between politics and entertainment. The notion of "service" foregrounded by Will Hays in his celebration of the Vitaphone, and elsewhere generally pervasive in the industry's public presentation of itself, precluded overt advocacy of any specific political candidate or position. Even the perception of such advocacy was something to be guarded against.

As the decade of the 1930s progressed, the film studios, however, would ultimately prove not immune to mobilizing on behalf of a political candidate or issue when it directly served their immediate financial interests. MGM would use staged "newsreels" as part of a coordinated, multiple-media campaign—much like the Creel Committee—against the gubernatorial candidacy of Upton Sinclair in 1933–34. And, of course, some studios deployed their resources at the service of FDR's New Deal in forms as varied as Paramount's Buddy Rogers short "New Deal Rhythm" and Warner Bros.' musical celebrations of FDR and the National Recovery Administration at the conclusion of *Footlight Parade* (1933). But overall, and at least amid the emergence of sound, the U.S. film industry sought to navigate the potential clash between politics and entertainment by avoiding the overtly political as part of the (commercial) service it provided.

Conclusion

Announcements heralding the advent of electrical-acoustic technology sought to frame emerging media in relation to consensual utopian desires. Sound media were made meaningful by rhetorically and practically mapping them onto the preexisting spheres of commerce, culture, and politics. Distinctions between the local and the national, the rural and the urban, the immigrant and the native-born provided powerful frameworks that shaped how technology was understood and experienced. A perceived need for national unification provided a governing logic that powerfully shaped the deployment of electrical sound technology even as the technology itself—particularly radio—made the notion of a shared national imaginary more salient, more important, and more desirable.

Acoustic media would promote democracy—but democracy understood as "consumer" democracy in the first instance and only secondarily cultural and political democracy. Consumer society could itself promise to provide longed-for social cohesion and, not surprisingly, the mass media of capitalism (embodied in increasingly powerful media conglomerates) readily participated in cultivating such a form of a commercial public sphere. As media economic relations were solidifying and the "meanings" of technology were still undergoing some negotiation, the nation underwent a national economic crisis that lent urgency and solidity to an overlapping set of definitions of mass media and the roles they should play in commerce, culture, and politics.

Emerging media and increasingly interrelated media providers facilitated a longed-for social cohesion, but one most often articulated in commodity form. The emerging political economy for radio synthesized the three spatial definitions of the medium, collapsing the conquest of distance into a national mode of address and codifying a system of network-affiliate relationships that subsumed the notion of local service as an adjunct to the dominance of commercial national radio networks. This process would ultimately exert a powerful conceptual inertia on all future attempts to imagine an alternative, more democratic, deployment of radio and indeed all electronic communication systems. Following a model arguably introduced by magazines and successfully embodied by commercial radio networks, the film industry introduced commercially sponsored films. Short subjects, both live and animated, sought with mixed critical success to conform to the conventions of entertainment while also promoting nationally distributed products. The wholly owned theater chain of a vertically integrated film company, supplemented by the widely dispersed independent theaters contracted to screen its films, became a "network" of affiliates, essentially enabling the "broadcast" of commercial messages on a national scale.

In the gap between utopian promises and actual existing practices, a rhetoric of service provided an often-powerful rhetorical shield through which media organizations sought to dissimulate for-profit activities as serving the public interest. While this rhetoric of service failed in the case of Hollywood's sponsored film experiment, it proved to be much more successful in supporting network radio and commercial sponsorship of broadcasting. The comparatively short life of sponsored films demonstrates that the advent of electrical sound media redefined the boundary between entertainment and commerce, solidifying both media identities and modes of consumer address. Radio embraced the direct marketing of commercials—even making it the object of running gags in some comedy programs—but the cinema expunged such direct commercial appeals, opting instead for the indirect advertising of product placement in films.

Emerging hierarchical and spatial understandings of Culture accompanied technological change, and the new sound media amplified ongoing debates about cultural value by offering the potential for broader dissemination of cultural forms of all types. Consensual definitions of "high culture"—whether doled out in small doses through sustaining network radio programming, aligned with a pedagogical function (Damrosch), or strategically used to announce new technology (Vitaphone's opera short subjects)—functioned as a powerful component of corporate media's claims about service. While structurally marginalized within the economics of emerging acoustic media (albeit an ongoing site of debate and rhetorical posturing), "high culture" attested to the benevolence of corporate media providers, serving as an ongoing assertion of good corporate citizenship. But while prestige and cultural capital might be garnered from bringing "Wagner to the Sticks," profits resided elsewhere. On the one hand, CBS president William Paley could claim the cultural capital associated with the CBS School of the Air, but on the other hand, he could assert that ultimately all broadcasting was in fact educational for it illustrated the popular tastes of the nation. Because of its commercial nature, broadcasting, and indeed all commercial media, guaranteed that the public received only what it wanted—consumer-citizens voted with their radio tuner and their leisure spending—and media corporations, ever responsive to consumer demands, merely provided for those needs. The modern media corporation inevitably functioned at the service of public interest, convenience, and necessity, or so the rhetoric of service claimed.

Politics, when understood as the nexus of public policy and power, fit uneasily into such a model for media. But *electoral* politics, increasingly configured according to a binary model, proved more readily assimilated by media providers.[130] Ultimately, radio's innovation of equal-time provisions, and the principles through which such provisions were enforced, amounted to a codification of political blandness—or an essentially exclusionary disciplining of which types of political speech were worthy of reproduction. An election-year choice could resemble yet another iteration of the choices consumers made each day, and political "content" took on some of the trappings of "entertainment." Commercial broadcasters, advertisers, and the film and phonograph industries sought to hail an appropriately "modern" consumer, who even in the midst of an economic crisis might be successfully urged to become precisely a citizen of the world of standardized, commercially produced goods, including "entertainment." Citizenship slipped ever closer to consumption, and the "citizen-consumer" as a subjectivity to inhabit began to place ever-increasing emphasis on the latter term.

Although ultimately overruled by the courts, the New Deal's National Recovery Act—with its advertising slogan and Blue Eagle brand ("We Do Our Part")—re-created the orchestrated multiple-media campaign of first the Creel Committee and, later, L-I-S-T-E-R-I-N-E. Point-of-sale signs emblazoned with the Blue Eagle joined solemn-toned radio tag lines and the extravagance of Warner Bros.' *Footlight Parade* in asserting a national consensus that linked producer with retailer with consumer, collapsed politics with entertainment, and consumption with citizenship in a shared attempt to overcome the growing financial catastrophe. For a brief time, the Blue Eagle—like the Gold Dust Twins before it—became a pervasive marker of ideological consensus even as it masked its hegemonic function behind the illusion of a shared national interest.

The process of mapping emerging sound media onto preexisting discourses and practices had a series of consequences. "Participation"—whether political or cultural—became increasingly to be experienced through listening. That is, participation moved ever closer toward consumption. That act of listening became discursively aligned with nationalism via the immigrant; what one listened to and how one listened determined whether an auditor was an *appropriate* participant in the "public sphere." Sound media fulfilled, to a large extent, the desire for a shared national imaginary, but that imaginary privileged commerce over both culture and politics. Entertainment as a category of experience purged and/ or otherwise carefully contained politics *because* it was potentially divisive and therefore inimical to the smooth functioning of commodity relations. Through their rhetoric of service, large for-profit, private media concerns cultivated and solidified an identity as "public" institutions.[131]

These shifts did not fully and inevitably determine the process whereby new media were grafted onto social relations. Instead the relationship should be understood as mutually reinforcing: emerging media can function conservatively by appearing to realize previously established consensual goals (to the exclusion of often competing, equally material social needs) even as they can also create conceptual and practical spaces in which new social realities might be imagined. New media are inevitably shaped by the dominant ideational and material relations of the social world in which they emerge. But new media also open up spaces within social relations—spaces in which residual or entirely new social visions might be (partially) realized. In the case of electrical sound technology, the growing power of large corporations within a consumption-based economy, together with a vision that government regulation might best support this growth and a related vision of citizens as *subject* to this consumer society and its promulgation of a shared national imaginary, powerfully contained and effectively marginalized any potential challenge enabled by technological change.

Five

Transcription Versus Signification

Competing Paradigms for Representing with Sound

You sing into a little hole year after year, and then you die.

—Bing Hornsby (Bing Crosby) in *The Big Broadcast*

The above line of dialogue, spoken in a 1932 film by one of the first and most successful multiple-media performers of the electrical-acoustic era, expresses a commonsense notion of media identity. The radio was a transient medium, characterized by live performances and direct address. While it was able to eclipse spatial limits, to unite urban centers with the hinterlands (or to unite advertiser with consumer), to bring cultural uplift and edification (or uplifting and edifying entertainment), the radio performance remained a transient, fleeting phenomenon, adrift in the ether and caught in the moment of its creation through the wonders of applied science and corporate-funded engineering. Implicit in Crosby's line is the contrast between the identity of radio and the corresponding capacities of the phonograph and cinema. While Crosby could croon into the microphones of NBC's Red Network, transmitting year after year, and then simply die, his performance could also live on, stored for posterity and continued consumer pleasure through the media of cinema and phonography.

Crosby's line (both what it says and what it implies) echoes widely shared generalizations as to the epistemological status of these mass media. But simply categorizing the radio as a technology that transmits the "live" and cinema and the phonograph as technologies that store the "recorded" neither accurately characterizes the uses of technology nor exhausts the issues surrounding these media's identities. Instead, the advent and initial diffusion of any new representational technology can prompt drastic reconsiderations of not only a medium's epistemological status but also its social function and the circumstances under which it is produced and consumed. Technological change in the mass media challenges patterns of thought and habits of practice, opening up conceptual and practical spaces in which dominant media identities are challenged either by emerging models or by the reemergence of residual identities previously foreclosed.

This chapter examines the issue of technology-in-use: the manner in which visions of technological possibility inform and interact with representational practices. Two competing conceptual paradigms—that is, systematic assumptions about the technology's identity and its relationship to sounds—characterized initial applications of electrical sound technology. Often, individual texts uneasily combined representations of sound produced according to each of these competing paradigms. Such hybrid texts throw into sharp relief the manner in which at moments of technological change, technology-in-use (like institutional arrangements, regulatory frameworks, patterns of consumption, and economic identities) becomes a site of contestation. Conflicting models for radio's content and representational practices vied for dominance simultaneous with the struggles over radio's regulatory and economic identities. In earlier chapters I describe the important role played by national transmissions of significant (and pseudo-) events across often ad hoc networks to constructions of radio during the early 1920s. These national broadcasting events were central to assertions that radio might best offer a national mode of address, and they became a powerful weapon in support of the economic and institutional array that was eventually classified as best serving the public's interest. But the representational practices used for these events and the conceptual model on which they were based clashed with competing models.

Throughout 1927, radio trade journals printed a lively debate on the appropriate content for radio.[1] Some argued against the broadcasting of live events and called instead for a shift in programming to more controlled content, more studio-originated programming, as part of a larger professionalization of radio performance. Some radio critics hailed programming like WGN's (in Chicago) 1927 series "Old Time Prize Fights" in which championship bouts from the previous decade were re-created in the broadcasting studio, pointing to the success of the announcer in creating the illusion that the bouts were actually happening.[2] Meanwhile, others argued: "The broadcast of actual events is best, such as football, church services, conventions, banquets, etc. The greatest thing about the radio is that it brings the world into our home. Our Victrola still plays the better music, orchestral, vocal, etc. But the radio is something *real, alive, actual,* bringing great minds, great voices, great personalities, great happenings, great thoughts, great news events, etc. to us all wherever we are."[3] Radio engineer Carl Dreher countered that broadcasters performed best when "allowed to run things for themselves" as independently existing events structured and performed for a copresent audience led to various qualities not conducive to broadcasting.[4]

More than an argument over radio's content, this debate embodied a clash between two conflicting models for the relationship between technology and sound. On the one hand, radio programming like WGN's "Old Time Prize Fights" posited aural representation as involving a silent space ("dead air") into which one assembled sounds so as to *construct* a representation. An alternative model assumed that aural representation was essentially a documentary task and that electrical acoustics was best used to *document* independently existing aural events.[5] This model, which I refer to as the *transcription* paradigm, imagined the technology-in-use as properly documenting preexisting acoustic events. Values and functions associated with this paradigm included accuracy, fidelity, preservation, and "presence." The other paradigm, which I refer to as the *signification* paradigm, conceptualized electrical acoustics as a tool for expression, for illusion, for greater freedom to expressively manipulate sounds.[6] These conflicting models for aural representation informed debates and representational protocols in radio and phonography throughout the late 1920s and into the 1930s. But the struggle to develop conventions of representation in the aftermath of technological change was most substantial within the cinema.

These conceptual paradigms do not operate on the phenomenal or ontological level but rather on the epistemological. They involve not the actual relationship between technology and the phenomenal world, for all sound recording involves a sampling or a representation, and recordings provide us with only a partial access to a never fully present original acoustic event.[7] Instead, the transcription and signification paradigms operate on the level of *perceived* relationships to a phenomenal world; they appear in the realms of rhetoric, discourse, or faith. But these conceptual frameworks, this faith, do have material effects.[8]

Sound Recording Practice Before the Electrical Era

Transcription of sounds constituted the dominant conceptual paradigm governing sound technology when electrical methods were introduced. It provided a preexisting set of models for recording and reproduction practice and an epistemological framework for understanding sound media widely shared by consumers and producers alike. The dominance of this paradigm was a product not of the technology itself but of the social construction of the technology-in-use; it resulted not from material qualities endemic to the technology, but rather from historical struggles to fix discursively the identity of technologies of sound recording (for example, the discursive construction of the phonograph as a musical instrument during the 1910s).[9]

The generalized assumption that the phonograph was considered a medium of transcription owes much to the phonograph industry's own methods of advertising products to consumers. Promotional rhetoric continually stressed, and announced new innovations in terms of, the promise of ever greater fidelity to an original event. The industry's announcements of the transition to electrical methods of recording further solidified this commonsense identity of the phonograph as a tool of transcribing the voice or music by often juxtaposing old "unnatural" methods of recording performance with new, electrical methods to demonstrate a more "natural" arrangement of musicians and instruments. But before electrical acoustics made possible the manipulation of sound phenomena with greater control, a model of sound recording as signification coexisted with the dominant paradigm of sound recording as transcription. Throughout the first two decades of the century, the three major gramophone and phonograph companies marketed recordings premised on a model of sound as signification long before the advent of electrical acoustics and its enhanced ability to manipulate sound.

The Columbia Phonograph Company's 1905 catalog of disk recordings reveals the coexistence of sound technology envisioned as both an instrument of transcription as well as of signification.[10] Much of the catalog lists recordings of musical performances (everything from bugle calls and yodeling to band music and grand opera records), and Columbia consistently promoted these recordings in terms of their fidelity to original performances. In the case of cornetist Jules Levy, recording technology allowed Levy's performances to transcend his death: "In their way, these records are inimitable and it is extremely gratifying that his last perfect work was done after the metal master had been invented; for through this permanent method of record-making, Mr. Levy's records can continue to be furnished though the great artist himself is no longer with us."[11] Sound recordings framed with recourse to the transcription paradigm, according to the rhetoric of Columbia, were also characterized by their "presence." Recordings of male vocal "quartettes [sic]" were so finely rendered that "the effect is so natural that no stretch of the imagination is required to bring the singers so strongly forward in spirit that their actual presence seems to be achieved."[12] Columbia used appeals to both authenticity and the scientific precision of the recording and reproduction apparatus when describing its Alpine Yodels recordings as the company's "expert record-maker in Berlin" sought out authentic "Tyrolese Yodelers." "The Alpine yodels, so elusive in their sudden but harmonious changes of tone quality that even the great Moscheles failed to transcribe them correctly in his Tyrolese Melodies, are now reproduced on the COLUMBIA RECORDS in exact counterpart of the original."[13] In promoting large portions of its catalog as transcriptions of musical performances, the Columbia Phonograph Company, like announcements of electrical-acoustic technology some twenty years later, foregrounded the ability of this earlier generation of (mechanical) recording to preserve authentic acoustic events in such a way that provided enough illusion of presence to sometimes be mistaken for the event itself.[14] In 1906 even "the great Moscheles," with his well-trained musical ear practiced in the art of transcription, was no match for the scientific objectivity of an acoustic transcription apparatus.

In addition to the transcription-based musical recordings, Columbia also simultaneously recorded a significant number of nonmusical performances and released records that mixed the spoken word with music. These recordings included political speeches, scenes from plays, vaudeville acts, and several collections of recordings designed so that when played in sequence they produced a minstrel show in the home. Such recordings included "Address by the Late President McKinley at the Pan-American Exposition," a "Dutch Dialect Series" (i.e., "Schultz on Christian Science" or "Schultz on Malaria"), recordings by famous vaudevillians Weber and Fields, and thirty-one different recordings in the "Uncle Josh Weathersby's

Laughing Stories" series. Theatrical scenes in the Columbia catalog in-cluded "The Transformation Scene from 'Dr. Jekyll and Mr. Hyde,'" "Ghost Scene from Hamlet," and "Flogging Scene from Uncle Tom's Cabin."[15] Whether dramatic enactments of discreet theatrical scenes, dialect com-edy, or vaudeville sketches, these records, like the musical recordings, tran-scribed performances known to the consumer or understood in terms of other (live) performance contexts and venues.

Besides the substantial portion of its catalog devoted to recordings based on a transcription paradigm, Columbia also released material premised on a paradigm of sound signification. Often categorized as "Descriptive Discs," these recordings involved the strategic orchestration of sounds so as to represent some illusory scene. "Capture of Forts at Port Arthur" and "In Cheyenne Joe's Cowboy Tavern (A Coney Island Scene)" exemplify this practice. Columbia described "Capture of Forts at Port Arthur" in the fol-lowing way: "A scene from one of the Russian forts, with cannonading, and shriek of shells. The Russian Band is heard playing the National Anthem. The Japanese approach, headed by their band playing their National Air, and take possession of the forts, amid loud cries of 'Banzi'"[16] The Chey-enne Joe's recording is described as

A realistic descriptive record depicting a scene in the famous Rocky Mountain tavern which was one of the features of the Pan-American Ex-position, and has since been removed to the Coney Island Bowery. The outside barker invites the passing ones to stop and enjoy the hospitality of the tavern. The cowboy orchestra plays a characteristic overture. Bill Brindle, a cowboy waiter sings a rollicking drinking song, in which the habitues of the tavern join heartily, expressing their approval by volleys of pistol shots and cowboy yells, given in typical western style. It is a de-cidedly novel and entertaining record.[17]

Like Columbia's transcriptions of performances, these "Descriptive Discs" were also presented in terms of their realism, but the realism of these carefully constructed audio spectacles differs from the fidelity and au-thenticity assigned to recordings of performances.[18] Recordings like "Uncle Josh at Delmonico's" or "Uncle Josh on a Fifth Avenue Bus" seem to have been framed as vaudeville performances, thereby encouraging consum-ers to listen to the record as a transcription of a preexisting performance event. But similar recordings by the same artist (Cal Stewart) in the same character (Uncle Josh), such as "Evening Time at Pumpkin Center Descriptive Scene," cue a different type of reception. While Columbia recordings premised and marketed according to a transcription paradigm promised to transport a performer into the home, "Descriptive Discs," and the signification paradigm on which they were based, promised to transport the listener to a different (fictional) space. Within the Columbia catalog the category "Descriptive Discs" seemingly refers to a type of acoustic text. But the recordings so categorized embody a range of performances from "song and dance with clogs" to the elaborate acoustic scenes described above. Like all generic labeling, "Descriptive Discs" indicated a way of listening, cueing consumers how to listen, how to experience a recording.[19] More importantly here, the category attests to a way of conceptualizing the act of recording. These records demonstrate that, although the transcription paradigm was the conceptual dominant long before the potential for acoustic manipulation made possible by electrical methods of handling sound, the signification paradigm informed some early recording industry practices.[20]

In his discussion of early phonography, James Lastra notes that nearly all written commentary about the phonograph stressed its status as an instrument of transcription, but that Edison and others also understood that "the process of inscription could similarly *create* sonic effects." Lastra continues, "Edison gradually weaned his performers and technical staff from the prejudice that inscription was the only properly phonographic act, instilling in its place a flexible and essentially 'diegetic' understanding of sound representation that created coherent but spatiotemporally nonliteral musical worlds."[21] Lastra describes Edison adapting recording practice and shaping the "pro-phonographic" performance to accord with the limits and requirements of the technology. "Performers and technicians alike . . . came to believe that sonic representation was not simply a matter of precisely transcribing completely prior and autonomous events." Ultimately then, Edison's rhetoric of "re-creations" masks a manipulation, or rather, the "original" musical event resides in *playing* the Edison record rather than in the historically prior, inevitably absent "pro-phonographic" performance. Importantly though, practices like Edison's Tone Tests or Victor's fidelity assertions provided powerful and widely accepted models of *understanding* the phonograph as a transcription technology.

Acoustic Representation by Electrical Means

The clash between transcription and signification paradigms during the innovation of electrical-acoustic technology is perhaps most dramatically demonstrated through rather extreme advocates of each position. Throughout the late 1920s and into the early 1930s, social scientists Milton Metfessel and Carl E. Seashore from the State University of Iowa introduced a new scientific method to fellow social scientists for the collection, measurement, quantification, and study of music and other sounds. Called "phonophotography," their method involved the collection and transcription of sound waves on motion picture film. As early as the fall of 1925, in a field study of black American folk music, Metfessel was "photographing" voices using a portable "phonophotographic" camera.[22] The photographically encoded sound waves could then be quantifiably measured and analyzed based on frequency, amplitude, duration, and wave form.[23] Beyond the collection, preservation, and scientific analysis of folk music, Metfessel and Seashore saw in the apparatus the ability to analyze the expression of emotion in music and to quantify aesthetic value with science. Describing a multivolume phonophotographic study of vibrato, Seashore claimed that because of this scientific advance in acoustic transcription, they were "now enabled to define, describe, measure and control such a subtle aspect of the expression of tender emotion as the slightest change in the character of a vibrato."[24] He also suggested that a phonophotographic study of the vibrato of well-known singers was "the basis for determining what we may call good practice, which is the first step in an effort to determine a large number of component factors that operate to make a thing beautiful."[25] Here, the potential of electrical-acoustic technology to transcribe sounds accurately and with scientific precision enabled the isolation and quantification of the component factors of aesthetic "beauty."

Contrasting their method with traditional musicological analysis based on the careful listening of a trained anthropologist to either the sound itself or to a phonograph or cylinder reproduction, Metfessel and Seashore claimed to have scientifically eliminated the subjective limitations of previous approaches. In a 1928 report of his work on black American folk music, Metfessel located two important sources of error or potential bias in then currently prevailing methods. First, even the "keenest" anthropological ear was ill-suited to "the American Negro vocal embellishments"

because of the "bias due to past musical experience which deafens the no-
tator to elements foreign to his own music."[26] Second, conventional methods of musical notation, designed to represent European music, neglected
factors that made folk music distinctive. "In Negro music, that part which
is characteristically Negro is not found in the stilted notes on the conventional five-line staff, but rather in the twists and slides between the lines."[27]
Although fascinating as an early implicit critique of ethnocentric anthropological procedures and the Eurocentric bases of conventional musical
notation, Metfessel's discussion of phonophotography is most interesting
in this context for the manner in which it constructs the electrical-acoustic
apparatus as a scientific instrument of transcription.

In 1927, Seashore located the origin of this method in attempts to certify
participants in the prevailing subjective approach:

> My personal interest in phonophotography dates from the time when the
> Smithsonian Institution sent its specialist in Indian music, Miss Frances
> Densmore, to our laboratory to have her ears certified with reference to
> the degree of reliability for the transcribing of phonograph records. It
> then occurred to me that it was possible to avoid depending upon the ear,
> which is quite inadequate for the purpose, and substitute a photographic
> method. This led to the developing of photographing of phonograph rec-
> ords, and that in turn to the direct photographing from a musician's per-
> formance. [28]

Trained though she was, Miss Densmore with her scientifically certified
ear, much like the "Great Moscheles" before her, was no match for the
objective documentation provided by the electrical-acoustic apparatus.
Metfessel's and Seashore's phonophotographic device and their descriptions of their methods exemplify a conceptual paradigm that viewed electrical acoustics as an apparatus of transcription. The values they assigned
to the apparatus hinged on its precision (of measurement), its (scientific)
objectivity, its neutral method of notation, and its ability to preserve an
acoustic event accurately.

While Metfessel and Seashore conceptualized electrical acoustics as a
scientific apparatus of transcription more faithful to detail than the human
ear, Seashore at least saw the potential of the apparatus to function in arguably the opposite way. "It is now only a matter of patient workmanship for

the future inventor to make a synthetic human voice automaton capable of speaking in languages, playing upon the whole gamut of emotions in vocal expression and even executing artistic effects not yet attainable through the voluntary performance of the singing or speaking artist."[29] For Seashore, electrical sound technology offered not only the potential for recording and analysis more detailed and accurate than human perception, it also provided for the future possibility of musical production and signification far in advance of human capacity.

This power of the electrical-acoustic apparatus to function as an instrument of signification was advocated by a voice speaking in realms far removed from the world of social science occupied by Seashore and Metfessel. While the Iowa social scientists sought, among other tasks, to study aesthetic practice scientifically, Harry Alan Potamkin sought to shape the future of that practice through his film and music criticism. Potamkin's critical work provides a vivid description of electrical acoustics conceptualized as an instrument of signification. Writing for various newspapers and such periodicals as the *Phonograph Monthly Review*, *Close-Up* (an international film journal), *Hound and Horn*, and *New Masses*, Potamkin provided a consistent voice for the nonmimetic application of electrical acoustics.[30] Contrasting the "sonal" or "sonorous film" with the "talking picture," Potamkin called for overt signification in the application of electrical sound technology and the rejection of its transcription possibilities. Invoking and refuting commonsense assertions that the camera provided the potential for illusion but the microphone provided factual duplication of the real, Potamkin valorized a series of sound strategies that would explore the signifying potentials of the new combined visual and acoustic apparatus.[31] These included the rejection of sound and image redundancy, the use of speech as abstract sound, the fragmentation of individual acoustic events or performances, and the incorporation into film soundtracks of both "found sounds" and artificially produced sounds.

In "The Phonograph and the Sonal Film," Potamkin argued that a true sound aesthetics hinged on a rejection of the mimetic potentials of the apparatus. Supporting the use by film studios of sound effects libraries (sounds prerecorded independently of the images they eventually accompanied), Potamkin advocated even further efforts to conceptualize film sound and film image as distinct representational technologies.

There is no need to get distinct sounds for identical images; standardization is not counter-quality. As a matter of fact, such standardization,
which is an elementary kind of conventionalization of sound, has already
been instituted by the film companies. That is a first rudimentary step in
the establishment of sound-engineering, and sound-aesthetics.[32]

For Potamkin, this first rudimentary step would soon be replaced by a manipulation of acoustic phenomena that avoided redundancy between sounds
and images. He called for the combination of a "sound from one source
with an image not issuing the sounds: in other words a non-coincidence of
sound" rather than the synchronization of sound with image.[33] The careful
orchestration of aural phenomena as signifying units would accomplish this
noncoincidence of sound. "There is no reason why that medium [the sync-
sound film] must blunder in the erroneous path of duplication. Aesthetic
organization is the very secret of health for any expression of man called
entertainment or art."[34] Frequently citing the manifesto on the "contrapun-
tal" use of cinema sound written by Soviet filmmakers Sergei Eisenstein,
Vsevolod Pudovkin, and Grigori Alexandrov that appeared in English translation in *Close-Up* in 1929, Potamkin called for a sound practice emphasizing "sound suggestiveness, as opposed to sound literalness."[35]

One of the key components of Potamkin's view of a "sound suggestiveness" involved the use of speech as abstract sound. He broke most specifically with the Soviet manifesto in advocating the use of speech, but according to a principle he called speech-as-utterance. "I oppose also the Russian
directors who accept the sonorous film but not the film of speech. Speech
can be stylized, harmonized and unified into an entity with the visual image."[36] "Speech as utterance is stylized speech," Potamkin argued. "Its basis lies in a variety of sources: explosive speech, uniform pitch, monotone,
etc."[37] For Potamkin, recorded speech offered to the cinema far more than
the ability to transcribe dialogue accurately and audibly. Instead, spoken
utterances offered to the filmmaker a range of signifying potentials including rhythm, pitch, and intonation, all of which were susceptible to aesthetic
manipulation at the service of larger ends. Speech could thus function as
an abstract sound independent of the linguistic content it offered.

Perhaps the apotheosis of Potamkin's aesthetic for the sound film came
in his valorization of multiple levels of acoustic manipulation and the extent to which virtually any sound could serve the larger aesthetic interests

of a filmmaker so inclined. He often held up as an example the acoustic manipulations of Alexandrov's *Romance sentimentale* (1928) (released in the United States by Paramount as *A Sentimental Romance*). In the film, Alexandrov sought to design his soundtrack, in part, according to acoustic equivalents for the principles of Soviet montage. The techniques he used, including running a recorded sound backwards, breaking the recording of a sustained sound into very brief fragments, cutting the recording of a siren into its component frames to create an ascent and descent in steps rather than a regular incline, and directly inscribing (writing) onto the soundtrack, all were valorized by Potamkin as potential directions profitably pursued by a practice that viewed soundtrack construction as signification.[38]

> Let us say a note is banged on the piano, impressed on the negative. Immediate cutting—and there are a variety of ways—will change the character of the sound and give it an absoluteness. That is to say, it will not be associated with the instrument from which it will have emanated. One may record a jazz-band and then play around with the sounds as impressed, and get thereby any number of possible arrangements. The same can be achieved with speech: it may be clipped, stretched, broken into stutters, made to lisp, joined with all sorts of sound combinations either in discriminate melange or in alternating, repeating motifs.[39]

Potamkin called for the fabrication of "natural" sounds, the acoustic manipulation of performance events with no respect for their duration, and the use of tone test records (for example, recordings in which a pure tone of 6,000 Hertz gradually glides down to 100 Hertz) originally designed to ascertain the frequency curve of electrical sound apparatus as the raw material (essentially found sound) for the filmmaker's manipulation.[40] His aesthetic mandate for the sound film demonstrates some of the furthest implications of a sound practice premised on electrical acoustics conceptualized as an apparatus of signification.

It is Potamkin's advocacy of the use of sounds that don't exist in nature that brings into sharpest relief the divergence between the two conflicting paradigms for the work of sound reproduction. Here, Potamkin conceptualizes the potentials of electrical sound technology in direct opposition to the paradigm embodied in Seashore's and Metfessel's social science discourse. Metfessel and Seashore assumed an identity between an original acoustic event and its electrical-acoustic transcription. Their scientific process hinged on an epistemological faith in the power of the apparatus to provide a detailed, dispassionate, and direct encryption of an event. For Potamkin, such an epistemological faith in the apparatus was an obstacle to the fulfillment of the true signifying potentials offered by electrical sound technology. To some extent, these divergent positions are directly related to the different discursive networks within which they operate. The various disciplines of social science require an epistemological faith in their meth-ods of observing, measuring, quantifying, or otherwise describing social phenomena. The development of electrical acoustics, or advances in other representational technologies, provided new, objective methods of obser-vation precisely because technology was assumed to function neutrally.[41] Although necessary to social science as a practice, this epistemological faith in the technological apparatus' ability to document phenomena, to supersede and improve upon human perception, was not unique to social sciences (as the discussion below will indicate). Potamkin's position was located within a discursive tradition of aesthetics that argued not for the duplication of reality but for the expressive transformation of phenomenal reality. The conflict embodied in these two divergent examples indicates the interpretive flexibility that results when a technological artifact, com-plete with the rhetorical baggage of science, enters into the realm of aes-thetics and clashes with equally imposing rhetorical traditions.

Potamkin was not alone in his overt advocacy of electrical acoustics con-ceptualized as an instrument of signification.[42] Nor were the polemical po-sitions advocated by Seashore and Metfessel on the one hand and Potamkin on the other isolated in realms (social science vs. aesthetic criticism) as distinct as these two examples might imply. Instead, the conflict between these two competing conceptual paradigms was vividly enacted through a variety of Hollywood sound practices and debates among Hollywood sound practitioners as to the appropriate representational conventions for the synchronous-sound film. This interpretive flexibility of the apparatus at the level of sound practice or technology-in-use also informed various conflict-ing practices within both the radio and the phonograph industries.

In 1923, when Lee De Forest predicted the potential applications of his Phonofilm sound-on-film system, he foresaw the use of the appara-tus conforming to both of the two extremes represented by Seashore and Metfessel on one side and Potamkin on the other. In applications such as educational films, recordings of great men for posterity, or recordings of both popular and high-culture performers, De Forest foresaw the Phonofilm successfully transcribing and preserving acoustic events. The Phonofilm could surpass the phonograph's ability to transcribe performances by offering longer recordings, images as well as sound, an improvement in "naturalness of tone quality," and the ability to be reproduced to far larger public audiences.[43] In a familiar rhetorical move that echoes throughout the history of sound recording, De Forest invoked the social value to future generations when "both the voice and the visage of the nation's leaders" would be preserved for posterity by the Phonofilm.[44] "How priceless now would be the film reproduction of Lincoln delivering his immortal Address at Gettysburg, or of Roosevelt as he stood before the Hippodrome audience at his last public appearance delivering a message to his countrymen, the inspiration of which has already been, how sadly, lost."[45] As Metfessel and Seashore would do some five years later, De Forest assigned to electrical acoustics the values of fidelity, presence, and preservation. Here, the technologically augmented body cheated death through its encryption.

But simultaneous with De Forest's description of transcription-based synchronous-sound cinema, he also outlined a series of applications based on the signification paradigm. Phonofilm could provide weekly newsreels not only with vocal identifications of scenes and situations (analogous, perhaps, to photographic captions) but also interpretations "by the voice of some well informed, entirely invisible speaker."[46] De Forest also foresaw poetic verbal description and music "interwoven" with film images to "thereby elevate to ennobling heights" scenic films and travelogues.[47] This use of sound as an *additive signifying* material for newsreels and travel films also characterized De Forest's predictions for the silent drama. The Phonofilm could provide standardized musical accompaniment regardless of exhibition venue and, through recorded voice-overs, eliminate intertitles. Synchronous spoken dialogue was to be reserved only to produce dramatic effects. "I can in fact picture some very dramatic effects which may be obtained where, perhaps, only one or two words or sentences spoken throughout the entire run of an otherwise silent drama, will grip the attention, and startle the imagination, as does the occasional introduction of a hand-tinted object in an otherwise monotonous black and white picture."[48] In general, De Forest's view of the dramatic signifying possibilities of the

Phonofilm for the screen drama required their strategic rather than pervasive use.

> Thus I claim that an entirely new form of screen drama can be worked
> out, taking advantage of the possibilities of introducing music and voice,
> and appropriate acoustic effects, not necessarily thruout [sic] the entire
> action, but here and there where the effects can be made much more
> startling, or theatrical if you will, or significant, than is possible by panto
> mime alone, no matter how cleverly such may be worked out.[49]

De Forest's list of Phonofilm applications conforming to both the transcription and signification paradigms demonstrates that the eventual uses
to which electrical acoustics would be put were not predetermined by either
of these conceptual models. Both of these paradigms could and did underwrite a variety of social uses for the technology as well as different representational practices. But the interpretive flexibility evident in De Forest's
predictions for the synchronous-sound film also characterized Hollywood
sound practices long after synchronous sound was introduced. Whereas
De Forest's Phonofilm predictions separated transcription and signification paradigms to distinct categories of film practice, the early years of
Hollywood sound practice involved a number of instances in which, in a
single text, transcription and signification paradigms for handling sound
dramatically clashed.

In order to exemplify this clash, I turn now to two case studies from
different Hollywood practices: the use of sound in early Vitaphone and
Movietone films, and the 1932 feature-length musical *The Big Broadcast*.
These case studies illustrate instances in which both of these conflicting
paradigms for sound representation inform individual films. The resulting
"hybrid texts" usefully point not only to confusion about and struggle over
representational practices but also to the ways in which the introduction
of electrical sound technology to the American film initiated a period of
competition among diverse theories of representation.

The notion of "hybrid texts" conflicts with some of the dominant methods of considering cinema's aesthetic history. Whereas André Bazin viewed
technological change as contributing to film's inexorable, asymptotic evolution toward greater realism, historians like David Bordwell view the advent
of synchronous-sound technology as posing little more than a temporary

obstacle to the hegemony of a particular aesthetic and economic system of production.[50] For Bordwell, synchronous-sound technology was simply adapted to the needs of the preexisting "classical" narrative mode. From such a perspective, textual hybrids constitute no more than a transitional stage. While Bordwell's account produces a coherent narrative of Hollywood's conversion to sound, it fails to recognize the ways in which technological change imposes the need to negotiate conflicting conceptions of the relationship between technology and the task of representing. Rather than being characteristic of a comparatively short "transitional" period, hybrid texts visibly and audibly embody larger ongoing clashes between representational practices, representational choices, and modes of *thinking* about representation that are brought to the surface, made visible and audible, by technological change.

Transcription Versus Signification in the Initial Vitaphone and Movietone Releases

In chapter 2 I demonstrate the ways in which initial announcements of electrical-acoustic technology often involved assertions about the greater fidelity offered by the new methods. These references to expanded fidelity differentiated the new technology from previous attempts to innovate synchronous-sound film (Kinetophone, Cameraphone, the initial incarnation of Phonofilm, etc.) and, importantly, from preexisting methods of mechanical recording and reproduction widely diffused through the phonograph. Thus, electrical acoustics was at first predominantly defined as a method of documenting or transcribing acoustic events. The dominant view of sound technology as properly functioning to transcribe acoustic phenomena exerted a powerful weight, a conceptual inertia, upon attempts to apply electrical acoustics to cinema.

Despite *Photoplay*'s October 1926 assertion that there were no current plans to innovate "talking pictures" (sync-sound would supply recorded orchestral accompaniment only), initial applications of electrical sound technology to the cinema relied heavily upon recording conceived as the transcription of acoustic events—particularly people speaking and performing.[51] Even initial recorded synchronous scores were publicly presented as virtual transcriptions that re-created the experience provided by the best orchestras in the largest urban motion picture theaters. The Fox

Film Corporation also initially used its Movietone sound-on-film system to document the remarks of international leaders and literary figures. Direct-address speech performances of government leaders (including King George V, King Alfonso XIII, the Prince of Wales, Lloyd George, President-elect Hoover, etc.) and literary personalities (for example, George Bernard Shaw and Arthur Conan Doyle) were released as autonomous short subjects or incorporated as segments into Fox's Movietone Newsreel. These highly discursive performances underscored the new technology's ability to document events. The representation of the speaker's body as the source of the sound posited an event through the synchronization of recorded voice with moving lips.

The earliest Vitaphone sound-on-disk shorts produced by Warner Bros. featured representations of both high- and middlebrow performance events. Like Fox's direct-address short subjects, the initial Vitaphone shorts, in historian Charles Wolfe's words, evidenced a "documentary impulse."[52] The conceptual paradigm of transcription as enacted in the Vitaphone shorts was well suited to a particular economy of film production and distribution. Unlike the production and release patterns of feature-length films, these recorded performances could circulate long after their initial release, constituting an interchangeable catalog of film recordings.[53]

But simultaneous with these earliest transcription-oriented applications of electrical acoustics, other practices explored the new technology's ability to construct illusory events. Artificially created synchronous events occurred during soundtracks otherwise consisting solely of recorded musical accompaniment. In the earliest Vitaphone features, specific sound effects would be reproduced in isolation and, importantly, in approximate synchronization with an appropriate image. Examples include the sound of a champagne cork popping, the only sound from an otherwise busy but silent nightclub kitchen in *The Singing Fool*, a knock on a door or the rattle of clashing sabres in *Don Juan*, a whip cracking across an otherwise silent slave's back and a scene in which a soldier in the trenches of World War I responds to an officer's command (communicated by a dialogue title: "And listen you eggs, be hardboiled") with a raspberry reproduced in rather startling synchronization (both in *Noah's Ark*). While such a highly selective approach to recording and reproducing sound effects gestured toward the creation of a realistic acoustic space, the intermittent nature of precisely such selection undermined a notion of verisimilitude. Similarly, while these pseudo-synchronous events might be heard as early attempts to narrate acoustically—that is, to mobilize the acoustic at the service of a narrating agency actively directing audience perception to important story information—too often the reproduced sounds serve no specific story function. Instead, the pseudo-synchronous acoustic effects of early Vitaphone features announce various technological possibilities, demonstrating the range of sounds that could be reproduced and, more importantly, performing the possibility of synchronization.[54]

Heterogeneous sound practices characterized Hollywood's initial forays into the application of synchronous-sound technology to feature-length films. Fully synchronous dialogue scenes were interpolated within films in which the soundtracks otherwise consisted of recordings of musical scores. Intermittently presented sound effects also attested to differing acoustic practices. Reproduced sounds sometimes seemed to approximate the accompanying depicted object, while at other times they resembled silent film accompaniment in that noises were produced through musical instruments.

Warner Bros.' film *Noah's Ark* indicates the simultaneous presence of multiple potential approaches to rendering sound. The film illustrates both the complexity of Hollywood's initial sound practices and the manner in which early recorded soundtracks announced the acoustic possibilities of electrical sound technology. *Noah's Ark* begins with a depiction of the construction of the Tower of Babel, acoustically accompanied by a musical score and both the pseudo-synchronous sounds of slaves being whipped and apparently synchronous construction sounds altogether out of keeping with the depicted events. The construction sounds demonstrate (approximate) synchronization but not fidelity to their depicted sources. The film's shift to a series of contemporary scenes begins with a dissolve to a close-up of a ticker tape accompanied by an appropriately recorded and reproduced sound effect. The ticker tape sound continues, unrealistically maintaining a consistent volume across several cuts and under expository intertitles. Here, the film initially anchors a sound effect to its source, justifying its presence, but the sound is soon visually disconnected from its origin and continued as if it were a component of an ongoing musical score. Sound effects become even more fully musical throughout *Noah's Ark* as cymbals stand in for, and are in fact synchronized with, the crash of lightning strikes. Throughout the film the approach to representing sound effects oscillates between attempts to render sounds that seem appropriate to a depicted source and sounds for which musical equivalents are used. For example, cymbals are also used to accompany the breaking of a bottle over a character's head. Crowd sounds are denoted through a murmur of voices, such as when a mob surrounds and mocks Noah as he constructs his ark. This particular acoustic strategy accompanies only long- and medium-long shots, as close-ups of the taunting crowd visually isolate the sources of single voices that are then presented through dialogue titles. Exterior shots of the Orient Express are "appropriately" accompanied by train sound effects, but when the scene and camera move to a location inside the train, the conventions of silent film accompaniment return through the use of different musical themes linked to different passengers in what otherwise is a nonsynchronous scene. The use of pseudo-synchronous sound effects returns with the crash of the train as cries of passengers momentarily become prominent on the soundtrack.

Early sync-sound films like *Noah's Ark* and other so-called "part-talkies" exhibited conflicting approaches to the task of acoustically representing fictional worlds. Like the Tower of Babel which it depicts, *Noah's Ark* features the overlapping of various representational strategies, threatening to render its acoustic world confusing if not altogether incoherent. The silencing of this Babel of representational possibilities would come to the Hollywood film in time, but for these initial sync-sound presentations, the presence of various, competing, and sometimes conflicting sound strategies served to announce to audiences some of the acoustic possibilities of the emerging sound film. Scenes reproducing dialogue attested to successful synchronization. The highly selective incorporation of sound effects in real or approximate synchronization with depicted sources attested both to the fact of synchronization and to the range of possible sounds that could be successfully reproduced. In these films, reproduced sound became a spectacle as early synchronous-sound films demonstrated the potentials of the emerging technology.

The Singing Fool, Warner Bros.' and Al Jolson's highly successful follow-up to *The Jazz Singer*, literally performs for its audience the process of conversion to synchronous sound. Mixing conventions of the silent film, Vitaphone's synchronous musical accompaniment, and the fully synchronous-sound film, *The Singing Fool*'s soundtrack within itself enacts Hollywood's multistage embrace of recorded sound. The film begins much like a silent film, or an early Vitaphone feature in which a recorded non-diegetic score accompanies on-screen action. Printed intertitles identify characters and provide expository information orienting the audience to the narrative's depicted time and space. But when a group of late-night revelers arrive at Blackie Joe's speakeasy, *The Singing Fool*'s approach to sound shifts to another register. Pseudo-synchronous sound effects accompany an image depicting the group knocking at the speakeasy's door. Even more notable, Warner Bros.' sound technicians manipulate the volume of the music emanating from inside the speakeasy each time a peephole in the front door is opened. As the group attempts to gain entry, attenuations in the volume of nightclub sounds accompany the doorkeeper's action of opening and closing the peephole. Pseudo-synchronous sound effects create an acoustic representation of space approximately analogous to what one might expect standing with the revelers. Visual perspective—with the camera positioned outside the speakeasy and with those seeking entry—combines with narrative information and realistic motivation (we know the club is busy and we would expect to hear more of the sound from inside when the peephole is opened) to underwrite a realistic approximation of actual acoustic space. The nonsynchronous soundtrack functions as if it had been recorded in sync with the image and thus two distinct representational systems (image-producing and sound-producing) function collaboratively so as to elide the distinction between them.

But any realistic acoustic effect thereby generated is quickly undermined when the camera's position moves to the other side of the door and the film's audience adopts the visual perspective of the doorkeeper. No consequent change in either volume or quality of sound coincides with the shift in camera position to inside the speakeasy. The simple and representationally logical cut to a camera position inside the club undermines the realism of the pseudo-synchronous events outside the speakeasy (knocks on the door and shifts in volume of crowd noise).

As the doorkeeper moves through the club to find Blackie Joe, the camera adopts his point of view. The soundtrack exclusively consists of music. Although diegetically motivated by the speakeasy's orchestra, the musical score only partially represents acoustic space. The soundtrack contains no other sounds, and the mobile point-of-view shot even includes rather startling moments of silently mouthed direct address. Like the pseudo-synchronous events intermittently present in many of the earliest Vitaphone features, sound effects (and in this case volume manipulations) are used in the opening moments of *The Singing Fool* less to create a consistent and coherent version of acoustic space than literally to announce some of the representational possibilities offered by the Vitaphone device. The apparently uncanny pseudo-synchronous sound effects remain particularly striking precisely because of their difference from the silent film conventions with which *The Singing Fool* begins and to which it rather suddenly returns.

Soon, a table approaches the camera, carried by an unseen employee, and as the table is set down, the waiter is revealed to be Al Jolson. Dancing in close-up, literally mugging for the camera, Jolson and his antics push *The Singing Fool* into the overtly performative, into the realm of the direct-address characteristic of many Vitaphone performance shorts. Jolson's arrival on-screen is indeed an "appearance," a revelation. Still conforming to silent film conventions, however, the film uses an intertitle to inform us of Jolson's fictional identity, policing our response to the star's appearance by reanchoring the moment of direct address back within the diegetic world.

Further narrative exposition and abrupt shifts in sound practice follow, including revelations that Jolson (character Al Stone, that is) is a really nice guy; he rescues a cigarette girl from a masher. Stone is not only oblivious to the cigarette girl's obvious admiration for him, but he is also madly in love with Molly, the speakeasy's singer. All this narrative exposition, establishing the central problematic that will become a primary focus of the film, is accomplished through pantomime augmented by expository and dialogue intertitles and suitably accompanied by a musical score potentially motivated by the speakeasy's offscreen band.

The musical score remains largely distinct from depicted narrative events, maintaining a consistent volume even when the action moves into adjoining rooms—that is, until Al and the film retire to Molly's dressing room. Diagetically motivated dance music abruptly becomes the silent film's melodramatic nondiegetic score. Conventions of silent film accompaniment continue as Al leaves Molly's dressing room and "fixes" two flat beers by rapidly pouring one into the other and back again. Here, the soundtrack provides an aural equivalent to the character's movements, musically mimicking the action. The acoustic distinction between two depicted spaces continues as the film cuts several times back and forth between the nightclub and Molly's dressing room. To this point the soundtrack of *The Singing Fool* has largely conformed to conventions of silent film accompaniment while intermittently incorporating several pseudo-synchronous acoustic events that *hint at* some of the possibilities of a fully synchronous soundtrack.

The soundtrack and consequently the film as a whole undergo an even more dramatic representational shift moments later when Al Stone, the singing waiter, moves to the center of the nightclub to perform. Here, fully thirteen minutes into the film, *The Singing Fool* (and Hollywood) again "convert" to sound. With the words, "Wait a minute, wait a minute," Jolson and the film slip into the fully synchronous. Striking in its abrupt transition, this shift is no more performative than Jolson's initial on-screen appearance. Only here the performer technologically announced is not only Jolson but also the Vitaphone apparatus itself.

Through its shifting use of sound, the first twenty-four minutes of *The Singing Fool* perform Hollywood's conversion to synchronous sound. The film begins by conforming to the narrative and acoustic conventions of the mid-1920s silent film, thus invoking a soon-to-be-eclipsed representational practice. Pseudo-synchronous knocks at the speakeasy door, easily accomplished by an alert live drummer with appropriate traps, give way to a pseudo-sync event made possible only with the Vitaphone—the attenuation of crowd noise to accompany the opening and closing of a peep hole. But the film quickly slips back into the acoustic protocols of the silent film. Then, in a startling acoustic announcement of technological potential, *The Singing Fool* becomes a synchronous-sound film.

These earliest synchronous-sound films arguably demonstrated a range of representational options, a variety of approaches to the incorporation of recorded sound into the American motion picture. Often emphasizing contrasts with older, preexisting modes of representing, these textual displays made sound into a kind of spectacle, but a spectacle heralding an inexorable evolutionary advance that promised eventually to collapse the difference between the real and its representation. The "pseudo-synchronous events" in the earliest Vitaphone features functioned to announce certain technological potentials of the apparatus, but more importantly in this context, they point to a working model among some Hollywood sound practitioners that viewed electrical acoustics as a tool for the construction of soundtracks at the service of a larger goal of signification.

At Warner Bros., electrical acoustics was applied to the cinema according to both the transcription and the signification paradigms. While the Vitaphone shorts transcribed performance events, the strategic inclusion of approximately synchronous sound effects (including instances of the human voice) within the recorded scores of Vitaphone features involved the application of a signification paradigm to soundtrack construction. Wolfe notes that two representational strategies were at work in these earliest Vitaphone films. "Vitaphone features treated the voice as an intermittent sound effect, post-synchronized to the image, never fully anchored amid the fluid relationship of orchestral score to depicted action. Vitaphone shorts, in contrast, privileged the recorded vocal utterance of a performer as a phenomenal event."[55]

The initial Vitaphone and Movietone releases demonstrate the prescience of Lee De Forest's 1923 predictions for the sound film. Transcription and signification paradigms simultaneously informed different representational practices. Movietone's audiovisual documents of political leaders and literary figures fulfilled De Forest's call for sound film recordings for posterity. The Vitaphone (and, to a lesser extent, Movietone) performance shorts, essentially modeled on phonograph recordings, similarly applied electrical acoustics as an instrument of transcription. Simultaneously though, the technology was used as an instrument of signification when the strategic use of recorded sounds augmented essentially silent feature films and new forms of audiovisual drama were considered. De Forest's predictions were doubly prescient in that, as he anticipated, the different approaches to sound practice characterized distinct categories of films (transcription for Vitaphone's shorts, signification applied to the feature-length scores). But as the next section will indicate, the clash between these conflicting paradigms for approaching sound soon crossed these boundaries and informed a variety of film types, sometimes uneasily coexisting in a single film.

The Big Broadcast: Conflicting Models for the Relationship Between Technology and Representation

Like the vast majority of American films, Paramount's *The Big Broadcast* (1932) occupies only a marginal place in conventional histories of American cinema. Perhaps most widely cited (and quickly dispensed with) as Bing Crosby's first feature-length starring role, *The Big Broadcast* fits uncomfortably into many prevailing historical approaches. As an exemplar of a "classical Hollywood cinema," *The Big Broadcast* is atypical at best; its director Frank Tuttle is an unlikely candidate for "auteur" status; as a musical, little recommends the film for sustained analysis in terms of its deployment of or challenge to generic conventions.[56] While the ostensible subject of the film, the fortuitous rescue of a financially troubled radio station, suggests that *The Big Broadcast* is relevant to discussions of radio-cinema connections, no such analyses exist. Its comparatively late date of release also generally eliminates the film from technological or economic accounts of Hollywood's conversion to synchronous sound. Conventional wisdom generally holds that by 1932 synchronous-sound film technology had been successfully diffused throughout the production and exhibition ends of the American film industry. But a closer consideration of *The Big Broadcast* demonstrates that the film bears striking traces of the process of technological change. Indeed, its diversity of representational strategies actually stages the aesthetic struggles that accompany such change. Through its use of various, often contradictory, methods of depicting performances and representing sound, the film exhibits the hybrid qualities indicative of the struggle over how electrical sound technology would be deployed in the cinema and elsewhere. *The Big Broadcast*, especially the final third of the film, throws into radical juxtaposition the clash between transcription and signification paradigms for electrical sound technology's relationship to acoustic events.[57]

Although *The Big Broadcast* was loosely based on William Ford Manley's play *Wild Waves*, this original source seems to have supplied little more than the suggestion of locating a comedy in a radio station.[58] Appearances by well-known radio performers and announcers provided one of the primary contemporary appeals of the film. The presence of well-known radio and popular recording performers is "announced" in the film's opening credits as a voice-over announcer declares, "Attention all stations . . . calling all stations. Clear the air lanes. Clear all air lanes for the big broadcast. The Big Broadcast." Next, a series of performers' photographs on an apparent publicity poster in turn come to life and present a few bars of signature songs or routines (in order: Bing Crosby, Kate Smith, the Boswell Sisters, Cab Calloway, the Mills Brothers, Arthur Tracy, Vincent Lopez and his Orchestra, and Burns and Allen). *Billboard* claimed "the radio names should mean something, but the film itself is not so hot an entertainment."[59] Apparently the radio names did mean something as *The Big Broadcast* was seventh in *Variety*'s list of top film grosses for 1932.[60]

The narrative of *The Big Broadcast* initially focuses on a financially troubled New York City radio station, WADX. The star of the station, the chronically late crooner Bing Hornsby (Bing Crosby), is dumped by his showgirl sweetheart Mona Lowe (Sharon Lynne) and befriends a heartsick Texan named Leslie McWhinney (Stuart Erwin). After being exposed to multiple renditions of the dirgelike song "Here Lies Love," the two new friends attempt suicide. Their attempt to asphyxiate themselves in Hornsby's kitchen is foiled by the timely intervention of Anita (Leila Hyams), a secretary at WADX and coincidentally the long-lost girl-from-back-home for whom Leslie pines. Leslie, who turns out to be an oil millionaire, resolves the narrative's initial tension when he buys not only the fiscally unsound station but also a national radio network. Thanks to its angel, WADX is on firm financial ground, but in the suicide aftermath a love triangle between the now-single Bing, the crooner-infatuated Anita, and the still heartsick Leslie provides a second narrative quandary. This too is eventually resolved in the film's final sequence when Mona returns to Bing and, after attempting to aurally stand in for the again tardy crooner during the network's inaugural "big broadcast," Leslie wins back Anita.

In its review, *Variety* labeled this story "childish" and noted that *The Big Broadcast* was "not a feature film but a succession of talking shorts." In the words of the reviewer, "That these specialties were dove-tailed as they were is not bad considering that somehow the entire works had to be hung onto some sort of a story structure, no matter how fragile."[61] The characterization of the film as a "succession of talking shorts" is largely true of the climactic sequence in which Leslie scours the city for a phonograph copy of Bing's "Please." Leslie searches for the recording in order that this transcription of Hornsby's performance might stand in for the actual presence of the absent crooner during the ongoing "big broadcast." Leslie's mishaps as he struggles to find the record are made all the more urgent as the film crosscuts between Leslie and the broadcast's series of performances that signal its inexorable advance toward the climactic "performance" by Hornsby.

The film as a whole involves a complex and heterogeneous approach to the combination of narrative and moments of performance. Throughout the film, differing strategies embed performance events within diegetic space and thereby justify their presence. These strategies exist on a continuum from performances firmly anchored within the diegesis to the uneasily interpolated "succession of talking shorts" to which the *Variety* reviewer referred.[62]

A scene of Bing Hornsby singing "Please," although a Crosby performance per se, is easily contained by its narrative function—the crooner rehearsing for the big broadcast. More typical of the future so-called "integrated" Hollywood musical, *The Big Broadcast* includes an impromptu, "spontaneous" performance by Hornsby/Crosby as his shoeshine boy provides rhythmic accompaniment for the song "Dinah." The incorporation of both of these performances follows established cinema conventions from the initial phase of Hollywood's "backstage" musicals.

Cab Calloway's first instrumental performance is similarly diegetically motivated as part of the daily broadcasting business of station WADX (Cab Calloway and his Girdleers performing on the "Grip-Tight Girdle Hour"). Although diegetically anchored, as it were, this Calloway performance is only marginally contained within the representational strategies initially established by the film. As the Calloway performance "heats up," visual consistency gives way to visual strategies that arise from the music itself. Rather than a naturalistic style of representation in which the visible and the audible work in concert, mutually reinforcing both each other and what we might call "diegetic integrity," the music begins to dominate and to motivate the visual representation. As the musical tempo and Calloway's signature gyrations intensify, the mise-en-scène becomes literally animated (a clarinet grows rubbery in the hands of its musician and a microphone leaps up and down on its stand). In terms of cinematography, the framing dizzyingly cants to the left and to the right. During the Calloway performance it is as if the visual world of the film had suddenly fallen prey to what conservative critics of the era labeled the narcotizing effects of jazz. The scene begins as a kind of audiovisual transcription of Calloway and his band performing, a transcription in which sound recording and visual strategies collaborate not only to document a performance but also to locate that event in a narrative world that we accept as consistent with our own. But as the performance progresses, both the narrative world (microphone and clarinet) and the visual representation of that world (cinematography) become affected by the literally infectious rhythms of Calloway's band.

The Big Broadcast's presentation of multiple versions of "Here Lies Love" provides by far the most elaborate attempt to interpolate performance into the narrative. First the accordion-clad Arthur Tracy initiates the song in a speakeasy where Bing and Leslie have just met. The film cuts to a rooftop nightclub where Vincent Lopez and his orchestra continue the song. After Lopez performs one verse of the song and the chorus, an elaborate camera movement reveals that the nightclub is adjacent to Hornsby's apartment and that, offscreen, Bing and Leslie have made their way to Bing's apartment. The final version of the song is performed spontaneously by Bing (accompanied by the now offscreen Lopez orchestra) as he and the morose Leslie contemplate suicide. The three versions of the song are performed continuously, with the film's soundtrack providing a transition from two distinct spaces while indicating (an only slightly implausible) temporal continuity. The song is reprised moments later as Arthur Tracy returns in the form of a ghostly apparition to accompany the final waking moments of Leslie and Bing's suicide attempt. The various performances of "Here Lies Love" function narratively to indicate a continuity of time, place, and action while simultaneously commenting thematically on the psychological states of the two despondent main characters. The performances by Tracy, Lopez, and Crosby are thus multiply integrated within the plot's presentation of story material.[63] These performances not only originate within the diegetic space of the film but also reinforce visual spatial and temporal cues while providing additional depth to the narrative by echoing the characters' inner psychological states.

Other performance moments are not so studiously integrated into the narrative of *The Big Broadcast*. News of Hornsby's termination by station WADX and Mona Lowe's marriage to a Wall Street financier (thereby jilting Hornsby) is spread through the radio station by three telephone operators (played by radio performers Major, Sharp, and Minor). This performance is far less integrated into the visual and narratological strategies of the film. It is transacted as an overt performance when first the operators and then the employees with whom they speak interact in rhyming dialogue.[64]

Although not immediately related to any of the major characters or central questions posed by the narrative, two radio-like performance moments featuring George Burns's comically frustrating encounters with Gracie Allen are loosely "narrativized" through their roles as station manager and receptionist. These performances feature the familiar combination of Allen's non sequiturs, malapropisms, and apparently aimless familial anecdotes with Burns's ever-increasing frustration—a performance style recognizable to the audience not only from Burns and Allen's radio appearances but also from their appearances in a series of popular short subjects. In each of these short films, Burns encounters Allen in some mundane, everyday capacity with a simple type of business to transact, only to have that task derailed by the comic encounter. That pattern is repeated here with little attempt to integrate these scenes into the narrative's chain of cause and effect.[65]

The techniques used in *The Big Broadcast* to represent music and comedy could be considered as cinematic adaptations of previous approaches to representing performance variously informed by traditions from vaudeville, musical theater, radio, or backstage film musicals. Some of these strategies anticipated conventionalized fixtures of Hollywood practice (so-called "integrated musicals") while others would be quickly marginalized (rhyming dialogue, Burns and Allen's vaudevillian schtick). *The Big Broadcast* thus might be seen as an illustration of some of Hollywood's approaches to the problem of integrating performance within narrative in the aftermath of the conversion to sound. But from some perspectives the integration of performance into narrative might not have been considered a "problem" at all, or rather, might have been formulated as a problem only intermittently. Notions of "interpolation" and "integration" privilege a specific approach to both producing and consuming films—an approach centered on narrative. Simple reflection on the circumstances of much of American film exhibition directs us to the recognition that the filmgoing event consisted of a decidedly nonintegrated aggregate of film styles. The evening's film program, especially in 1932, typically could include a newsreel, an animated film, live-action entertainment shorts of varying styles and genres, and a feature film perhaps further augmented by some form of live performance. Such a mix of film styles, modes of address, and types of performance suggests that perhaps for exhibitors and audiences, integration was not an issue.

If we ignore this exhibition context and consider feature-length films in isolation of the texts that accompanied them, further evidence indicates that integration may not be the most salient issue. Historian Alan Williams suggests that the *lack* of integration of performance into narrative is in fact an *integral* component of the film musical. Williams provocatively argues that it is precisely the type of music in musicals—the popular song—that authorizes and indeed mandates a nonintegrated approach to performance and narrative. "The musical film's alternation between music and diegesis seems both permitted and required by the strong boundaries that popular songs set for themselves. Music and narrative are set up as two terms of a paradigm, and the relative rarity of film operas as compared to film musicals suggests, also, that marketable effects obtain somehow from the alternation."[66] Anecdotal evidence indicates that Williams' perspective was shared by some period film critics. For example, many of the reviews of film musicals appearing in *Photoplay* during the early sync-sound era indicate little concern for whether or not, in the words of the *Variety* reviewer, performance "dove-tailed" with the narrative.[67] The notion that musical performance and narrative were in fact appropriately two distinct realms also seems to have partially underwritten some of Hollywood's production practices in the early 1930s. Credits for some early musicals indicate a division of labor, with directing credits assigned to two studio employees—one for the musical numbers and another for the story.[68]

While *The Big Broadcast* could be viewed as a veritable catalog of potential strategies for interpolating performance within narratives, such a focus on integration may prove less fruitful than other approaches that locate the film in a non-media-specific struggle to develop an identity for electrical sound technology. A focus on interpolation of performances emphasizes *The Big Broadcast*'s difference both from conventional narrative film per se (however understood) and from a teleologically posited and (more) fully integrated narrative musical film. But period critical and production practices, as well as the work of Alan Williams and Martin Rubin, at least partially accept the nonintegration of performance into narrative as either characteristic of the film musical and/or part of a larger tradition of "aggregate" performance practices. If we replace a focus on the relationship between performance and narrative with instead a focus on the relationship between representations and sound technology, such a view directs us toward similarities between *The Big Broadcast* and other preceding, simultaneous, and subsequent hybrid practices across several media forms.

Variety's characterization of *The Big Broadcast* as "a succession of talking shorts" most accurately refers to the final thirty minutes of screen time during which Leslie's quest for a Bing Hornsby recording is crosscut with the ongoing "big broadcast." It is this segment of the film that exhibits the most striking contrast between the transcription and signification paradigms as conflicting models for the recording and representation of sound. Six discrete performance events are presented in their entirety, separated by cutaways to Leslie's quest for the Hornsby recording. The performers are, in order, the Mills Brothers, Vincent Lopez and his Orchestra, Donald Novis, the Boswell Sisters, Kate Smith, and Cab Calloway. The following well-known radio announcers appear on-screen as well: Don Ball, James Wallington, Norman Brokenshire, and William Brenton. With the exception of one moment during Leslie's quest (his telephone call to the station to warn them of his impending arrival with "Bing," actually a phonograph recording of Bing's voice), the series of scenes that make up his journey through the city are accompanied entirely by nonsynchronous, nondiegetic sound effects and music. The origin of the nondiegetic score is only intermittently and marginally grounded within the narrative space through the occasional presence of loudspeakers supposedly broadcasting the performances originating in the studios of WADX.[69] But the absence of dialogue (even during Leslie's encounters with first a music store owner and then a policeman), the absence of sounds with any spatial signature conforming to that of the image, and the presence of clearly nondiegetic sound effects "synchronized" to narratively significant objects and events, lend these scenes depicting Leslie the quality of a comic pantomime *accompanied* by a discrete, albeit carefully orchestrated, series of sounds. The scenes appear and sound as if the image and the soundtrack were separately produced, in an additive relation, thereby attesting to some of the material practices of the film's production. Even during those marginally diegeticized moments when the conspicuous presence of a loudspeaker horn rationalizes the presence of a musical score, the sound clearly originates in a space distinct from that of the depicted action.

The uncanny relation between sound and image in the depictions of Leslie's quest is striking because of its sharp contrast with the preceding sequences of the film.[70] The disjunction between sound and image characteristic of Leslie's travels through the city conflicts even more dramatically with the method of presenting sound during the six performance events. As the "big broadcast" unfolds, two contradictory approaches to film sound collide, enacting the conflict between two conceptual models for the relationship between electrical-acoustic technology and sound events.

The representation of musical performances in the final third of *The Big Broadcast* is determined by the performances themselves. Each performance event is presented in its entirety, without interruption, and with sound and a depiction of the sound's source continually in synchronization. Only a single brief cutaway to a shot of people listening during the

Boswell Sisters' number and the final bars of the Vincent Lopez performance (which function as an aural bridge to the space and representational strategies of Leslie's quest) provide exceptions to the principle of synchronization. Pro-filmic performance events, their duration, and their production of sound determine representation strategies. Framing is often dependent on the content of the performance. Long shots are used to capture the participation of the entire Lopez orchestra during its novelty number or the full-body physical movements of Cab Calloway during his rendition of "Minnie the Moocher." Medium shots of a single performer focus attention on the "trumpet" solo in the Mills Brothers' routine or the emotive facial expressions and overwrought hand gestures of Kate Smith. For those performances that do not provide overt cues for image strategy, the film is at some pains to provide visual accompaniment, as when the Boswell Sisters are framed by an open piano lid or when Donald Novis stands as rigid and stiff as the microphone into which he sings. In each performance segment, the presence of the WADX microphone within the image justifies the soundtrack's consistent acoustic properties (i.e., uniform volume, absence of reverberation, consistent spatial signature) despite shifts in shot scale or camera angle. Throughout the six numbers, performances are, in effect, transcribed—with no apparent manipulation of sound and relatively unobtrusive visual accompaniment.

The sharp contrast between the sound strategies of these recorded performances and the nonsynchronous relationship of sound and image in the final section of *The Big Broadcast* point toward two distinct approaches to the representation of sound phenomena, two conceptualizations of how to represent sound—in effect, two competing theories for the relationship between electrical acoustics and the task of recording. The visual and aural representation of the radio broadcast's performance events grow out of a transcription paradigm for the relationship between sound technology and acoustic events. Here the desire to collect and preserve a performance event motivates the strategies used for sound recording, image choice, and the relationship between the two. Sound (and image) strategies function at the service of a preexisting event, respecting its duration and documenting its integrity, its wholeness as an event. Sound recording is conceptualized as the ordered transcription of a preexisting world.

In contrast, the film's depictions of Leslie's quest for the Hornsby recording evidence an approach to sound recording based on the signification paradigm. Rather than the transcription or documentation of a preexisting event, the work of sound recording is conceptualized as the ordered arrangement of diverse sound phenomena of different origins to create the illusion of a new "reality." In effect, the sound reproduction is conceived as a blank space—silence or "dead air"—into which one then arranges aural phenomena. Although far from realistic in the way we might normally use the word, the sounds of Leslie's quest do maintain an internal logic and consistency, albeit with little resemblance to our experience of phenomenal reality.

While the protocol for handling sound during *The Big Broadcast*'s final series of performance events is historically based on dominant conceptions of the gramophone and the phonograph and their ability to document and preserve acoustic events, the strategies for constructing the sounds of Leslie's quest are based instead on a nonmimetic protocol for sound construction evident in some silent cinema accompaniment and in the theories and practice of predominantly European advocates of a sound strategy not normally associated with American cinema (for example, René Clair, Sergei Eisenstein, Dziga Vertov, Grigori Alexandrov, and others). The final third of *The Big Broadcast* throws into sharp relief Hollywood's equivalents of the conceptualizations of electrical acoustics as embodied in Seashore's and Metfessel's "phonophotography" and Potamkin's vision of the "sonorous film."

The opposition staged in *The Big Broadcast* between two conflicting conceptual models for the relationship between electrical-acoustic technology and sound—transcription versus signification—is not just the aesthetic choice of a single director or studio. Instead it is the embodiment of a major clash within aesthetic, technical, and industrial realms central to the diffusion of electrical acoustics. What is at stake in textual hybrids like *The Big Broadcast* and the early Vitaphone and Movietone films is nothing less than the proper use of a new technology. This clash was also enacted in larger theoretical and practical debates among Hollywood's sound practitioners. Those debates illustrate that far from being rare, marginalized instances, the texts I have labeled "hybrids" are symptomatic embodiments of a more pervasive struggle over proper protocols for technology-in-use.

Hollywood's Sound Practice Debates

Throughout Hollywood's conversion to synchronous sound, technicians, engineers, and other film industry participants publicly struggled to shape the methods by which electrical-acoustic technology would be applied to the cinema. Published in film industry trade papers and technical journals, these debates most often focused in general on different definitions of the appropriate relationship between soundtrack and image and between technology and sound. In recent years, these debates have drawn significant scholarly scrutiny because the issues raised (including the number and position of microphones, when and how to mix the outputs from several microphones, the relationship of sound volume and amount of reverberation to shot scale, acceptable uses of post-synchronous mixing, etc.) involve more than mere technical details but instead offer a privileged insight into the development of realistic conventions and standardized protocols for representation. But these debates also vividly demonstrate the important roles played by transcription and signification as competing paradigms for understanding the relationship between technology and sound. These conflicting paradigms not only shaped the daily activities of Hollywood sound practice but also shaped the conceptual frameworks sound practitioners relied upon to develop precisely what would become Hollywood's sound practice.

Historians like James Lastra and Rick Altman have productively reassessed these debates in light of two competing models for handling sound.[71] Lastra finds in Hollywood's discussion of sound practices a conflict between a "phonographic" (or "perceptual fidelity") model and a "telephonic" (or "intelligibility") model in which "the former sets as its goal the perfectly faithful reproduction of a spatiotemporally specific musical performance" while "the latter, like writing, [takes as its goal] intelligibility or legibility at the expense of material specificity, if necessary."[72] On the one hand, the "phonographic" (or "perceptual fidelity") model sought to re-create acoustically the specific conditions of the production of a sound. The "telephonic" model, on the other hand, advocated the intelligibility of dialogue as the primary function of sound practice and involved such procedures as relatively close microphone placement, dialogue recording with more or less stable sound levels independent of shot scale, and the minimalization of reverberation as an indicator of realistic sound space. In Lastra's terms the "phonographic" model strove to re-create "*perceptually* specific information," while the "telephonic" model sought to encode and represent "*narratively* important information."[73]

Altman sees in these same debates a similar conflict between competing systems of handling sound. Initially contrasting "recording" with "construction," Altman later uses the terms "realism" versus "intelligibility" to characterize the divergent methods advocated by Hollywood's sound practitioners.[74] In positing the historical antecedents of these divergent sound practices, Altman suggests that radio influenced the initial documentary quality of cinema sound.[75] Radio's methods of sound representation were never entirely uniform. *Some* radio practices did conform to Altman's "realism" model (broadcasts of independently existing public address or performance events, for example), and therefore both radio and some early sound cinema drew on the preexisting transcription paradigm for conceptualizing technology's relationship to sound.[76] Altman also asserts that the cinema's emphasis on intelligibility (its "intelligibility system") was modeled on or borrowed from theatrical and telephonic precedents.[77] While clearly the telephone was socially constructed as an instrument that provided an intelligible version of the human voice, it was developed predominantly as a transcription technology, a documentary technology, distinguished from the preexisting telegraph by its ability to transmit across distances a recognizable voice. Within Hollywood sound practice, however, the "intelligibility system" was clearly a product of electrical-acoustic technology understood as an instrument of *signification*, as the discussion below will illustrate. In general, the technologies and protocols for twentieth-century sound representation reveal a shifting series of emphases between practices shaped by transcription and by signification paradigms across multiple sound media. Rather than a specific identity for any given sound medium—i.e., the radio as a documentary medium, the phonograph as a medium of transcription—each medium evidences the simultaneous presence of *both* conflicting paradigms, albeit in different relationships and to different degrees varying across time.[78]

Participants in Hollywood's sound practice debates (with backgrounds in film, acoustic engineering, radio, etc.) navigated these preexisting paradigms by grounding their various arguments on a number of recurring assumptions about the function of representational technologies. Some advocated a particular technique because of its ability to mimic human perception. Here, technology-in-use was configured as a surrogate for the audience. Others located the task of technology as a self-effacing fidelity to the physical world. Still others eschewed recourse to either the corporeal or phenomenal and instead understood emerging technology in economic terms, conceptualizing its function as either serving narrative requirements or the more ubiquitous category of entertainment. Hollywood's sound practice debates demonstrate that differing understandings of electrical-acoustic technology led to conflicting conceptions of the technology's representational function and thereby competing sound recording practices.[79]

Sound practitioners who took the human body and human perception as the model upon which the technology-in-use was to be based conceived electrical acoustics as properly a mimetic apparatus. Hollywood's conversion to sound posed the problem of approximating binaural perception through a monaural system.[80] This led both to various methods of technologically *simulating* binaural perception and to further research and development designed to move the electrical-acoustic apparatus toward a truly binaural system. By the time of Hollywood's wholesale conversion to synchronous sound, the technology existed that could provide theater audiences with a truly binaural experience. Recording on two separate channels and reproducing sound through headphones ("telephones," in the radio terminology of the period) would have been the logical outcome of developing sound technology on the model of the human body. With two recording and reproducing channels linked to the auditor-spectator's two ears, electrical acoustics could have provided film audiences with a binaural experience. But this solution was quickly discounted by sound practitioners for both technical and economic reasons, namely the added cost to exhibitors already burdened by the expense of converting theaters to synchronous sound.[81] Further, such a binaural acoustic system would have posed a drastic contrast to preexisting theatrical filmgoing experiences. Would audiences watch and listen to a live theatrical presentation and then at some cue from the theater reach beneath their seats and put on their telephone headsets? Would they sing along, following an on-screen bouncing ball, aurally isolated by their personal acoustic reproduction device, not even able to hear the sound of their own voice? And what of the construction of the cinema as a live, group-based social experience?

Binaural sound recording and its reproduction through individual headphones could deploy electrical acoustics as an approximation of human perception, but a variety of economic and social reasons obviated its implementation. Instead, Hollywood's sound practitioners and acoustic engineers pursued two independent solutions to this problem. Harvey Fletcher and other engineers at Bell Labs developed a binaural sound system, demonstrating it to the Society of Motion Picture Engineers in 1934.[82] But while Fletcher's system showed promise for eventually approximating the experience of binaural hearing, it didn't address the problems of the monaural system already diffused throughout the American film industry. Instead, a variety of voices in Hollywood's sound practice debates proposed methods of microphone placement, recording, mixing (or rather, avoiding mixing), and set design to simulate monaurally the perceptual experience of binaural hearing.

Other Hollywood sound practitioners conceptualized the problem of conversion to synchronous sound and with it the identity of the electrical-acoustic apparatus in different ways. For some, the transition to sound provided not simply a step toward the greater technological approximation of human perception but instead a new and powerful method of expression independent of mimetic objectives. While few participants in Hollywood's sound practice debates went as far as Harry Alan Potamkin in advocating the rejection of sync-sound's mimetic possibilities, some participants in these debates argued for the need to privilege the intelligibility of the voice to the exclusion of potentially realistic acoustic markers. These sound practitioners also often advocated techniques of sound signification at the service of the film image with little concern for respecting the notion of an original acoustic event. Others went even further in the direction advocated by Potamkin by overtly viewing electrical acoustics as an instrument of signification, calling for the exploration of its signifying possibilities independent of the image.

One of the central issues of Hollywood's sound practice debates, and the issue most frequently emphasized by contemporary historians, involved the relationship between soundtrack and image in terms of spatial cues. Numerous contributors to the debates remarked on the unrealistic result when changes in shot scale (i.e., a cut from a long shot to a close-up) were not accompanied by corresponding shifts in sound quality (i.e., decrease in volume, decrease in the ratio of direct to reflected sound). John Cass of the RCA Photophone Gramercy Studio located the source of such conflicting spatial cues in the practice of mixing the output from multiple microphones into a single sound designed to "attain intelligibility of speech and good quality of speech and music."[83] "The resultant blend of sound," Cass maintained, "may not be said to represent any given point of audition, but is the sound which would be heard by a man with five or six very long ears extending in various directions."[84] Framing the problem in terms of a grotesque spectator's body, Cass posited human perception as the standard against which sound practices were to be judged.

Cass was far from alone in bemoaning a sound practice that featured continuous close miking and consistent sound qualities independent of shot scale. While in March 1930 Porter Evans, chief engineer of the Eastern Studios of Warner Bros.–Vitaphone, largely dismissed the disjunction described by Cass as characteristic of the "early days of sound recording," Joe Coffman of Audio-Cinema in February of the same year denigrated the ongoing practice and blamed it on the influence of radio technicians recruited by Hollywood.[85] Although frequently reprinted and sometimes accepted at face value by contemporary historians, Coffman's accusations about radio engineers is belied not only by the multiple practices that characterized radio engineers' approaches to sound but also by the contributions of radio engineers to these debates.[86]

Former radio engineer Carl Dreher, vice president of RCA Photophone Co., clearly identified as a problem the disjunction between spatial cues provided by sound and accompanying image in February of 1929. "When characters go backstage after a close-up there is no corresponding diminution in the level of their speech. This is something that should be taken care of in recording but usually isn't."[87] Calling for the industrywide elimination of this practice and the greater standardization of technique, Dreher further located the problem in the use of multiple microphones to record a scene so that "when an actor turns away from the audience there is not the change in his voice one would usually expect."[88] Instead a second microphone "was used to pick him up when he turned away and the recording expert neglected to bring down the gain control somewhat on his transmitter to take care of its direction with respect to the future audience."[89] Dreher proposed a solution for this kind of "faulty" recording in February 1929 when he suggested that film projectionists compensate for such an unrealistic recording practice by riding gain (increasing and decreasing volume of reproduction) with a control located in the theater audience. For a time, such theater volume control systems were incorporated into some projection installations, but this film projection practice ran counter to the desire for increased standardization among film producers.[90] Although initially aligning himself with Cass, Dreher with his often apparently shifting position in the sound practice debates demonstrates the vast disagreement among sound practitioners as they struggled to standardize sound protocols and conventions. Only ten months after identifying the absence of sound perspective as a problem, Dreher in a different publication advocated close miking—a practice that would eliminate the sound perspective effects for which he had earlier called.[91]

Perhaps the most strident critic of close miking independent of shot scale was ERPI's J. P. Maxfield. In a series of articles, Maxfield clearly articulated the failure of what Lastra has called "telephone" models of sound recording to approximate the reality of human perception.[92] Unlike Dreher, who in 1929 suggested imposing spatial effects through manipulating volume during the exhibition of films, Maxfield proposed a system that carefully *recorded* spatial signature through not only volume control but also control of the relationship between direct and reflected sound. Like many of the participants in the sound practice debates, Maxfield conceptually premised and rhetorically presented his model in terms of the human body and human perception. Taking as his goal the technological reproduction of the experience of binaural hearing through electrical acoustics' monaural system, Maxfield advocated the use of a single microphone for each camera position, thereby allowing (with careful attention to microphone placement) a sound recording that, through volume but more importantly through the relationship between direct and reflected sound, approximated the spatial conditions denoted in a given shot. Microphone placement would depend on the position of the camera and the focal length of its lens. Maxfield's system avoided the use of a mixer during a take so that once appropriate volume levels for the action were set, no manipulation during a scene would change the consistent relationship between the viewer addressed by the image and the auditor addressed by the sound.

Maxfield's careful calculation of "appropriate" microphone positions ran counter to those who suggested more arbitrary methods of microphone placement.[93] Dreher, for example, claimed that "with even the shrewdest technical personnel on the job, microphone placing is a matter of cut and try."[94] As late as 1931, Albert W. DeSart, technical director of sound at Paramount-Publix Studios, in an essay that immediately followed Maxfield's in a publication of the Academy of Motion Picture Arts and Sciences, described a complicated studio setup and concluded: "This particular arrangement was, like any other arrangement, an arbitrary one; there are a large number of ways of arriving at equally good results in the

making of sound pictures, and no director and technical crew are likely to
use exactly the same set-up as another staff would."[95] Even advocates (with
Maxfield) of careful attention to sound perspective offered less than precise solutions. Coffman, for example, suggested that any director or sound
technician could locate the best microphone position by simply covering
one ear (thereby mimicking the monaural recording system) and moving
around the set during rehearsal.[96]

Maxfield's system of microphone placement and recording practice
also ran counter to those more concerned with intelligibility than approximating human perception of the depicted scene. Referring implicitly to
Maxfield, Porter Evans wrote,

> It has been suggested many times that the microphones be placed ap-
> proximately the same distance from the action as the camera. Some
> people claim that satisfactory results can be obtained in this way if the
> sets are given the proper acoustical characteristics. It has been our ex-
> perience, however, that when the microphone is placed at any distance
> from the action, the noise level is greatly increased by comparison with
> the sound, and a great deal of detail in the speech or music is lost, and a
> blurring of effect is introduced.[97]

The clash between positions respectively advocated by Maxfield and
Evans has been taken by some historians, most eloquently Altman and
Lastra, as a conflict between two distinct models for recording practice.
One could also identify a similar conflict between the various positions
articulated at one time or another by Carl Dreher. Dreher first identified
the absence of perspective matching as a problem, then advocated close,
"cut and try" miking which would largely eliminate aural spatial markers,
and finally concluded in June 1931 by acknowledging this tension. Noting that "the principal characteristics of good recording" included both
intelligibility of dialogue and "acoustic fidelity to the original rendition,"
Dreher argued in 1931 that "since the reproduction of sound is an artificial
process, it is necessary to use artificial devices in order to obtain the most
desirable effects."[98] This tension then, between sound practices privileging
dialogue versus those that privilege acoustic fidelity to an original sound
event, was for Dreher by 1931 less a conflict between mutually exclusive
models than two goals that equally might well be accomplished by artificial

means. While on the surface the conflict Altman and Lastra find in these Hollywood sound practice debates between "telephone" versus "fidelity" (or "construction" versus "realism") models for sound seems to echo the two conceptual categories I develop (transcription vs. signification), in actuality I take my cue from Dreher and the observation that both the positions advocated by Evans and Maxfield involve the use of sound technology for the purpose of signification.

Signifying Fidelity: Hollywood's Changing Sound Practice and a Third Conceptual Paradigm

Maxfield's system of guaranteeing sound perspective analogous to shot scale prevented the disjunctive perceptual experience characterized by Cass's description of a grotesque, multi-eared body. Maxfield's model, especially in terms of sound perspective and sound/image matching, appears to coincide with a paradigm of sound recording practice conceived of as the technological transcription of an acoustic event. Indeed recent historians like Altman and Lastra have found in Maxfield a champion for the use of "realism" instead of "construction" (or "fidelity" instead of "telephony"). But the conception of sound technology's function as properly *signification* underwrites Maxfield's system as well. Because electrical-acoustic technology's status as a monaural system clashed with the professed goal of repeating the experience of binaural hearing, Maxfield advocated several types of compensatory acoustic manipulation. In Maxfield's view, monaural recording and reproduction decreased the audience's ability to consciously discriminate between sounds of varying levels of importance. Therefore, in recording practice it was necessary to limit both the volume and reverberation of sounds categorized as unimportant (i.e., irrelevant to narrative or other signifying needs) to lower than their naturally occurring levels. This manipulation was necessary, in Maxfield's words, "to satisfactorily create the illusion of reality."[99] Recording practice thus involved the creation of a hierarchy of sounds so as to compensate for the technology's inability to reproduce binaurally. As Maxfield stated,

> With the loss of the sense of direction, which accompanies the use of monaural hearing, this conscious discrimination becomes much more difficult, and the incidental noises occurring in a scene, as well as any

reverberation which may be present, are apparently increased to such an extent that they unduly intrude themselves on the hearer's notice. It is, therefore, necessary to hold the reverberation, including these noises, down to a lower loudness than normal, if a scene recorded monaurally is to create a satisfactory illusion of reality when listened to binaurally.[100]

Here, limiting the volume of sounds categorized as extraneous and limiting some degree of reverberation (this necessary extra acoustic "deadness" was conveniently provided by the absent ceiling and fourth wall of most studio sets) amounted to manipulating acoustic phenomena neither to document an acoustic event nor at the service of signification (as understood by Potamkin, for example) but rather to *simulate* the perception of an event. Despite a rhetoric grounded in appeals to the human body and perceptual experience, even "realistic" approaches like Maxfield's simultaneously preferred the strategic use of sound technology for the purpose of signification. Maxfield advocated manipulation so as to signify bodily experience, to signify fidelity to the perceptual experience of an acoustic event.

Maxfield was not alone in advocating a kind of fidelity to acoustic phenomena through carefully mobilized procedures of signification and, in fact, if anything can characterize in general the range of sound practices developed by Hollywood in the aftermath of technological change it is this notion of *signification of fidelity*. Later in the decade, when RCA's Harry Olson and Frank Massa advocated the use of directional recording and reproduction for the sake of realism, they too described a series of signifying procedures at the service of connoting fidelity to acoustic phenomena.[101] Building on advocates of scale matching like Maxfield and Cass, Olson and Massa assumed from the outset that the space indicated by a sound should be consistent with that depicted by the accompanying image. Like Maxfield's, their system of realistic acoustic representation sought to compensate for the currently available (and, in their view, acoustically impaired) monaural sound systems. In calling for procedures that "established the center of gravity" of action through the strategic suppression of extraneous sounds, Olson and Massa's use of directional microphones technologically structured acoustic events so as to approximate and encode human perceptual processes. The directional microphone Olson and Massa had at their disposal (the velocity microphone) was a recent engineering development incorporating into the design of sound collection itself the directional

qualities previously supplied to microphones through accessories like baffles and sound condensers.[102] To some extent, Olson and Massa successfully redefined the apparent clash between sound practitioners advocating
fidelity versus those privileging intelligibility of the voice. For Olson and
Massa, Hollywood's emphasis on intelligibility was itself an acoustic fidelity to human perception of an original event, since in binaural hearing the
individual, like the directional velocity microphone, is able to attenuate extraneous sounds.

Throughout these sound practice debates, the various positions advocated are not necessarily as distinct as they might at first appear. While
advocates of speech intelligibility to the exclusion of acoustic markers of
realistic space argued for sound technology as a signifying rather than transcribing apparatus, advocates of realistic sound perspective also in practice
called for *strategic signification* at the service of simulating the human perception of auditory phenomena. In his 1931 contribution to these debates,
Dreher noted that when reproducing dialogue at volume levels greater than
the original performance (a practice necessary for theatrical film exhibition) it was sometimes necessary to drop out some of the lower frequencies
to insure clarity of dialogue. "This is an artificial device to preserve intelligibility; properly used, however, it may enhance the illusion of natural
sound reproduction for the audience."[103] Even while bemoaning the less
than realistic practice used by Hollywood, Cass announced the importance
of signification to recording practice in the title of his article: "The Illusion
of Sound and Picture." Maxfield, Olson, and Massa, and their use of the
human perceptual experience as the standard against which sound practices were to be judged, called for a series of procedures designed to create
acoustic illusions of reality. The sound practice debates thus reveal less two
competing models for the function of electrical-acoustic apparatus than the
development of a third conceptual paradigm that, while often embracing
the rhetoric of fidelity characteristic of transcription practices, relied upon
recording procedures mobilized at the service of signification. The conflict
between transcription versus signification models of handling sound was,
in effect, healed by a new paradigm in which the standards invoked for
measuring fidelity often shifted from a preexisting event to *perception* of
that event. While some sound practitioners overtly called for technology
to serve narrative or entertainment ends, advocates of so-called "fidelity"
or "realism" models of sound practice were themselves calling for technol-

ogy to serve signification. Both "telephone" and "fidelity" models involved signification; the difference was largely one of rhetoric. Maxfield, Cass, Olson, and Massa disguised the fact of signification through a rhetoric that invoked the human body and human perceptual processes as an objective standard.

Simultaneous with Maxfield's rhetoric of a documentary model of sound recording, numerous Hollywood production practices relied on the strategic use of technology at the service of manipulating acoustic events. These manipulations occurred both during recording and after an acoustic record had been collected. Microphone baffles and concave sound collectors attached to microphones focused microphone pickup to limit "extraneous" sounds. These accessories limited the amount of reflected sound reaching the microphone and, in effect, simulated the perceptual focusing characteristic of binaural hearing and/or supplied the "presence" to match close-up images when "the use of a long-shot camera precludes the possibility of getting a microphone close enough to the actor to produce the desired close-up sound."[104] Other types of microphone baffles together with various kinds of electrical filters could disproportionately record various sound frequencies. Often used to eliminate the "boominess" of recorded dialogue due to the reinforcement of lower voice frequencies by qualities of a set, these devices allowed greater flexibility in sound recording at the service of aesthetic rather than documentary goals. As L. E. Clark, technical director of sound at Pathé Studios wrote in 1931, "To be specific, a narrow bathroom or kitchen set is generally boomy. In certain cases it is allowable to reproduce exactly what the microphone hears. In others, artistic requirements demand a close match of voice quality with what has gone before. In this case, since the principal frequencies affected are below 200 cycles, a filter, attenuating below this value, will often work wonders."[105]

It is particularly difficult to sustain a characterization of Hollywood sound practices as documentary or as conforming to a transcription paradigm at the time of Maxfield's interventions, given the strategies for acoustic manipulation of previously recorded sounds that were then commonly practiced. The use of electrical filters to simulate the quality of the voice when augmented by telephones, old gramophones, or radios might be characterized as electrical-acoustic manipulation at the service of "realism," but such realism was accomplished through the strategic use of sound technology's signifying potential.[106] Similarly, the use of sound-looping technology could be categorized as "realistic," providing the sounds of real street scenes, a realistic acoustic ambience, under studio-produced images. But the technological bases of looping, the repetition of a series of prerecorded sounds, and period characterizations of the process as an acoustic equivalent to process photography, point toward the larger conception of technology increasingly underwriting Hollywood's sound practices during the 1930s—the ability of electrical acoustics to signify fidelity to (non)existent events.[107]

Whether to improve sound takes (with high- or low-frequency filters), add previously recorded sound effects and music, or overtly manipulate prerecorded sounds (adding reverberation and distance effects, or, for example, slowing down a recorded rifle shot to approximate the sounds of a distant cannon or even thunder), post-synchronous dubbing technology provided a wide range of options for sound signification.[108] One period commentator, Kenneth Morgan, recording manager at ERPI, estimated that fully 50 to 100 percent of all film sound was rerecorded in some way by 1931.[109] While some sound practitioners advocated that dubbing could be used to alter either volume or frequency, others argued that it was undesirable to manipulate or modify in any way the frequency characteristics of original recordings.[110] Loyalty to conceptions of an original acoustic event provided fewer constraints on the wholesale use of the signifying potentials of dubbing than other factors. Public reactions to revelations of voice doubling prompted some limitations on the possibilities technology offered for acoustic manipulation. Because the initial announcement and construction of electrical acoustics stressed the increased fidelity it offered and conformed to a transcription paradigm for the relationship between technology and acoustic events, careful discursive management was necessary when the public was presented with the possibility of voice doubling and other instances of technology mobilized for an overt signifying function. The report by the Progress Committee of the Society of Motion Picture Engineers for February 1930 noted that, "There has been further progress in voice doubling so that it can on occasion be done to a very high degree of perfection, but a public reaction against it has developed in those cases where it has become known. The practice is, therefore, being very rapidly discontinued."[111] Whether discontinued or not, the *possibility* of voice doubling was officially incorporated into the range of protocols available to Hollywood sound practitioners. In 1930 the practice was first codified in the "Players' Standard Contract," the industrywide document specifying employment conditions between freelance actors and producers.[112] Public perception and the Society of Motion Picture Engineers aside, the Producers' and Actors' Branches as well as the Academy of Motion Picture Arts and Sciences as a whole had clearly endorsed the practice early on.[113]

The SMPE's "Progress Report" also noted that a procedure similar to voice doubling was being used increasingly. The voice and image of the same actor would be recorded but at different times. The voice, recorded first to control acoustic conditions, would then be played back during filming of the image: the two distinct recordings were eventually combined through dubbing technology.[114] Some historians attribute to *The Broadway Melody* (MGM, 1929) the initial use of this form of prerecording, supposedly innovated by Douglas Shearer for the film's culminating Technicolor number, "The Wedding of the Painted Doll."[115] Period reports claimed that such a procedure characterized the production of entire revue films, such as Universal's *King of Jazz* (1930).[116] Prerecording became an acceptable procedure within Hollywood sound practices, albeit largely limited to representations of musical performances. The examples of voice doubling and prerecording musical performances point toward further ways in which Hollywood sound practice quickly embraced some of the possibilities offered by the signification paradigm for technology's relationship to acoustic events.

At issue in the sound practice debates was thus less a conflict between two competing paradigms for the cinematic use of electrical acoustics than a rhetorical conflict, a struggle between competing methods of talking about and, more importantly, thinking about technology-in-use (i.e., how do practitioners of sound think about the relationship between technology and their representational tasks). This rhetorical conflict threatens to obscure the extent to which Hollywood's sound practices increasingly conformed to a new paradigm for technology-in-use, what I have labeled *signifying fidelity*.[117] Although held up as perhaps the paradigm for "documentary" or "fidelity" models of recording sound, Joseph Maxfield's carefully calculated model of sound recording and reproduction clearly involved the deployment of electrical acoustics as a technology of strategic signification. The rhetoric of Maxfield's presentation of his model (or for that matter Olson and Massa, or Cass) emphasized the technology's relationship to the human body, to human perception as a standard against which recording practice was to be judged. With such a standard, a monaural system would always be deficient. Although wrapped in a rhetoric of fidelity to actually existing acoustic events, the task for Maxfield and for Olson and Massa was the strategic use of electrical acoustics so as to simulate, that is to signify, binaural perception.

Despite those, like Maxfield (or movie audiences scandalized by revelations of voice doubling), who invoked a rhetoric of sound technology as an instrument of transcription, the practices used by Hollywood's sound technicians moved toward this third conceptual paradigm. Edward W. Kellogg, from General Electric's Research Laboratories, anticipated in 1930 the eventual attitude of Hollywood's practitioners toward a tension between sound technology envisioned as an instrument of transcription or as a tool of signification. "While we are accustomed to thinking of the purpose of electrical sound equipment as that of faithfully imitating sounds which have been produced by voices or musical instruments, there is no reason why its function should be so limited."[118] Kellogg's use of the word "accustomed" points to the dominance of a transcription paradigm for conceptualizing the function of sound technology before the advent of electrical methods. This dominant model functioned conservatively to shape initial Vitaphone and Movietone synchronous-sound films. But as Hollywood sound practice adopted procedures designed to signify fidelity, the initial conflict between competing conceptual paradigms was tenuously resolved with the development of this new paradigm.

The documentary quality of the earliest sound-on-film and sound-on-disk shorts denoted the "mimetic objectivity" of the combined electrical-acoustic and cinematic apparatus. This demonstration of synchronization built on the preexisting social construction of (and consequent public knowledge about) acoustic technology—for example, the social construction of the gramophone and phonograph and fidelity-based announcements of technological change. At the time of those announcements, signification models were marginal to the practice and social understanding of recorded sound. The scandal of voice doubling or the disjunction caused by scale mismatching (aural spatial cues clashing with visual spatial cues) threatened to undermine epistemological faith in electrical acoustics as a transcription apparatus. These textual practices and public revelations of voice doubling reveal the work of sound practice as an act of signification.

During the sound practice debates, the film industry faced several options: it could embrace signification as a conceptual model for the work of sound production; it could revise cinema sound practice to make it conform to transcription models (i.e., through the use of a binaural headphone-based system of reproduction, or through massive capital investment in a binaural system like Fletcher's); it could rhetorically reinscribe sound practice as transcription while moving toward strategic signification practices as the operative conceptual model. Hollywood embraced the last solution. The emergent conceptual paradigm of *signifying fidelity* underwrote, and was in turn reinforced by, technological advances that were the product of application-driven research and development: the ascendancy of sound-on-film as the dominant recording and reproducing system; the innovation of microphone booms; the development of electrical filters and looping machines; the use of collectors attached to microphones and then the introduction of the directional velocity microphone; the refinement of mixing procedures and the rise of postsynchronous-sound production.[119]

The sound practice debates demonstrate that despite being technologically eclipsed and largely eclipsed in practice, the transcription model for sound technology persisted rhetorically. It persisted in the ways that some sound engineers framed their arguments for various sound strategies. Even advocates of what Lastra and Altman call a "fidelity" or documentary model of recording sound readily acknowledged the need to manipulate acoustic events in some ways. Although openly acknowledging the impossibility of conforming to the model of human binaural perception, these sound practitioners clung to a rhetoric of fidelity to an original acoustic event. The rhetorical resilience of this eclipsed model indicates how the initial social construction of sound technology as an instrument of documentation, transcription, or encryption persisted as an epistemological residue attached to later developments associated with the technology-in-use.

Conclusion

With the application of electrical-acoustic technology to American cinema, a model conceptualizing the work of recording as properly signification reemerged and competed with the dominant transcription paradigm underwriting sound recording and reproduction. With Hollywood's embrace of a third emergent paradigm—signifying fidelity—the conception of sound technology's transcription function was not completely elided. Instead, transcription models of sound continued to exist in various practices and, importantly, in various discourses surrounding sound technology. This residual conception of technology-in-use continued in some documentations of musical performance, largely underwriting much of the rhetoric if not the actual practices of the phonograph industry.[120]

The application of electrical sound technology to sound recording and reproduction expanded the frequency range of reproduced sounds and to that extent enhanced fidelity. The phonograph industry publicized its conversion to electrical methods as an "epochal advance" allowing more "realistic" recording and more "natural" studio methods. Period accounts cited the more natural arrangement of musicians and their use of conventional instruments as contributing to more "realistic" results since the recording studio now more closely approximated the conditions under which musicians normally performed.[121] But the new technology also introduced multiple ways in which recording engineers could intervene in the recording process. Typically two or more microphones were used for studio recording, thereby providing the recording engineer with an opportunity to mix the acoustic outputs from several positions. Studio performances were not "natural" in some pretechnological way but were instead performances for-the-sake-of-recording—a performance directed not at an audience surrogate, nor a future absent audience somewhere in the domestic sphere, but a performance directed multiply toward the apparatus itself and practical understandings of its potentials and limitations. Electrical methods provided recording engineers with an expanded arsenal of tools through which they might technologically mediate sound so as to signify fidelity to an original event. Engineers adjusted the shape of recording studios and manipulated acoustic damping materials to control the amount of reverberation recorded in the studio. Some period accounts equated this capture of "atmosphere or room tone" with a greater sense of realism, in that recorded reverberation signified "presence."[122] Volume indicators and monitoring loudspeakers or headphones allowed recording engineers to adjust volume during recording, and they used acoustic filters to eliminate low frequencies that could cause the recording stylus to cut from one groove into the next. Each of these sites of potential intervention by the recording engineer contributed to the reconceptualization of recording as an active process rather than as the technological transcription of an independently existing acoustic event. Mixing the output from several microphones actively structured the resulting recording as a hierarchically arranged electronic version related to but not identical with an original studio performance. The elimination through filters of some audible frequencies further modified the resulting sound, and the addition or diminution of recorded reverberation was openly acknowledged by engineers as an act of signification.

Throughout the late 1920s and into the 1930s, radio's representational practices continued to embody both the transcription and signification models for the relationship between technology and sound. The faithful transmission of electrically transcribed events so essential to the rise of radio networks continued to inform radio's representational practices. Remote broadcasts of musical performances, political events, and other public address forums continued to provide a portion of radio's content long after the rise and institutionalization of radio as a network-dominated, predominantly commercial-based medium. But as radio developed throughout the 1930s, the construction of illusory "events" grew in importance, especially through the increased hegemony of serial, fictional programming. Conventions of radio drama and much comedy programming (particularly the use of sound effects and music) increasingly relied not on the transcription of independently existing acoustic performances but on a conception of the radio medium as an empty space to be filled with sounds. Radio's development of the convention of a live studio audience presents an interesting synthesis of these two conflicting models. The construction of an acoustic event (including the production of sound effects and "nondiegetic" music) is performed in front of an audience. The broadcast transmission is arguably a transcription of a performance, the documentation of an event, but one that would not exist independently of its transmission.

Hollywood's sound practice debates point toward the ways in which technology as an instrument of transcribing an acoustic event or documenting phenomenal reality, although largely marginalized in practice, returns discursively again and again throughout the history of sound technologies. This return of a largely eclipsed model for sound technology's function speaks to the epistemological weight attributed to notions of sound technology's mimetic objectivity and to the conceptual conservatism that follows the tenuous fixing of any technology's identity. Although the innovation and diffusion of electrical acoustics brought about a paradigm shift in sound practices whereby selected techniques of signification displaced as the conceptual dominant transcription models of sound work, the residual definition of the technology's identity as properly transcription continued to underwrite certain representational practices, to introduce later technological innovations, to help sound practitioners make sense of their representational tasks, and to shape the social experience of listening to electrically reproduced sound.[123]

Films like *The Big Broadcast*, which echoed at the level of practice the discursive struggles of Hollywood's sound debates, bring into relief the clash between conflicting conceptual models for technology-in-use. *The Big Broadcast* embodies the conflict between sound practice conceptualized as the authentication and display of independently existing performance versus the signification of narrative business in an illusory nonrealistic space. This tension between authentication versus illusion is literalized in the film's climactic scene. Leslie signifies Hornsby's illusory presence in the broadcasting studio by playing a phonograph record over the air. But the badly warped record reveals the work of signification, the absence at the heart of the recorded, at the heart of signification. Leslie himself then steps to the microphone to provide an authentic, live, but woefully inadequate version of Hornsby's signature song "Please." He fails to fool anyone, but Anita, hearing Leslie's technologically mediated vocal struggles as if she were actually hearing him for the first time, rewards him with a kiss and her renewed devotion. In the film's culminating moment, the real Bing Hornsby arrives, rescuing the "big broadcast" by finishing the song. In a final gesture of transcribed synchronization, Bing Crosby sings directly to the film's audience, reasserting the apparatus' mimetic potential and rescuing the audience's epistemological faith in the sound film as an instrument of transcription.

Conclusions/Reverberations

The forces at play in the technological change described in the preceding five chapters reverberate across the history of twentieth-century media and beyond. Our contemporary moment is characterized by the erosion of taken-for-granted notions of media specificity. Technological change and not-unrelated economic developments challenge commonsense assumptions about media forms and practices. Historian William Uricchio has suggested that the process of digitization and convergence are redefining "our present as a moment of media in transition."[1] The situation is not entirely unlike that surrounding the innovation of electrical methods of augmenting sound. Electrical acoustics prompted a reimagining of the phonograph and more particularly the cinema, the establishment of new representational practices, the development of new economic relationships, and the interrogation of conventional notions of media identity. Radio's development in the form of point-to-mass broadcasting similarly required the construction of a medium's identity, the determination of its representational conventions, its social role, and the economic principles that made it "profitable." Uricchio further notes that in the midst of our contemporary media transition, "the working agenda for historians can quite productively make use of those earlier transition moments when related forms of instability threw into question media ontologies."[2] Any history is necessarily implicated in its present moment of construction, and I would suggest that this is particularly true of histories of media undertaken in the midst of contemporary dramatic media change. Further, to paraphrase Hayden White, any history implies that it might have been told otherwise. So, by way of providing a conclusion that stubbornly resists imposing closure, I have isolated a number of themes developed elsewhere in the book and have begun to sketch out ways in which they appear to resonate powerfully with subsequent and ongoing media trends.

Utopian Visions

Anyone reading chapter 2 will recognize ways in which the often grandiose promises used to announce electrical acoustics have been frequently repeated to promote subsequent technological changes. Virtually any new form of mass media has been hailed as the instrumental fulfillment of a perpetually deferred participatory public sphere. Each iteration attests to the continuing rhetorical power of such a fiction. First television, then cable television, and more recently the Internet have promised to make broadly available culture, education, news, and thereby ultimately to facilitate individuals' inclusion in a broader social fabric and informed participation in public affairs. Innovators of media technology continue to invoke the promise of a more egalitarian social formation achieved through the latest technological wonder. As outlined throughout the book, these utopian promises can cohere to the technology after it is diffused, shaping public perceptions of media and popular understandings of specific media practices.

Our contemporary moment takes for granted the notion of "spin"—we assume *a priori* that all messages reaching us through their various mediations are shaped. Yet at the same time, we culturally long to experience the unmediated, the direct, and the transparent. Rhetorical presentations of blogging during the last presidential election cycle, for example, focused on the absence of mediation, as if the practice provided the direct voices of the people, even as blogging itself relies constitutively on a technology of mass mediation. Bloggers were credited with instigating and perpetuating (Dan) "Rathergate" and the censuring of former Senate majority leader Trent Lott. Individual citizens have become watchdogs of the political process—stepping in for a popularly discredited mass media that now seems transparently self-interested. Technology momentarily delivered on its promise of empowerment as citizens became news gatherers and directly intervened in the political sphere. But these instances of revealing careless reporting by CBS or holding a politician accountable for racially insensitive remarks take on an importance in excess of their intervention into news practices or public affairs. They reaffirm yet again the possibility that technology might deliver on the inclusive promises that surround it. Like New York governor Al Smith circumventing a recalcitrant legislative branch and using radio broadcasting in 1925 to take his case for a tax cut directly to the people, these bloggers, and importantly popular knowledge about their actions, loudly reaffirm some of the faith vested in media technologies.

Compulsory Repurchasing

At a far earlier stage in this project, I arrived for a one-year teaching assignment as a sabbatical replacement at a small New England liberal arts college only to discover that, much to my chagrin, the school had just "deaccessioned" its 16mm film collection by donating it to the state's library system. When I discovered that I couldn't get the films back (employees at the state library system didn't know where the films had ended up), I dutifully followed my instructions and used the department's budget to acquire laser disk copies of films that the school had just discarded on 16mm. Often, the same titles sat on my office shelves in both VHS and Betamax formats. I often wonder now, what became of that school's collection of laser disks. Does the harried technician who responded to my research needs and came up with a 78 rpm phonograph (from his parents' garage) now tinker away well into the night trying to keep a single laser disk player running (no doubt using parts scavenged from a stack of broken machines)? Or have all the laser disks been "deaccessioned" as the 16mm films once were? Have the laser disks been sold on e-Bay to collectors? Donated to a state library system that didn't want them? Hauled to the dumpster and then to a landfill? Aren't the DVDs now on that office's shelves themselves on the verge of obsolescence?

The application of electrical methods to the phonograph arguably inaugurated the process of compulsory repurchasing characteristic of each substantial technological innovation in the mass media. U.S. media developments subsequent to the successful diffusion of electrical-acoustic media continued a process of innovation in which the new (or the better, the more convenient, or that which produced more and "better" sound) introduced competing formats and a process of compulsory repurchasing. Successful innovations, even in the short term, demanded of consumers a repurchasing of ever newer and better devices to replace those that had become outmoded. Electrically recorded 78 rpm phonograph records would, of course, be displaced by 33 rpm long-playing disks and 45 rpm formats. Audiotape seemed to render the disk outmoded through, first, reel-to-reel tape, and then 8-track and audio cassettes, all of which were first displaced by CDs and now by technologies like the iPod. Domestic 16mm film lingered despite the innovation of, first, 8mm film and then super 8mm film with, alternately, phonograph recordings for sound accompaniment, then optical and even magnetic sound-on-film technologies. Radio added FM and then became TV, which became cable TV, satellite radio and TV, TV with stereo sound of various formats and now (or very soon) HDTV.

Not coincidentally, subsequent twentieth-century media innovations have often involved the dissimulation of the machine described in chapter 3. Cable and satellite distribution of television signals eliminated the need for consumers to physically adjust—that is, to physically interact with—their television sets. It is a sign of one's age to recall the location and operation of "vertical hold" adjustments on TV sets or knobs that allowed one to fine-tune the tint and hue of color television reception. Today, consumer interaction with domestic entertainment technology is largely accomplished at a distance—via a remote, the keypad of which provides an ease and automation of the consumer's interface with the device. Where once one had to physically lift the tone arm of a phonograph and move it through space to reach a desired song (literally looking at the grooves in the recording to identify where songs began and ended), today the artifact of recorded sound increasingly has no observable, physical quality. Listening to a desired song merely requires the push of a button, a keystroke, a mouse-click.

More convenient? Most assuredly. Better? Well, perhaps. But the comparative benefits of digital versus analog are not the issue here. Instead, I'm suggesting that the shift to electrical methods of augmenting sound represents a crucial stage in a larger process through which domestic media technology became increasingly opaque. Technological processes have progressively receded from consumer sight and control. This dissimulation of the labor of technology takes its place alongside the progressive, attempted dissimulation of all labor within corporate capitalism. The containment of the phonograph within a cabinet, the physical design of radio receivers, the innovation of single-control radio tuning, and the introduction of automatic record changers (all during the 1910s and the 1920s) are steps in a larger process through which technological principles and processes become increasingly opaque. Progress, convenience, efficiency—at least as they have come to be commonly defined—all hinge on this dissimulation of labor, this elision of technology, the sealing of technology's black box. The illusion, or rather the structured containment of consumer control, lingers in the form of the remote control, which in many ways fulfills the onanistic promises of the Aeolian-Vocalion's transformation of phonograph enthusiasts into performers of sound.

Transcription Versus Signification

The innovation of electrical methods of augmenting sound introduced previously unheard-of possibilities for the manipulation of acoustic phenomena. Although innovators announced electrical acoustics with grand claims about enhanced fidelity in sound reproduction, the material practices of representing with sound moved inexorably away from the transcription of acoustic events. Instead, as I have started to detail in chapter 5, sound-representation practices moved more fully to a signification paradigm. Subsequent developments in acoustic technology have only exacerbated and enhanced this process. Research and development within Hollywood and by sound technology manufacturers provided filmmakers with the ever-increasing ability to creatively manipulate sounds. Later developments like, first, magnetic tape and, later, multitrack recording moved sound practice further from the documentation of preexisting events. More recent technological innovations and shifts in sound practice, chronicled by Paul Théberge and Timothy D. Taylor, have further extended the steps toward the creative manipulation of sound undertaken within Hollywood following its conversion to sound.[3] The visual techniques of the modernist avant-garde film are now simply a mouse-click away on most domestic computers, and with the introduction of consumer audio software (most recently, GarageBand), technology that might enable Harry Alan Potamkin's vision of the "sonorous film" is widely diffused, if not democratized. But this is not to suggest that "transcription" as a method of understanding the work of sound recording has been fully eclipsed.

Each audio innovation subsequent to the introduction of electrical acoustics has been announced with broad claims regarding enhanced fidelity to original performances. The belief, indeed the faith, that technological innovation can further close the gap between the real and its representation continues to have persuasive popular appeal, even in the face of continued indications to the contrary.

Concentration of Ownership

The concentration of media ownership in the hands of a few large corporations that grew out of the innovation of electrical methods for mediating sounds received some critical attention during the late 1920s and early 1930s. Chicago's labor-owned radio station WCFL warned its listeners and the readers of its magazine about the threat posed by "the radio trust" to free speech, democratic ideals, and the public's ownership of the ether. This criticism of radio broadcasting during the early 1930s often uncannily echoes contemporary critiques of corporate control of the mass media. In a 1934 publication, James Rorty noted:

> In brief, the broadcasting power of the nation is in the hands of a few powerful corporations, closely connected with power trusts and public utilities corporations. That means, not merely that time-on-the-air cannot be had, except from relatively insignificant independent stations, for criticisms and attacks upon the policies and performances of these corporations, but that the *status quo* of business and finance in general is protected from such attacks. This is insured by the community of interests, established through interlocking directorates, financial control, etc., to which appeal can always be made when a dissenting minority attempts to attack financially important individuals or corporations.[4]

Today, one can have the uncanny experience of holding Rorty's book in one hand and, for example, Robert McChesney and John Nichols' *Our Media Not Theirs* in the other.[5] The arguments are strikingly similar, the stakes seem equally as high. McChesney and Nichols' critiques of corporate media dominance now seem a bit stronger than Rorty's because of the weight of evidence culled from decades of corporate control of U.S. media dating back precisely to Rorty's moment. Media ownership in the 1930s was not as fully or as systematically concentrated as it is today, but to a large extent the rise of vertically and horizontally integrated media conglomerates that is justifiably of such concern to today's media activists originated in the period surrounding Hollywood's conversion to sound and other applications of electrical sound technology.

Marketplace Democracy and Economic Alternatives

Although the period surrounding the introduction of electrical acoustics led to an intensified concentration of ownership across mass media forms, the U.S. culture industry did not become monolithic. Within each medium small manufacturers, theater owners, and/or broadcasters continued an often precarious economic existence. For the U.S. system and its ideology of consumer choice to work, there had to remain some illusion of a market place that was indeed free. In his 1934 critique of U.S. broadcasting, Rorty quoted economist Walton Hamilton to characterize the ideological implications of corporate broadcasting: "Business succeeds rather better than the State in imposing restraints upon individuals, because its imperatives are disguised as choices."[6] While concentration of media ownership did not eliminate consumer "choice," the dominance of mass media by a handful of for-profit institutions necessarily imposed limitations on the extent and type of such choices available.

One of the abiding fictions of American life in the twentieth century was and is the notion that the market is an articulation, an enactment, of Democracy. U.S. media industries, like all firms entrenched in consumer culture, have long promoted the notion of consumer sovereignty. In his analysis of interwar advertising, Roland Marchand describes the "Parable of the Democracy of Goods," through which the U.S. advertising industry depicted a version of social reality where equality was achieved through the consumption of goods.[7] A similar principle underwrote the rhetoric surrounding emerging sound technologies and the notion of "service" promulgated by media conglomerates. The cinema, the phonograph, but most particularly the radio (because of the provision of "free" programming), promised consumers equal access to "quality" entertainment irrespective of social status. Such consumption also promised a participation of sorts. Consumers voted with their dollars or by tuning their radio dials, and the entertainment industry promised to listen and respond to consumer demands. Participation in the form of purchasing, or indeed of listening, collapsed citizenship with consumption.

In the late 1920s and early 1930s, labor-owned WCFL, socialist-operated WEVD (New York), and a few other media organizations vocalized and embodied in practice a critique of so-called marketplace democracy and the corporate domination of media. Our contemporary moment arguably includes a broader range of media directly opposed to that domination. Perpetually embattled (for there is no other way within the current political economy of mass media), a radio network like Pacifica joins dispersed public-access television activists (sometimes coordinated through Deep Dish TV or the Alliance for Community Media), various practitioners of LPFM radio, the international proliferation of "indymedia" centers, and a host of sometimes short-lived, always raucous, grass-roots alternatives to the corporate media status quo.

Intermedia Commodities, "Synergy," and the Commodity Logic of the Multimedia Conglomerate

In a broad overview of the music industry in the twentieth century, Reebee Garofalo characterizes the goals of 1980s–1990s transnational music companies as cultivating and exploiting "multiple revenue streams," including not only record sales but also revenue from licensing music to advertisers, movie tie-ins, and cross-media marketing.[8] He further notes the key role played by technology in this shift in business practices (specifically satellite and cable television, CDs, and the promise of, eventually, streaming audio on the Internet). As I've described in chapter 1, such a process also characterized the late 1920s and early 1930s and the business practices of U.S. entertainment companies in the aftermath of the diffusion of electrical acoustics. The integrated, transmedia marketing of Amos 'n' Andy that accompanied the national release of *Check and Double Check* in 1930, or Warner Bros.' "placement" of popular songs that they controlled in feature films, cartoons, phonograph recordings, and sheet music might seem almost historically quaint in light of the standard set by the contemporary "high concept" motion picture. The vertically and horizontally integrated entertainment firms of the early 1930s could not also include theme parks, television, home video, and so on in their "revenue streams." But like today's "high concept" media product, the synergistic marketing of entertainment commodities in the early era of electrical sound media was the logical and inevitable outcome of a particular economy of mass media dominated by multimedia conglomerates.

Politics and Entertainment

While the process through which American politics adapted to mass media clearly begins before the period covered in this book, the innovation and deployment of radio broadcasting (and to a lesser extent the sound film) played powerful roles in moving political presentations ever closer to the entertainment commodity. American politics adapted to emerging media, seeking to close the gap between the political process and the entertainment spectacle. In 1932, when Herbert Hoover took to the airwaves and unintentionally disrupted Ed Wynn's Texaco "Fire Chief" Hour, he erred in a way that has grown unthinkable today. Political conventions—unlike the Democratic Party's two-week marathon of 1924—are carefully managed so as to be mass-mediated. In 1928 political parties began to adapt their conventions to the emerging representational practices of radio (*and* in light of the 1924 national broadcasts). Today, keynote speakers address the floor and the nation (or that portion of it watching) within the temporal confines of "prime time" in carefully orchestrated events that have been designed to be broadcast. These calculated political performances become most notable, most newsworthy, when the careful preparations fail (e.g., when the balloons fail to drop on cue, as happened at the Democratic National Convention in 2004).

The collapsing (or at least the blurring) of the boundary between politics and entertainment is perhaps no more evident than in presidential election debates or public policy presentations that opt for a "town-hall" format (e.g., George W. Bush discussing his plan to privatize Social Security). In a highly theatrical enactment of "direct" interaction with "the people," citizens rise in turn (often nervously and hence "authentically"), read their preapproved question from a note card, and hurriedly return to their seats to await an answer that seems every bit as prepared as the hairstyle and necktie selection of politicians. The mise-en-scène of the town square, the grange hall, or the shop floor is here invoked under the glare of television lights and an inflated sense of seriousness recalling, but most clearly not reproducing, a fictional historic moment when men vying for public office addressed the people directly from the rear of train cars or the crush of a crowd. While, first, Western Electric's public address system and, later, radio broadcasting allowed politicians to address ever-widening circles of citizens, the electrical-acoustic expansion of the audience threatened a loss—the loss of intimacy, of direct connection. U.S. political culture has developed a series of compensatory strategies to address such a loss, including today's "town-hall" format. The "intimacy" of FDR's fireside chats was cut from the same bolt of rhetorical cloth as Jimmy Carter's cardigan sweater, "Dutch" Reagan's blue jeans, or John Kerry's hunting jacket.

Much of our shared contemporary understanding of the twentieth century revolves around a series of mediated, theatrical moments, endlessly recycled in mass-mediated accounts. Consider popular memory of the Cold War. In a broadcast speech from Missouri, Winston Churchill named for us the "Iron Curtain." Nixon went toe to toe with Khrushchev in a faux American kitchen. And while Khrushchev banged his shoe on the desk at the UN, Adlai Stevenson would later sit at a remarkably similar desk and instruct the Soviet representative not to wait for the translation but to answer the question *now*. When Ronald Reagan theatrically proclaimed, "Mr. Gorbachev, tear down this wall!" the televisual moment bore all of the same theatrical drama as when Kennedy declared, "Ich bin ein Berliner." But consider other moments: Ike warned us of the threat posed by a vast military industrial complex; CNN broadcast to the world the image (from China in 1989, after the Tianemen Square massacre) of an anonymous man standing in front of a tank (weren't there virtually identical images from Prague in 1968?); U.S. presidential politics changed forever with the Nixon-Kennedy debate; and John Dean's hyperbolic testimony about a "cancer on the Presidency," although widely scorned at the moment of its utterance, has come to signify "Watergate." What these and many other readily recalled moments of U.S. and international political history have in common are both their endless recycling in retrospective accounts of particular moments (their incorporation into a mediated mythos of the twentieth century) and the structural ease with which they were so incorporated. Each of these mediated events consciously played to the medium of its moment. With increasing intensity since the innovation of ever-new tools of mass mediation, politics have taken on, indeed embraced, the conventions of the melodramatic spectacle. After all, isn't FDR's statement about "fear itself" the first legitimate sound-byte?

Although I have begun to outline above a series of resonances that reverberate and echo between the historical innovation of electrical acoustics and subsequent media developments, this is not to suggest that the process of technological change is ultimately predetermined by events originating in the 1920s and early 1930s. Instead, as this book has argued, the process of technological innovation is always and inevitably the site of contestation and struggle. That contemporary events resound with the past attests to continuities between the past and the present that shaped technological innovation then as well as now. Among those continuities are the power of corporate capitalism to influence and in many ways control technological change, the longing for and recurring faith in media technology's ability to unite disparate individuals despite their differences, and the assumption that such a (inter)national unification, such a shared imaginary, is indeed desirable, the continuing pervasive fiction of a participatory public sphere, and the ideological erasure that equates capitalism with democracy.

Just as dominant ideology always bears within itself a series of gaps and fissures, so too does corporate control of mass media and technological innovation ultimately leave open a series of spaces that alternatives might inhabit and from which they indeed speak. Hegemony seeks to disguise internal contradictions and mutually contradictory beliefs. Its success might be measured in the absence or marginalization of alternative visions. Corporate media's hegemony can be measured (historically and in our present moment) in the degree to which such alternatives remain marginalized and essentially contained. In an era when large corporate scandals are represented (that is, *mediated*) as essentially aberrations attributed to individual avarice (Ken Lay at Enron)[9] or as proof that existing structures of oversight will, in fact, eventually catch up with "poor accounting practices" (we need only strengthen such oversight and reform such practices), there appears to be little conceptual ground from which a rigorous critique of corporate domination might gain some purchase. Such ground must be created. While technology—particularly media technology—most assuredly will not, and indeed cannot, in and of itself create such ground or fulfill the promise of a more egalitarian social formation, it may indeed play a role.

But such a view is fully complicit with the recurring notion that technological change offers us access to those grand, utopian promises.

Notes

Introduction

1. Alan Williams, "Historical and Theoretical Issues in the Coming of Recorded Sound to the Cinema," in Rick Altman, ed., *Sound Theory/Sound Practice* (New York: Routledge, 1992), 129.

2. Electrical acoustics was applied to other consumer products and preexisting communication forms, including public address systems and technological augmentation for the hearing-impaired. I address predominantly those configurations in which the apparatus was applied to and interpolated within mass media commodity relations.

3. Douglas Gomery, *The Coming of Sound* (New York: Routledge, 2005), 115.

4. Richard A. Schwarzlose, "Technology and the Individual: The Impact of Innovation on Communication," in Catherine L. Covert and John D. Stevens, eds., *Mass Media Between the Wars: Perceptions of Cultural Tension, 1918–1941* (Syracuse, N.Y.: Syracuse UP, 1984), 101.

5. Oliver Read and Walter L. Welch, *From Tin Foil to Stereo: Evolution of the Phonograph*, 2d ed. (Indianapolis, Ind.: Howard W. Sams, 1976), 237.

6. Nancy J. Ogren, *The Jazz Revolution: Twenties America and the Meaning of Jazz* (New York: Oxford UP, 1989), 98.

7. Erik Barnouw, *A Tower in Babel: A History of Broadcasting in the United States*, vol. 1—to 1933 (New York: Oxford UP, 1966); Erik Barnouw, *The Golden Web: A*

History of Broadcasting in the United States, vol. 2—1933 to 1953 (New York: Oxford UP, 1968); Erik Barnouw, *The Image Empire: A History of Broadcasting in the United States*, vol.3—from 1953 (New York: Oxford UP, 1970); Hugh G. J. Aitken, *Syntony and Spark: The Origin of Radio* (Princeton: Princeton UP, 1985); Hugh G. J. Aitken, *The Continuous Wave: Technology and American Radio, 1900–1932* (Princeton: Princeton UP, 1985); Susan J. Douglas, *Inventing American Broadcasting, 1899–1922* (Baltimore: Johns Hopkins UP, 1987); Robert W. McChesney, *Telecommunications, Mass Media, and Democracy: The Battle for the Control of U.S. Broadcasting, 1928–1935* (New York: Oxford UP, 1993); Susan Smulyan, *Selling Radio: The Commercialization of American Broadcasting, 1920–1934* (Washington, D.C.: Smithsonian Institution Press, 1994); Michele Hilmes, *Radio Voices: American Broadcasting, 1922–1952* (Minneapolis: U of Minnesota P, 1997); Thomas Streeter, *Selling the Air: A Critique of the Policy of Commercial Broadcasting in the United States* (Chicago: U of Chicago P, 1996); Douglas B. Craig, *Fireside Politics: Radio and Political Culture in the United States, 1920–1940* (Baltimore: Johns Hopkins UP, 2000); Susan J. Douglas, *Listening In: Radio and the American Imagination* (New York: Times Books, 1999); Catherine Covert, "'We May Hear Too Much': American Sensibility and the Response to Radio, 1919–1924," in Covert and Stevens, eds., *Mass Media Between the Wars*, 199–220; Frank A. Biocca, "The Pursuit of Sound: Radio, Perception, and Utopia in the Early Twentieth Century," *Media, Culture, and Society* 10 (1988): 61–79; Frank Biocca, "Media and Perceptual Shifts: Early Radio and the Clash of Musical Cultures," *Journal of Popular Culture* 24.2 (Fall 1990): 1–15; Mary Mander, "The Public Debate About Broadcasting in the Twenties: An Interpretive History," *Journal of Broadcasting* 28.2 (Spring 1984): 167–85; Leonard S. Reich, "Research, Patents, and the Struggle to Control Radio: A Study of Big Business and the Uses of Industrial Research," *Business History Review* 51 (Summer 1977): 208–35; Hugh G. J. Aitken, "Allocating the Spectrum: The Origins of Radio Regulation," *Technology and Culture* 35.4 (Oct. 1994): 686–716; Hugh Richard Slotten, "Radio Engineers, the Federal Radio Commission, and the Social Shaping of Broadcast Technology: Creating 'Radio Paradise,'" *Technology and Culture* 36.4 (Oct. 1995): 950–86.

8. Read and Welch, *From Tin Foil to Stereo*; Roland Gelatt, *The Fabulous Phonograph: From Edison to Stereo* (New York: Appleton-Century, 1954; 3d ed., 1965); Andre Millard, *America on Record: A History of Recorded Sound* (New York: Cambridge UP, 1995); Michael Chanan, *Repeated Takes: A Short History of Recording and Its Effects on Music* (New York: Verso, 1995); William Howland Kenney, *Recorded Music in American Life: The Phonograph and Popular Memory, 1890–1945* (New York: Oxford UP, 1999); Lisa Gitelman, *Scripts, Grooves, and Writing Machines:*

Representing Technology in the Edison Era (Stanford, Calif.: Stanford UP, 1999); Marsha Siefert, "Aesthetics, Technology, and the Capitalization of Culture: How the Talking Machine Became a Musical Instrument," *Science in Context* 8.2 (1995): 417–49; Marsha Siefert, "The Audience at Home: The Early Recording Industry and the Marketing of Musical Taste," in James S. Ettema and D. Charles Whitney, eds., *Audiencemaking: How the Media Create the Audience* (Thousand Oaks, Calif.: Sage, 1994), 186–214; Emily Thompson, "Machines, Music, and the Quest for Fidelity: Marketing the Edison Phonograph in America, 1877–1925," *Musical Quarterly* 79.1 (Spring 1995): 131–71; Lisa Gitelman, "Unexpected Pleasures: Phonographs and Cultural Identities in America, 1895–1915," in Ron Eglash, Jennifer L. Croissant, Giovanna Di Chiro, and Rayvon Fouche, eds., *Appropriating Technology: Vernacular Science and Social Power* (Minneapolis: U of Minnesota P, 2004), 331–44; Lisa Gitelman, "How Users Define New Media: A History of the Amusement Phonograph," in David Thorburn and Henry Jenkins, eds., *Rethinking Media Change: The Aesthetics of Transition* (Cambridge: MIT Press, 2003), 61–79.

9. Harry M. Geduld, *The Birth of the Talkies: From Edison to Jolson* (Bloomington: Indiana UP, 1975); J. Douglas Gomery, "The Coming of Sound to the American Cinema: A History of the Transformation of an Industry" (Ph.D., U of Wisconsin—Madison, 1975): as indicated in note 3 above, Gomery has revisited his work and published it in book form (Gomery, *The Coming of Sound*); James Lastra, *Sound Technology and the American Cinema: Perception, Representation, Modernity* (New York: Columbia UP, 2000); Donald Crafton, *The Talkies: American Cinema's Transition to Sound, 1926–1931* (New York: Scribner's, 1997).

10. Michele Hilmes, *Hollywood and Broadcasting: From Radio to Cable* (Urbana: U of Illinois P, 1990). Richard Jewell describes Hollywood's use of radio as a promotional outlet in the 1930s, focusing largely on RKO and the resentment among exhibitors about any potential or actual alliance between the two industries. Jewell recounts some of the debates within the film industry concerning radio's effect on cinema attendance. His essay is perhaps most interesting for the questions he poses in his conclusion—questions concerning technological and aesthetic cross-fertilization between the two media and the impact of the radio-film combination on American consumerism. Richard Jewell, "Hollywood and Radio: Competition and Partnership in the 1930s," *Historical Journal of Film, Radio, and Television* 4.2 (1984): 125–41.

Gregory Waller has begun to address the relationships between radio and film by examining radio performers' appearances in rural Kentucky movie theaters during the 1930s. Gregory A. Waller, "Hillbilly Music and Will Rogers: Small-town

Picture Shows in the 1930s," in Melvyn Stokes and Richard Maltby, eds., *American Movie Audiences: From the Turn of the Century to the Early Sound Era* (London: BFI, 1999), 164–79.

11. See, for example, Chanan, *Repeated Takes*, 54–79; Millard, *America on Record*, 136–75; and Crafton, *The Talkies*, 23–61.

12. Emily Thompson, *The Soundscape of Modernity: Architectural Acoustics and the Culture of Listening in America, 1900–1933* (Cambridge: MIT Press, 2002); Jonathan Sterne, *The Audible Past: Cultural Origins of Sound Reproduction* (Durham, N.C.: Duke UP, 2003).

13. The school where I earned my Ph.D. now separates film studies (in Comparative Literature) from media studies (in Communications Studies), which is itself separated from mass communications (a separate department with Journalism). I have taught identical courses in a Communication Studies Department, a free-standing Film Studies Department, a Theater Department and, most recently, an English Department. Rarely have radio or recorded sound comfortably "fit" into the preexisting curriculum. In our libraries, PN1991 (radio) is nearby PN1995 (film), but neither is particularly close to HE8698 (also radio) or ML1055 (recorded sound).

14. Gordon S. Mitchell "Applying Public-Address Equipment to Indoor Use," *Projection Engineering* 3.8 (Aug. 1931): 9; Western Electric, "How the Sounds Get Into Your Record by the Electrical Process," *Phonograph Monthly Review* 1 (Oct. 1926): 6.

15. Trevor J. Pinch and Wiebe E. Bijker, "The Social Construction of Facts and Artifacts: Or, How the Sociology of Science and the Sociology of Technology Might Benefit Each Other," in Wiebe E. Bijker, Thomas P. Hughes, and Trevor J. Pinch, eds., *The Social Construction of Technological Systems: New Directions in the Sociology and History of Technology* (Cambridge: MIT Press, 1987), 17–50. The historical model developed in this essay is elaborated upon in Wiebe E. Bijker, *Of Bicycles, Bakelites, and Bulbs: Toward a Theory of Sociotechnical Change* (Cambridge: MIT Press, 1997).

16. Pinch and Bijker, "Social Construction," 40. Although Pinch and Bijker use the term "interpretative," throughout the remainder of this text I will opt for the term *interpretive*.

17. Pinch and Bijker, "Social Construction," 46.

18. Rick Altman, *Silent Film Sound* (New York: Columbia UP, 2004); see esp. 15–23. Although this book was not available to me during the preparation of my manuscript, I certainly have been influenced by Altman's other work and by our numerous conversations over many years.

19. Altman, *Silent Film Sound*, 22.

20. See, for example, Kenneth Haltman, "Reaching Out to Touch Someone? Reflections on a 1923 Candlestick Telephone," *Technology in Society* 12.3 (1990): 333–54; Arthur P. Harrison, Jr., "Single-Control Tuning: An Analysis of an Innovation," *Technology and Culture* 20.2 (Apr. 1979): 296–321.

21. See, for example, Paul Théberge, "The 'Sound' of Music: Technological Rationalization and the Production of Popular Music," *New Formations* 8 (Summer 1989): 99–111.

22. See, for example, Patrick L. Ogle, "Technological and Aesthetic Influences Upon the Development of Deep Focus Cinematography in the United States," *Screen* 13.1 (Spring 1972): 81–108; Peter Baxter, "On the History and Ideology of Film Lighting," *Screen* 16.3 (Autumn 1975): 83–106; Philip Beck, "Technology as Commodity and Representation: Cinema Stereo in the Fifties," *Wide Angle* 7.3 (1985): 62–73.

23. See, for example, Lana F. Rakow, "Women and the Telephone: The Gendering of a Communications Technology," in Cheris Kramarae, ed., *Technology and Women's Voices: Keeping in Touch* (New York: Routledge & Kegan Paul, 1988), 207–28.

1. Technological Innovation and the Consolidation of Corporate Power

1. The origins and history of this program, as well as its relationship to what some saw as the assimilationist possibilities of radio, have been explored extensively elsewhere: Melvin Patrick Ely, *The Adventures of Amos 'n' Andy: A Social History of an American Phenomenon*, (New York; Free Press, 1991); Michele Hilmes, "Invisible Men: *Amos 'n' Andy* and the Roots of Broadcast Discourse," *Critical Studies in Mass Communication* 10 (1993): 301–21.

2. RKO claimed that the countrywide premiere was based on the flood of letters from the duo's radio fans who demanded to see the film as soon as New York audiences. "Record Tieups on Amos 'n' Andy," *Exhibitors Herald-World*, Oct. 11, 1930, 26.

3. *Exhibitors Herald-World*, Oct. 11, 1930, 10.

4. The origins and implications of this patent pool are described in some detail in W. Rupert MacLaurin, *Invention and Innovation in the Radio Industry* (New York: Macmillan, 1949); and Hugh G. J. Aitken, *The Continuous Wave: Technology and American Radio, 1900–1932* (Princeton: Princeton UP, 1985).

5. Such patent pools might be thought of as examples of what Trevor Pinch and Wiebe Bijker refer to as "closure mechanisms," procedures designed to close down the interpretive flexibility of a technological artifact and enforce consensus about

its meaning and design. By apportioning the profitable potential of a technology among its participants, a patent pool seeks to codify the identity of a technology.

6. U.S. Federal Communications Commission, *Proposed Report Telephone Investigation* (Pursuant to Public Resolution No. 8, 74th Congress) (Washington, D.C.: GPO, 1938), 232 (hereafter referred to as the Walker Report).

7. MacLaurin, *Invention and Innovation*, 27.

8. The agreement also included patents held by the United Fruit Company and two smaller wireless companies.

9. It wasn't until January of 1927 that the first commercial radio-telephone service between London and New York was established and 1929 before ship-to-shore wireless telephony operated commercially. Walker Report, 440–41; also see MacLaurin, *Invention and Innovation*, 157.

10. In a 1915 article in *Electrical World*, Carty declared, "The results of long-distance tests show clearly that the function of the wireless telephone is primarily to reach inaccessible places where wires cannot be strung. It will act mainly as an extension of the wire system and a feeder to it." Quoted in MacLaurin, *Invention and Innovation*, 92–93.

11. MacLaurin, *Invention and Innovation*, 113. Susan Douglas suggests that the nature of press coverage of radio during the 1915–1922 period—essentially neglecting ongoing developments and then responding primarily to the sudden rise of broadcasting in 1922—ultimately served the emerging corporate control of the medium. "This ahistorical stance made radio seem an autonomous force, so grand, complex, and potentially unwieldy that only large corporations with their vast resources and experience in efficiency and management could possibly tame it." Susan J. Douglas, *Inventing American Broadcasting: 1899–1922* (Baltimore: Johns Hopkins UP, 1987), 304.

12. The 1919–20 wireless patent pool left unclear which organizations held the rights to build and to license radio transmitters. AT&T interpreted the agreement as providing them with the exclusive rights to this market. It determined that all broadcasters should therefore become AT&T licensees. At a royalty rate of $4 per watt of power with a maximum fee of $3,000 and a minimum of $500, AT&T began in 1923 an aggressive campaign to force broadcasting stations to sign up for licenses. Under the terms of the license agreements, AT&T sought to reserve for itself the rights to sell radio broadcast time. Selling broadcast time, in effect toll broadcasting, was adopted by AT&T's New York station WEAF (established in 1923) and defined radio's commercial use as akin to that of the telephone. The commodity offered to the public was a service, a technology through which they could speak,

not to one but to many. For details on AT&T's attempts to license all radio transmitters and its refusal to furnish wired interconnections between nonlicensed stations, see Walker Report, 458–60.

13. Susan Douglas's *Inventing American Broadcasting* provides important details on the means through which corporations maintained control over technology during the transition from wireless telephony to broadcasting (315–22).

14. David F. Noble, *America by Design: Science, Technology, and the Rise of Corporate Capitalism* (New York: Oxford UP, 1977), 93.

15. Walker Report, 235–36.

16. Walker Report, 231.

17. Ibid., 237.

18. Ibid., 480.

19. Walker Report, 464–65. A similar practice of exclusive licensing helped Guglielmo Marconi gain international dominance in the field of wireless telegraphy. Marconi's initial licenses for shipboard use of wireless both leased equipment and contractually provided a trained operator who was allowed to communicate only with other Marconi licensees.

20. Douglas, *Inventing American Broadcasting*; Hugh G. J. Aitken, *Syntony and Spark: The Origin of Radio* (Princeton: Princeton UP, 1985), and Aitken, *The Continuous Wave*.

21. Under this arrangement, Marsh Laboratories recorded talks intended to inspire salesmen, pronunciation recordings, at least two recordings for the Industrial Workers of the World, and even a multisided "History of Greenhouse Structures." Such an eclectic list of recordings attests to Marsh's marginal status in the phonograph industry and his company's sale of recording and duplication services to private concerns suggests the limited capital available to Marsh Laboratories. In part because of this marginal status, Marsh more aggressively sought to experiment with technological innovations of his own design.

22. Martin Bryan, "Orlando R. Marsh: Forgotten Pioneer," *New Amberola Graphic* 71 (Jan. 1990): 3–14, and Bryan, "Orlando Marsh Revisited," *New Amberola Graphic* 72 (April 1990): 3–5; Dennis E. Farrara, "A Jesse Crawford Retrospective: The Marsh and Victor Recordings (1924–1933)," *New Amberola Graphic* 88 (Spring 1994): 15–16.

23. By the late 1920s, Marsh Laboratories had adapted to radio and offered a transcription service that recorded and duplicated programs on phonograph disks to be distributed to local stations, including a series of programs for Amos 'n' Andy and bandleader Ted Weems. By the early 1930s, Marsh Laboratories offered this electri-

cal transcription service to any client, providing support throughout the radio production process (developing program ideas, writing program continuity, selecting and rehearsing talent, directing as well as recording and distribution). Despite the organization's marginal status within post-1925 developments of applied electrical acoustics, Marsh Laboratories' promotional material in 1931 still carried the slogan "The Originators of Electrical Recording." Bryan, "Orlando R. Marsh," 3–14.

24. The copresence of such dramatically different visions of radio broadcasting within a single organization conforms to Pinch and Bijker's notion of interpretive flexibility and powerfully underscores that the ultimate shape a media form takes is the product of contestation and struggle.

25. This section appeared in a different form in Steve Wurtzler, "AT&T 'Invents' Public Access Broadcasting in 1923: A Foreclosed Model for American Radio," in Susan Squier, ed., *Communities of the Air: Radio Century, Radio Culture* (Durham, N.C.: Duke UP, 2003), 39–62.

26. The fact that radio's audience had yet to be "broadly conceived" characterized, in part, the medium's interpretive flexibility in February of 1923.

27. Quoted in William Peck Banning, *Commercial Broadcasting Pioneer: The WEAF Experiment, 1922–1926* (Cambridge: Harvard UP, 1946), 116.

28. Banning, *Commercial Broadcasting Pioneer*, 116.

29. Craft, quoted in ibid., 118. A report on the AT&T / AIEE long-distance public address demonstration framed the event in terms of both politics and a national mode of address. "Will the Entire Nation Listen to the Next President's Inaugural Address," *Radio Broadcast* 3 (May 1923): 13–14.

30. See, for example, Ralph B. Howell and Nathaniel B. Dial, "Senators Howell and Dial on Broadcasting the Senate Proceedings," *Wireless Age* 12 (Aug. 1924): 30–31, 79–80. Period rhetoric surrounding the broadcast of political conventions is exemplified by William A. Hurd, "How Broadcasting June Conventions Affects the November Elections," *Wireless Age* 12 (Aug. 1924): 18–22, 57.

31. AT&T's 1923 radio plan is described in the Walker Report, 455–58.

32. For an account of station WCFL, see Nathan Godfried, *WCFL: Chicago's Voice of Labor, 1926–1978* (Urbana: U of Illinois P, 1997). Also see Derek Vaillant's history of Chicago radio and Lizabeth Cohen's history of Chicago-area workers for descriptions of both station WCFL and local, ethnic-oriented radio programming: Derek W. Vaillant, "Sounds of Whiteness: Local Radio, Racial Formation, and Public Culture in Chicago, 1921–1935," *American Quarterly* 54.1 (Mar. 2002): 25–66; Lizabeth Cohen, *Making a New Deal: Industrial Workers in Chicago, 1919–1939* (New York: Cambridge UP, 1990), 129–43.

33. Hilmes, "Invisible Men," 313. Also see J. M. McKibbin, Jr., "The New Way to Make Americans," *Radio Broadcast* 2 (Jan. 1923): 238–39.

34. Susan Smulyan, "Radio Advertising to Women in Twenties America: 'A Latchkey to Every Home'" *Historical Journal of Film, Radio, and Television* 13 (1992): 299–314.

35. On radio and women's political participation, see Christine Frederick, "Women, Politics, and Radio," *Wireless Age* 12 (Oct. 1924): 36–37, 76, 78. More common in the period, however, were articles on radio as a tool for a more efficient home economics, such as Hortense Lee, "Cooking Eggs—Via Radio," *Wireless Age* 11 (Dec. 1923): 24–25. Interestingly, Frederick was also a powerful advocate of applying the principles of scientific management to home economics through emerging domestic technologies, yet some of her radio-related writing stressed women's participation in *public* realms.

36. Despite its interpretation of the wireless patent pool, AT&T was *not* solely entitled to construct and to license broadcasting stations. With an arbitration referee's decision in November 1924, the notion that AT&T might monopolize American radio through building and licensing transmitters came to an abrupt end.

37. This is similar to the strategy pursued by AT&T / Western Electric / ERPI in successfully introducing electrical acoustics to the U.S. film industry.

38. The Bell System vigorously pursued this strategy independent of its 1923 radio plan. In February of 1923, *Radio Broadcast* noted that WEAF (owned and operated by AT&T) enjoyed exclusive use of long-distance telephone lines for remote transmissions. Other stations had to use acoustically inferior telegraph lines. The article called for an agreement whereby all stations could make use of the "highest grade line available." "Fewer and Better Stations," *Radio Broadcast* 2 (Feb. 1923): 272.

39. Stuart Ewen, *PR! A Social History of Spin* (New York: Basic Books, 1996), 85–101, 192–96; Milton L. Mueller, Jr., *Universal Service: Competition, Interconnection, and Monopoly in the Making of the American Telephone System* (Cambridge: MIT Press, 1997).

40. AT&T was on the verge of deciding that while ownership of a broadcasting facility provided certain long-range benefits (it could supplement research and development), the day-to-day operation of a radio station was inconsistent with its established corporate identity and its existing areas of expertise. At least this is the position taken in Banning's retrospective account of AT&T's operation of WEAF.

41. Read and Welch continue to provide the best gloss of the theory of matched impedance. Oliver Read and Walter L. Welch, *From Tin Foil to Stereo: Evolution of the Phonograph*, 2d ed. (Indianapolis, Ind.: Howard W. Sams, 1976).

42. Robert W. Baumbach, *Look for the Dog: An Illustrated Guide to Victor Talking Machines, 1901–1929* (Woodland Hills, Calif.: Stationary X-Press, 1981), 149.

43. Western Electric, "How the Sounds Get into Your Record by the Electrical Process," *Phonograph Monthly Review* 1 (Oct. 1926): 5–6. Arguably, the electrical system of reproduction in its earliest stages of development provided less fidelity than that which was available through an acoustic method of reproduction relying on the folded horn. But electrical reproduction of phonograph recordings did provide greater volume and, importantly, greater *consumer control* over volume. The folded acoustic horn required the consumer to open or close doors located in front of it to raise or lower the volume of reproduced sound, but the consumer operating an electrical phonograph could essentially increase or decrease volume with the turn of a knob. An account of the Western Electric system written by J. P. Maxfield for the 1926 edition of the *Encyclopedia Britannica* is reprinted in Thomas Rhodes's column, "Life in the Orthophonic Age," *New Amberola Graphic* 79 (Jan. 1992): 13–17.

44. Denmark's Valdemar Poulsen developed a functioning electrical system of phonograph recording and reproduction and a practical method of magnetic recording. Poulsen's efforts are briefly described in Edward W. Kellogg, "History of Sound Motion Pictures," in Raymond Fielding, ed., *A Technological History of Motion Pictures* (Berkeley: U of California P, 1967), 174–220.

45. Benjamin L. Aldridge, *The Victor Talking Machine Company* (New York: RCA Sales Corp., 1964), 87.

46. Some evidence indicates that Edison undertook similar research into electrical phonograph methods, although he apparently objected to the potential distortion introduced into phonography through electrical means. Read and. Welch, *From Tin Foil to Stereo*, 271–72.

47. Elmer C. Nelson, "Brunswick Electrical Recording," *Phonograph Monthly Review* 1 (Oct. 1926): 20.

48. N. E. Branch, "What Is the Panatrope?" *Phonograph Monthly Review* 1 (Oct. 1926): 26; Elmer C. Nelson, "Electrical Reproduction," *Phonograph Monthly Review* 1 (June 1927): 387.

49. *Bell Laboratories Record* 3.3 (Nov. 1926): 82. After visiting the Victor Talking Machine Company's Camden plant in August of 1926, Maxfield left the staff at Bell Laboratories to become the manager of development and research at Victor on November 1.

50. Baumbach, *Look for the Dog*, 150.

51. Andre Millard notes that the phonograph industry suffered from disappointing sales in both 1923 and 1924. Victor's sales were down 60% in 1924 and sale

of Edison phonographs had declined by 50%. Millard reports that the phonograph industry attributed the decline to the influence of radio. Andre Millard, *America on Record: A History of Recorded Sound* (New York: Cambridge UP, 1995), 138.

52. I use the phrase "compulsory repurchasing" to refer to the ways in which technological change encourages consumers to (re)purchase new software with each "upgrade" in their acoustic hardware. With each format change ("long-playing" 33 1/3 rpm disks, magnetic reel-to-reel tape, 8-track tape, cassette, compact disc, etc.) consumers were actively encouraged not only to purchase new hardware but also to repurchase new versions of their recording collections.

53. Baumbach, *Look for the Dog*, 181. In introducing new phonographs, Brunswick followed a path opposite to that of Victor, releasing first its all-electric Panatrope phonograph and then, in advertisements from January 1927, referring to an as yet unnamed new phonograph specifically designed to reproduce the music of electrical recordings but without the Panatrope's electrical equipment. The machines were to be priced less than the Panatrope and were intended to compete with Victor's Orthophonic line. Brunswick-Balke-Collender, "Before Buying Any Musical Instrument or Radio Hear the Brunswick Panatrope," *Phonograph Monthly Review* 1 (Jan. 1927): 199.

Defining its system as a competitor to both Western Electric's electrical recording and Victor's well-publicized, nonelectric Orthophonic method of reproduction, Brunswick claimed to be the first to "synchronize electrical recording with electrical reproduction." In assertions that never identified the Orthophonic Victrola by name, Brunswick cited the superiority of its cone loudspeakers in that they produced vibrations acting directly on the air without a "period of confinement" in an overtone- and other distortion-producing horn. "What is the Panatrope?" 26; Nelson, "Electrical Reproduction," 387.

54. In 1925, Victor introduced the folded horn through three differently sized and priced Orthophonic Victrolas and then supplemented this line in 1926 with a series of all-electric models, under the general name of Electrola. The public seemed to favor the Orthophonics over the Electrolas. In 1926, the first full year in which both machines were available to consumers, Victor sold over 260,000 Orthophonic Victrolas to some 8,000 Electrolas. The apparent public preference for the Orthophonics no doubt involved more than fidelity, as the electrical devices cost significantly more than the all-mechanical Orthophonics. The Credenza, the first top-of-the-line Orthophonic, initially retailed for $275 while, by comparison, the Cromwell, one of the earliest all-electric Victor phonographs, retailed for $450 ("Sales by Class of Product," reprinted in Baumbach, *Look for the Dog*, 8). Although

302
Notes

extremely successful, the Orthophonic Victrola ultimately proved to be a transitional device, easing the conversion of domestic spaces to electro-acoustics by allowing the successful reproduction of electrical-process recordings on a redesigned acoustic machine.

55. Shifts in Victor's product lines, including its introduction of a cone loudspeaker and a phonograph combined with a radio receiver, can be tracked through Baumbach, *Look for the Dog*.

56. This combination approach to sound reproduction, an electrically amplified signal reproduced through an Orthophonic horn, produced only slightly greater volume than the Orthophonic horn alone. Using the Orthophonic horn to amplify a radio signal did, however, allow the consumer greater control over volume, but the amplifier apparently added an appreciable distortion and hum. Baumbach, *Look for the Dog*, 213.

57. "Sales by Class of Product," reprinted in Baumbach, *Look for the Dog*, 8.

58. See the Special Issue of *Film History* 11.4 (1999): "Global Experiments in Early Synchronous Sounds," ed. Richard Abel and Rick Altman. Also see Rick Altman, *Silent Film Sound* (New York: Columbia UP, 2004).

59. Merritt Crawford, "Some Accomplishments of Eugene Augustin Lauste— Pioneer Sound-Film Inventor," *JSMPE* 16 (Jan. 1931): 105–109; Merritt Crawford, "Pioneer Experiments of Eugene Lauste in Recording Sound," *JSMPE* (Oct. 1931): 632–44; Paul C. Spehr, "Eugene Augustin Lauste: A Biographical Chronology," *Film History* 11 (1999): 18–38; John Hiller, "Eugene Augustin Lauste: Apparatus in the Smithsonian," *Film History* 11 (1999): 39–44.

60. John B. McCullough, "Joseph T. Tykociner: Pioneer in Sound Recording," in Fielding, ed., *A Technological History*, 221; J. Tykocinski-Tykociner, "Photographic Recording and Photoelectric Reproduction of Sound," *Transactions of SMPE* 16 (May 1923): 90–119; Joseph E. Aiken, "Technical Notes and Reminiscences on the Presentation of Tykociner's Sound Picture Contributions," in Fielding, ed., *A Technological History*, 222.

Lauste and Tykociner were just two among many inventors who approached with varying degrees of success the innovation of synchronous-sound films. Valdemar Poulsen (Denmark), Oskar Messter (Germany), as well as Hans Vogt, Josef Engl, and Joseph Massolle (also Germany, working together to innovate the successful Tri-Ergon system), achieved notable results in applying the principles and components of electrical acoustics to synchronous-sound film. See Albert Narath, "Oskar Messter and His Work," in Fielding, ed., *A Technological History*, 109–17 (see esp. 115); Kellogg, "History of Sound Motion Pictures," 174–220; Douglas Gomery,

"Tri-Ergon, Tobis-Klangfilm, and the Coming of Sound," *Cinema Journal* 16.1 (Fall 1976): 51–61; Hans Chr. Wohlrab, "Highlights of the History of Sound Recording on Film in Europe," *SMPTE Journal* 85 (1976): 531–33.

61. Richard Koszarski, *An Evening's Entertainment: The Age of the Silent Feature Picture, 1915–1928* (Berkeley: U of California P, 1990).

62. Historian Sheldon Hochheiser cites the institutional memory of earlier failed systems: "The Kinetophone's failure was widely known to both the public and the trade, and served as the clinching evidence, in the minds of Hollywood moguls, that sound motion pictures would not work." Sheldon Hochheiser, "AT&T and the Development of Sound Motion-Picture Technology," in Mary Lea Bandy, ed., *The Dawn of Sound* (New York: Museum of Modern Art, 1989), 23.

63. I place quotation marks around "introduced" because, of course, Warner Bros. did nothing of the kind. Selected theaters had for some time been screening, for example, the electrically recorded short films produced by Lee De Forest's Phonofilm process. And numerous nontheatrical public demonstrations of electrically recorded soundtracks preceded the Vitaphone event. The issue of primacy—much like the superfluous debate over which radio station was the "first" to broadcast—tends to elide the larger significance of such events, namely, the publicity and prestige garnered by the Vitaphone premiere and its rhetorical significance as an announcement of technological change. See chapter 2 for an analysis of the techniques used to announce technological change. For a more detailed narrative of Hollywood's conversion to sync-sound, see Donald Crafton, *The Talkies: American Cinema's Transition to Sound, 1926–1931* (Berkeley: U of California P, 1997), and esp. Douglas Gomery, *The Coming of Sound: A History* (New York: Routledge, 2005).

64. MGM, for example, initially relied upon the sheet music library assembled by the Capital Theatre to "compose" recorded musical accompaniment for feature-length films. Soundtrack construction thus mimicked live accompaniment, drawing on the same musical resources and using the same compositional methods as those of the "best" large urban theaters. See David Mendoza, "The Theme Song," *American Hebrew* (Mar. 15, 1929): 124.

65. Charles Wolfe, "On the Track of the Vitaphone Short," in Bandy, ed., *The Dawn of Sound*, 38.

66. See U.S. Bureau of Labor Statistics, "Effects of Technological Changes Upon Employment in the Motion-Picture Theaters of Washington, D.C." *Monthly Labor Review* 33.5 (Nov. 1931): 1005–1018. The experiences of musicians and their response are chronicled in James P. Kraft, *Stage to Studio: Musicians and the Sound Revolution, 1890–1950* (Baltimore: Johns Hopkins UP, 1996).

67. Promotional rhetoric surrounding the Vitaphone reinforced this conservatism by invoking many of the rhetorical promises that had surrounded the innovation of the phonograph. Like Victor's promise of Caruso in every home via the phonograph, Vitaphone's promoters claimed that the most accomplished musicians and vocalists, as well as the greatest Broadway and vaudeville performers, would grace the small-town and neighborhood motion picture theater, thus providing all Americans equal access to the finest entertainment. This connection between recorded film sound and "musical democracy" was central to initial pronouncements about Hollywood's adoption of electrical acoustics. See, for example, Harry M. Warner, "Future Developments," in Joseph P. Kennedy, ed., *The Story of the Films* (New York: A. W. Shaw, 1927), 335.

68. Case's system used a light modulated by electrical currents to encode voices and other sounds picked up by a microphone onto a strip of film moving through a modified Bell and Howell film camera. The variable-density soundtrack was reconverted into electrical impulses upon projection through the use of a photocell.

69. While some period accounts credit the Fox Movietone system with greater frequency response than the Vitaphone, Movietone was prone to distortion through wow and flutter, and even under the best projection conditions Movietone's soundtracks initially contained more noise than their disk-based competitors. In addition, the presence of a Movietone soundtrack on a release print changed the aspect ratio of the projected image, necessitating that projectionists install a special aperture plate to restore the Movietone film to a standard shape of the projected film image. For a detailed account of Movietone, see E. I. Sponable, "Some Technical Aspects of the Movietone," *Transactions of the SMPE* 21.31 (1927): 458–74.

70. As Hollywood converted to electrical acoustics, the technical refinements to the Movietone sound-on-film system accomplished by the engineers at Fox-Case were incorporated into Western Electric's own sound-on-film system. The Fox-Case Corporation was not in the business of manufacturing and marketing equipment for film producers and theaters, although they did supply equipment to Fox's newsreel crews. Western Electric, on the other hand, did supply the industry as a whole and were granted access to the research and development undertaken by engineers within Fox-Case.

71. The Tri-Ergon patents formed the basis of prolonged litigation and for a time appeared to threaten the legitimacy of the film industry's licenses with Western Electric and RCA. Ultimately though, the U.S. Supreme Court overturned two lower court rulings and found Fox's patent position invalid on March 4, 1935. Had the Court upheld the lower courts' rulings, Fox's American-Tri-Ergon would have

been in position to bring suits asking for treble damages against virtually all manufacturers and consumers of sound-on-film apparatus.

72. Fox outlined his position in Upton Sinclair, *Upton Sinclair Presents William Fox* (Los Angeles: Upton Sinclair, 1933).

73. "Joseph Maxfield Interview (Apr. 16, 1973)," IEEE History Center (*see www. ieee.org/organizations/history_center/oral_histories/transcripts*/maxfield.html), accessed Feb. 21, 2000. Vitaphone commercially introduced AT&T's efforts to apply electrical principles to film sound, but experimental efforts within the Bell System had also simultaneously pursued a different system design. The sound-on-disk effort under the leadership of J. P. Maxfield competed internally with attempts to record sound-on-film under the direction of I. B. Crandall. This alternative sound-on-film design eventually superseded the Vitaphone sound-on-disk system, but initially at least, the Bell System focused more directly on sound-on-disk since this research proved mutually compatible with the development of a commercial system of electrical phonographs. Stanley Watkins, "Madam, Will You Talk?" *Bell Laboratories Record* 24.8 (Aug. 1946): 295.

74. As in other applications of electrical acoustics, Hoxie's system used a microphone to transform sounds into electrical signals. These signals then activated a vibrating mirror onto which a light source was shone. The vibrating mirror varied the amount of light reaching a strip of film, thereby encoding photographically a version of the original sound. The photographic sound recording was reconverted into electrical signals by running it between a source of illumination and a photoelectric cell. After proper amplification, a loudspeaker converted these electrical signals back into sound. Hoxie's system resembled in principle Case's method of recording sound on film, except whereas the Fox-Case system used electrical signals to modulate the intensity of a lamp, Hoxie's system used a consistent source of illumination and electrical signals instead modulated the ability of light to reach the celluloid negative. When Western Electric released a commercial version of its sound-on-film recording system in 1927, it relied on a light valve that, like Hoxie's method, encoded a photographic trace of sound through a shutter of stretched metallic ribbon modulated by electrical signals, thereby varying the amount of light exposed on moving film.

75. Kellogg, "History of Sound Motion Pictures," 182.

76. Douglas Gomery, "Failure and Success: Vocafilm and RCA Photophone Innovate Sound," *Film Reader* 2 (1977): 213–21.

77. Paramount Famous Lasky, United Artists, Metro-Goldwyn Pictures, and First National, as well as the smaller Christie Film Company and Hal Roach Studios, signed contracts in May and June. Universal signed with ERPI in July, and Columbia followed in September.

78. Despite the coercion to adopt Western Electric's sound-projection equipment, many independent exhibitors initially outfitted their theaters with equipment manufactured by recently established firms. Some independents found the cost of Western Electric's equipment prohibitive, while others balked at Western Electric's mandatory service contracts. Some theater owners sought equipment from RCA or smaller competitors because of the inability of Western Electric's manufacturing facilities and ERPI's installation engineers to meet theaters' demands for sound-projection apparatus. Douglas Gomery notes that around January 1927 approximately one new sound-film system was advertised to the industry per month, but that number increased to approximately one new system per week by May of 1928. Gomery, "Failure and Success."

79. Moody's Manual of Investments and Security Rating Service, 1929 (New York: Moody's Investment Service, 1929), 584–90.

80. Hoover, quoted in Thomas Streeter, *Selling the Air: A Critique of the Policy of Commercial Broadcasting in the United States* (Chicago: U of Chicago P, 1996), 88.

81. Spectrum overcrowding shaped much of the research and development agenda for this period of broadcasting. Early broadcasters desired efficient and inexpensive transmitters, but as their numbers grew, they sought increasing power levels to reach wider audiences or to overpower competitors broadcasting on the same frequency. The resultant "noise" necessitated some form of technological refinement, and innovators explored methods of transmitting on more stable frequencies, directional transmission, and still greater power. Radio listeners sought first more powerful and efficient radio components (tubes, condensers, circuits, and antennas) and then similar qualities in pre-built sets. The "noise" of spectrum overcrowding also led to home receivers with increased sensitivity in tuning and combinations of greater signal amplification and efficient loudspeaker technology. Other researchers sought to apply new principles to broadcasting's various components in order to circumvent the patent pool vested in RCA.

82. Streeter, *Selling the Air*, 90. My analysis of radio regulation also draws on Hugh Richard Slotten, "Radio Engineers, the Federal Radio Commission, and the Social Shaping of Broadcast Technology: Creating 'Radio Paradise,'" *Technology and Culture* 36.4 (Oct. 1995): 950–86; and Hugh G. J. Aitken, Jr., "Allocating the Spectrum: The Origins of Radio Regulation," *Technology and Culture* 35.4 (Oct. 1994): 686–716.

83. For a 1929 account of spectrum allocation, see Robert H. Marriott, "United States Radio Broadcasting Development," *PIRE* 17.8 (Aug. 1929): 1395–1439, esp. 1405–1406. Marriott served as an engineering consultant to the Federal Radio Commission, shaping the very developments he described.

84. We can retrospectively identify early radio regulation as "highly political," although few voices of the day saw it as such—a testament to the power of an emerging corporate-capitalist hegemony and this particular rhetorical screen.

85. Streeter, *Selling the Air*, 95.

86. See Aitken, "Allocating the Spectrum," 699–706. At the time, the station was jointly owned by the Zenith Radio Corporation and the newspaper *Chicago Herald Examiner*. Zenith was under the leadership of Eugene F. McDonald, president of the National Association of Broadcasters, who was well versed in current broadcasting policy and the potential legal limitations on the Department of Commerce's licensing power. Marvin R. Bensman provides an account of the case in "The Zenith-WJAZ Case and the Chaos of 1926–27," *Journal of Broadcasting* 14.4 (Fall 1970): 423–40.

87. The *Chicago Tribune* tried to use the state court system to protect the frequency assignment of its station WGN during the 1926 lapse in federal regulatory authority. Louise M. Benjamin, "The Precedent That Almost Was: A 1926 Court Effort to Regulate Radio," *Journalism Quarterly* 67.3 (Autumn 1990): 578–85.

88. Robert W. McChesney, *Telecommunications, Mass Media, and Democracy: The Battle for the Control of U.S. Broadcasting, 1928–1935* (New York: Oxford UP, 1993).

89. These issues are developed in more detail in chapters 3 and 4.

90. Douglas Gomery summarizes film industry developments following the transition to sound. Gomery, *The Coming of Sound*, 115–27.

91. While the vertical and horizontal integration of RCA facilitated cross-media promotion and the production of intermedia commodities like *Check and Double Check*, the scope of RCA's holdings sometimes caused conflicts between divisions within the larger organization. Alexander Russo describes one such conflict surrounding the recording and distribution of broadcast programming on disks. While the NBC radio network decried the practice and advocated live, wired transmission of programs, another division of RCA, RCA-Victor, produced these "transcriptions." Alexander Russo, "Defensive Transcriptions: Radio Networks, Sound-on-Disc Recording, and the Meaning of Live Broadcasting," *The Velvet Light Trap* 54 (Fall 2004): 4–17.

92. Douglas Gomery, *The Hollywood Studio System* (New York: St. Martin's, 1986): on Warner Bros., see 104, 109–10; on Loew's/MGM, see 58; on Paramount, see 30–31. Paramount's purchase of CBS stock is discussed in William Boddy, " 'Spread Like a Monster Blanket Over the Country': CBS and Television, 1929–1933," in Annete Kuhn and Jackie Stacey, eds., *Screen Histories: A Screen Reader* (New York: Oxford UP, 1998), 129–38; and Jonathan Buchsbaum, "Zukor Buys Protection: The Paramount Stock Purchase of 1929," *Cine-tracts* 2.3–4 (1979): 49–62.

93. Lewis A. Erenberg, *Steppin' Out: New York Nightlife and the Transformation of American Culture, 1890–1930* (Chicago: U of Chicago P, 1981); see esp. ch. 5.

94. Barney Dougall, "Theme Songs, I Love You: A Tremulous Tribute to a New National Ballad Which the Talking Films Brought to America," *Vanity Fair* 33 (Nov. 1929): 67.

95. See, for example, "Song Hits in Current Films," *Motion Picture Herald* (Mar. 7, 1931): 26–27.

96. "Theatre Publicity Punch Urged to Sell Public Robbins Music," *The Loew-Down* 1.8 (July 18, 1930): 4. The twice-monthly internal publication subsequently continued to remind Loew's managers of the corporate connection between the theater chain and Robbins music, all within MGM. (Copies of *The Loew-Down* are located in the archive of the Theater Historical Society, Elmhurst, Illinois.)

97. Gomery, *The Coming of Sound*, 152–53. This dominance would continue into the 1930s. Susan Douglas notes that by one account, in 1938 eight of the top fifteen music publishers were owned or controlled by Hollywood studios, and only 18 percent of the top songs on the radio came from houses not affiliated with Hollywood. Susan J. Douglas, *Listening In: Radio and the American Imagination* (New York: Times Books, 1999), 152.

98. "Theatres Develop Radio on Co-operative Basis," *The Loew-Down* 1.5 (May 19, 1930): 4.

99. "Canton Scores Knockout in Radio Fight," *The Loew-Down* 1.7 (July 7, 1930): 1–2.

2. Announcing Technological Change

1. "A Reporter's Visit to the Boston Telephone Exchange," *Scientific American* 56 (Feb. 12, 1887): 106.

2. "A Reporter's Visit," 106.

3. Ibid.

4. Ibid.

5. Stuart Ewen, *PR! A Social History of Spin* (New York: Basic Books, 1996).

6. "A Reporter's Visit," 106.

7. Tom Gunning, "Re-Newing Old Technologies: Astonishment, Second Nature, and the Uncanny in Technology from the Previous Turn-of-the-Century," in David Thorburn and Henry Jenkins, eds., *Rethinking Media Change: The Aesthetics of Transition* (Cambridge: MIT Press, 2003), 40–41, 42. I explore similar issues in an essay on the curatorial practices surrounding recent museum displays of archaic technology in "The Space Between the Object and the Label:

Exhibiting Restored Vitaphone Films and Technology." *Film History* 5 (1993): 275–88.

8. Gunning, "Re-Newing Old Technologies," 45.

9. In what follows, I trace ways in which public performances of emerging technologies sought to incorporate media into already-existing ideas of progress, convenience, the modern, and other widely held and reproduced values. While the "announcements" I describe below certainly could not ultimately guarantee consumers' relationships to emerging media forms, like William Boddy's recent work on personal video recorders in the late 1990s, I see such "ephemeral industry discourses" as some of the "tools with which the public makes sense of technological change." William Boddy, "Redefining the Home Screen: Technological Convergence as Trauma and Business Plan," in Thorburn and Jenkins, eds., *Rethinking Media Change*, 191–200.

10. "Victor Auditorium Orthophonic Victrola," *Phonograph Monthly Review* 1 (Mar. 1927): 250–51. I am not here (or subsequently) suggesting that press coverage of such a public performance offers us transparent, objective accounts of events. They, of course, reveal a particular perspective on that performance, but they function most powerfully as period articulations *about* the performance, extending (and rarely contesting) the rhetorical shaping of technology undertaken in such public demonstrations.

11. "Victor Auditorium Orthophonic Victrola," 250.

12. Brief reports of these demonstrations appeared in the trade publication *Talking Machine World* throughout 1927. See, for example, "Auditorium Victrola Installed in Omaha," *Talking Machine World* (Jan. 15, 1927): 74; "Auditorium Model Victrola Aids New Orleans Trade," *Talking Machine World* (Feb. 15, 1927): 38; "Auditorium Orthophonic Stimulates Music Trade in Buffalo Territory," *Talking Machine World* (Feb. 15, 1927): 120; "Automatic Orthophonic Shown to Indianapolis Trade," *Talking Machine World* (Mar. 1927): 127; "To Feature Orthophonic," *Talking Machine World* (June 1927): 98; "Auditorium Victrola Is Installed in Atlantic City," *Talking Machine World* (June 1927): 108; "Auditorium Orthophonic Installed in Willow Grove," *Talking Machine World* (July 1927): 145.

13. Benjamin L. Aldridge, *The Victor Talking Machine Company* (New York: RCA Sales Corp., 1964), 89–94.

14. "Paris Pays to See Columbia at Work," *Phonograph Monthly Review* 3 (Mar. 1929): 192.

15. Rick Kennedy, *Jelly Roll, Bix, and Hoagy: Gennett Studios and the Birth of Recorded Jazz* (Bloomington: Indiana UP, 1994), 85.

16. James Lastra locates nineteenth-century edifying spectacles that promoted emerging technologies in relation to the "operational aesthetic" described by Neil Harris. James Lastra, *Sound Technology and the American Cinema: Perception, Representation, Modernity* (New York: Columbia UP, 2000), 67–71. Lastra notes that "systematic demystification played an important role in legitimating these devices, both as inoffensive forms of entertainment and as technologies capable of making particular epistemological claims" (68).

17. Transcribed from a 1906 version recorded by Edison artist Len Spencer. Len Spencer, "I Am the Edison Phonograph" (1906), Smithsonian Institution, *Great Speeches of the 20th Century*, vol. 2 (Santa Monica, Calif.: Rhino Records, 1991). Similar recordings were available for demonstration purposes in the late nineteenth century. They continued to be used most often in demonstrations performed by Edison dealers into the twentieth century.

18. Sullivan, as noted in Charles Musser, with the collaboration of Carol Nelson, *High-Class Moving Pictures: Lyman H. Howe and the Forgotten Era of Traveling Exhibition, 1880–1920* (Princeton: Princeton UP, 1991), 34–35.

19. "Talking 'Movies,'" *Outlook* 103 (Mar. 8, 1913): 517.

20. Isaac F. Marcosson, "The Coming of the Talking Picture," *Munsey's* 38 (Mar. 1913): 956.

21. Bailey Millard, "Pictures That Talk," *Technical World* 19.1 (Mar. 1913): 20.

22. Edison, quoted in Millard, "Pictures That Talk," 21.

23. For a detailed discussion of the Tone Tests and their importance to the phonograph industry, see Emily Thompson, "Machines, Music, and the Quest for Fidelity: Marketing the Edison Phonograph in America, 1877–1925," *Musical Quarterly* 79.1 (Spring 1995): 148–69. She reports this estimated attendance on page 153. Edison's Tone Tests continued well into 1927, overlapping with the phonograph industry's conversion to electrical methods. *Talking Machine World* reported in May of 1927 the conclusion of the spring season of Tone Tests, noting that over 500 such "concerts" had been presented during that season ("Spring Season of Edison Tone-Tests Is Completed," *Talking Machine World* [May 1927]: 34a). These demonstrations promoted mechanical phonographs and recordings as Edison did not release an electrical machine until later in 1927, the Edisonic. Jonathan Sterne takes note of Tone Tests and suggests that their most important rhetorical function was to assert a correspondence between the reproduction of a recording and an "original" acoustic event. Such an equivalency was by no means obvious and had in fact to be discursively constructed. Jonathan Sterne,

The Audible Past: Cultural Origins of Sound Reproduction (Durham, N.C.: Duke UP, 2003), 261–65.

24. Edison dealer quoted by Virginia Powell, Edison performer, in a letter to Edison. Reprinted in Raymond Wile, "Virginia Powell in Kansas (An Edison Artist on the Tone Test Trail)," *New Amberola Graphic* 88 (Spring 1994): 4.

25. For a reprint of a "Tone Test" program from Olympia, Washington, on March 22, 1921, see *New Amberola Graphic* 73 (July 1990): 8. The program describes the appearance of Arthur Collins and Byron G. Harlan as well as saxophonist William Reed. The evening's entertainment mixed popular songs, instrumental performances, and a recording by soprano Anna Case (with live accompaniment provided by Reed). Mixing the live and the recorded demonstrated the Diamond Disc's performance through a number of categories of the recorded repertoire.

26. For a more complete description of the mise-en-scène of these performances, see "The Tone Test and Its Stage Setting," from *Edison Diamond Points*, Nov. 1920, reprinted in Thompson, "Machines, Music," 150. Like the Sullivan Brothers' late-nineteenth-century phonograph performances, that mise-en-scène included a framed portrait of Thomas Edison.

27. Thompson, "Machines, Music," 156–60.

28. American Telephone and Telegraph Co., "A Wonder of Wonders," *American Magazine* 80.6 (Dec. 1915): 71.

29. William Peck Banning, *Commercial Broadcasting Pioneer: The WEAF Experiment, 1922–1926* (Cambridge: Harvard UP, 1946), 199. The Amateur Radio Relay League undertook similar transcontinental demonstrations in 1916 and 1917. Here again, corporate innovators followed preexisting patterns to introduce and to shape public perceptions of emerging media. Susan J. Douglas, *Inventing American Broadcasting, 1899–1922* (Baltimore: Johns Hopkins UP, 1987), 296–97.

30. Banning, *Commercial Broadcasting Pioneer*, 248–52. A second National Defense Test Day, held on July 4, 1925, similarly demonstrated selected potentials of networked radio stations (ibid., 273).

31. I. W. Green and J. P. Maxfield, "Public Address Systems," *Transactions of the AIEE* 42 (Apr. 1923): 357–58.

32. Craft's remarks are selectively reproduced in "The Vitaphone Tells Tales of Itself," *Bell Laboratories Record* 3.4 (Dec. 1926): 126–28.

33. "The Vitaphone Tells Tales of Itself," 128.

34. Ibid.

35. Ibid., 126; see also *Bell Laboratories Record* 3.2 (Oct. 1926): 57.

36. "In the Month's News," *Bell Laboratories Record* 4.1 (Mar. 1927): 247.

37. "In the Month's News," *Bell Laboratories Record* 4.2 (Apr. 1927): 289; "News Notes," *Bell Laboratories Record* 4.3 (May 1927): 332; "News of the Month," *Bell Laboratories Record* 4.4 (June 1927): 376.

38. "D. & R. News and Notes," *Bell Laboratories Record* 3.5 (Jan. 1927): 169.

39. "D. & R. News and Notes," *Bell Laboratories Record* 3.6 (Feb. 1927): 215–16. The Bell System's use of talking pictures as part of this larger combined public relations and pedagogical project should be viewed in light of the organization's use of silent film technology in public performances throughout the late 1910s and early 1920s. Some of these efforts are detailed in the Federal Communication Commission's 1938 report on the Bell System. See U.S. Federal Communications Commission, *Proposed Report Telephone Investigation* (Washington, D.C.: GPO, 1938), ch. 17 ("Public Relations"), esp. 569–73. The report concluded in part, "Visual education through motion-pictures has been a powerful influence in aiding the Bell System to sell its ideology to the youth of the country" (569). Also see John Mills, "The Silent Drama of Telephony," *Bell Laboratories Record* 3.2 (Oct. 1926): 39–45.

40. While realistic representational strategies seek to elide the processes through which films are produced, a host of simultaneous discursive practices surrounding film and filmgoing, *including* behind-the-scenes accounts and the other methods of announcing technological change described in this chapter, foreground that which realistic practices seek to repress.

41. "Bringing Sound to the Screen," *Photoplay* (Oct. 1926), in Miles Kreuger, ed., *The Movie Musical from Vitaphone to "42nd Street"* (New York: Dover, 1975), 2–3.

42. Elsewhere in the same issue, in a review of the Vitaphone "Tannhäuser" short subject, *Photoplay* invoked the claim echoed widely in the period that the synchronous-sound film (and indeed all manifestations of the electrical-acoustic apparatus) offered to consumers something more than an enhanced entertainment experience: "[The Vitaphone "Tannhäuser" short is] a musical education for the novice, as close-ups of the various sections of the orchestra, judiciously cut into the film, give a casual idea of some of the intricacies of the Wagner orchestration." Review from *Photoplay* (Oct. 1926), in Kreuger, ed., *The Movie Musical*, 3.

43. "How Talkies Are Made," *Photoplay* (Nov. 1928), in Kreuger, ed., *The Movie Musical*, 8–9. The two years between these examples of consumer pedagogy illustrate that technological change and corporate innovators' attempts to shape public perceptions were ongoing.

44. A. P. Peck, "Giving a Voice to Motion Pictures," *Scientific American* 136 (June 1927): 378–79; A. P. Peck, "Sounds Recorded on 'Movie' Film," *Scientific American* 137 (Sept. 1927): 234–36; "The Movie 'Theater' Up-To-Date," *Scientific American* 137 (Dec. 1927): 516–17.

45. Orrin E. Dunlap, Jr., "When the President Broadcasts," *Scientific American* 137 (July 1927): 36–38; Joseph P. Maxfield, "Electrical Research Applied to the Phonograph," *Scientific American* 134 (Feb. 1926): 104–05. Also see Joseph P. Maxfield, "Electrical Phonograph Recording," *Scientific Monthly* 22 (Jan. 1926): 71–79.

46. Western Electric Company, "How the Sounds Get into Your Record by the Electrical Process," *Phonograph Monthly Review* 1 (Oct. 1926): 4–6.

47. Peter P. Decker, "Echoes from the Okeh Recording Studio," *Phonograph Monthly Review* 2 (Dec. 1927): 86–88; Nathaniel Shilkret, "Modern Electrical Methods of Recording," *Phonograph Monthly Review* 1 (June 1927): 382.

48. Elmer C. Nelson, "Brunswick Electrical Recording," *Phonograph Monthly Review* 1 (Oct. 1926): 19–21; Elmer C. Nelson, "Electrical Reproduction," *Phonograph Monthly Review* 1 (June 1927): 386–87, 90; N. E. Branch, "What Is the Panatrope," *Phonograph Monthly Review* 1 (Oct. 1926): 26–27.

49. Brunswick-Balke-Collender Co., "Before Buying Any Musical Instrument or Radio Hear the Brunswick Panatrope," *Phonograph Monthly Review* 1 (Jan. 1927): 199; Columbia Phonograph Company, "'Like Life Itself': The New Viva-tonal Columbia," *Phonograph Monthly Review* 1 (Mar. 1927): 253.

50. Jerome Beatty, "The Sound Investment," *Saturday Evening Post* 201 (Mar. 9, 1929): 19, 129.

51. Beatty, "The Sound Investment," 8–9.

52. Peggy Wood, "See and Hear," *Saturday Evening Post* 202 (July 20, 1929): 20–21.

53. Beatty, "The Sound Investment," 19.

54. Western Electric, "What Makes the Picture Talk?" *Saturday Evening Post* 202 (July 13, 1929): 111.

55. Western Electric, "Making a Sound Picture with Western Electric Equipment," *Photoplay* (July 1929), in Kreuger, ed., *The Movie Musical*, 32.

56. "Making a Sound Picture with Western Electric Equipment."

57. Western Electric would intensify this branding as they faced increasing competition from the RCA Photophone and other sound-film systems. During the early 1930s they used an advertising campaign with the slogan "Look for the Sign" to

suggest that theaters outfitted with Western Electric equipment guaranteed better sound reproduction. (See note 72.)

58. Masonite Corporation, "Presdwood Stars Again," *Saturday Evening Post* 201 (Apr. 6, 1929): 175.

59. P. Lorillard Co., "Please pardon my frown . . . but some one in the studio just coughed . . . *and spoiled our love scene,*" *Life* (Apr. 12, 1929): 40; P. Lorillard Co., "He coughed . . . the Villain! and the love scene had to be taken all over!" *Life* (Mar. 29, 1929): 32.

60. "Please pardon my frown . . . "

61. The Bellamy testimonial names her latest release *Mother Knows Best,* and the Barthelmess advertisement cites "the First National–Vitaphone masterpiece *Weary River* in which Mr. Barthelmess adds to his laurels with a voice of rare dramatic quality." Both ads also refer to the Old Gold–Paul Whiteman Hour broadcast Tuesday evenings on the CBS network.

62. *Projection Engineering* 2.3 (Mar. 1930): 12.

63. Elaine Ogden,"Inside the Monitor Room," *Photoplay* (July 1930), in Kreuger, ed., *The Movie Musical,* 196.

64. Western Electric, "The 'Rule of Thumb' Is Over," *Scientific American* (Dec. 1926): 447.

65. At issue here is less the actual practices of invention than how those practices are represented to the public. Some historians have begun to investigate invention before and during this period. On this issue, see Paul Israel, *From Machine Shop to Industrial Laboratory: Telegraphy and the Changing Context of American Invention, 1830–1920* (Baltimore: Johns Hopkins UP, 1992); Leonard S. Reich, "Edison, Coolidge, and Langmuir: Evolving Approaches to American Industrial Research," *Journal of Economic History* 47 (June 1987): 341–51.

66. "The Test," *American Magazine* 80 (Dec. 1915): 60–61.

67. "What Edison Likes in Music," *Country Gentleman* 86 (May 21, 1921): 21.

68. Alan L. Benson, "Edison's Dream of New Music," *Cosmopolitan* 54 (May 1913): 797.

69. For a brief, insightful analysis of Edison's cultivation of a personal mythology, see Portia Dadley, "The Garden of Edison: Invention and the American Imagination," in Francis Spufford and Jenny Uglow, eds., *Cultural Babbage: Technology, Time, and Invention* (London: Faber and Faber, 1996), 81–98. Dadley includes an analysis of the cultural functions of tales from Edison's boyhood.

70. "Edison's Golden Day," *Literary Digest* 103 (Nov. 2, 1929): 10.

71. "Edison's Golden Day," 10.

72. Western Electric, "Look for This Sign: People Are Learning That There's a Difference In SOUND QUALITY," *Photoplay* (Feb. 1930): 95. Also see the following Western Electric advertisements, all of which instructed consumers to look for theaters displaying the "Western Electric equipped" sign: "Follow This SIGN and Hear Talking Pictures That Sound Natural," *Life* 95 (Feb. 28, 1930): 27; "Select Your Theatre by EAR TEST," *Life* 95 (Apr. 25, 1930): 1; "Which Theatre Tonight? Let the EAR TEST Decide," *Life* 95 (June 27, 1930): 25. For the prominence of Western Electric's brand in local theater promotions, see the advertisement for the Lyric Theatre of Atkinson, Nebraska, reprinted in *Motion Picture Herald* (Mar. 14, 1931): 102.

73. Western Electric, "Sound Pictures, a Product of the Telephone," *Saturday Evening Post* (Apr. 6, 1929): 138–39.

74. Western Electric, "Great Names Pioneering in This Great New Art," *Life* 93 (May 10, 1929): 4.

75. "Organized Common Sense," from the Telephone Almanac for 1933, reprinted in *Bell Laboratories Record* 11.5 (Jan. 1933): 156. For a similar presentation from earlier in the period, see "The Service of Knowledge," *Literary Digest* (Aug. 16, 1924): 55.

76. Attempts to engineer public perceptions of the Bell System and its affiliated companies including Western Electric have been detailed both in the 1930s by the Federal Communication Commission's telephone investigation and more recently by historian Stuart Ewen, *Proposed Report Telephone Investigation*, 563 passim; Ewen, *PR!*, 84–101, 192–97.

77. Ewen, *PR!*, 94. For a discussion of AT&T's use of female operators to cultivate a corporate identity, see Lana F. Rakow, "Women and the Telephone: The Gendering of a Communications Technology," in Cheris Kramarae, ed., *Technology and Women's Voices: Keeping in Touch* (New York: Routledge and Kegan Paul, 1988), 207–28. This embodiment of the corporation through the telephone operator continued into the 1930s; see American Telephone and Telegraph Co., "Friend and Neighbor," *Bell Laboratories Record* 11.12 (Aug. 1933): inside back cover.

78. "Sound Pictures, a Product of the Telephone," 138.

79. Although numerous individuals and stations might be credited with the first broadcast, I am referring to the commonplace assertion that Frank Conrad inaugurated broadcasting from his Pittsburgh garage with Westinghouse-funded station KDKA by transmitting the 1920 election returns. That broadcasting arguably began with a political event seems particularly appropriate given the argument I develop below.

80. Maurice Henle, "A Non-Partisan Political Medium," *Wireless Age* 10 (Dec. 1922): 27.

81. William A. Hurd, "How Broadcasting June Conventions Affects the November Elections," *Wireless Age* 11 (Aug. 1924): 18.

82. "William Jennings Bryan Tells Us Radio Will Help Bring World Peace," *Wireless Age* 11 (May 1924): 17.

83. "Senator Royal S. Copeland: Radio Will Improve Politics," *Wireless Age* 11 (July 1924): 19.

84. Christine Frederick, "Women, Politics and Radio," *Wireless Age* 12 (Oct. 1924): 36–37, 76, 78.

85. Edwin F. Ladd, "The Harmonizing Effect of Radio on Our Social Order," *Wireless Age* 12 (Oct. 1924): 20.

86. Hurd, "How Broadcasting," 18.

87. Ibid., 21. Not to be outdone, fellow radio industry giant RCA through spokesman Gen. James G. Harbord echoed this industrywide chant. "Broadcasting will give the people cleaner politics and more effective self-expression." Proclaiming the end of the "old style stump speaker," "the old fashioned ranter" who relied on "semaphore gestures" and "tremulo in his voice" for "emotional and hypnotic effect," Harbord attributed to radio the rise of "the keen, clear-voiced, self-contained, logical speaker" (ibid., 57).

88. Dwight F. Davis, "The Foundation Laid by Radio for Our Social Edifice," *Wireless Age* 12 (Oct. 1924): 21, 62; James J. Davis, "Promoting International Understanding," *Wireless Age* 12 (Oct. 1924): 19, 62.

89. Hubert Work, "The Beginning of a New Political Era," *Wireless Age* 12 (Oct. 1924): 18. See also the statements by eight national political figures in "How Broadcasting Political Speeches Affects Our Politic Body," *Wireless Age* 12 (Oct. 1924): 24–25.

90. Roland Marchand, "Corporate Imagery and Popular Education: World's Fairs and Expositions in the United States, 1893–1940," in David E. Nye and Carl Pederson, eds., *Consumption and American Culture* (Amsterdam: VU UP, 1991), 18–33.

91. Marchand, "Corporate Imagery," 19.

92. Ibid., 20.

93. "The Reproduction of Orchestral Music in Auditory Perspective," *Bell Laboratories Record* 11.9 (May 1933): 254–61.

94. "Music and Science," *Bell Laboratories Record* 11.10 (June 1933): 314.

95. "The Bell System Exhibit at the Century of Progress Exposition," *Bell Laboratories Record* 11.11 (July 1933): 340.

96. R. F. Mallina, "Seeing Sound at the Chicago Exposition," *Bell Laboratories Record* 11.12 (Aug. 1933): 361–64; C. N. Hickman, "Delayed Speech," *Bell Laboratories Record* 11.10 (June 1933): 308–10; Harvey Fletcher, "An Acoustic Illusion Telephonically Achieved," *Bell Laboratories Record* 11.10 (June 1933): 286–89; "The Bell System Exhibit at the Century of Progress Exposition," 340–48.

97. Robert W. Rydell, *World of Fairs: The Century-of-Progress Expositions* (Chicago: U of Chicago P, 1993) 98–99.

98. Nelly Oudshoorn and Trevor Pinch, eds., *How Users Matter: The Co-Construction of Users and Technologies* (Cambridge: MIT Press, 2003); Trevor Pinch and Frank Trocco, *Analog Days: The Invention and Impact of the Moog Synthesizer* (Cambridge: Harvard UP, 2002); and Ron Eglash, Jennifer L. Croissant, Giovanna Di Chico, and Rayvon Fouche, eds., *Appropriating Technology: Vernacular Science and Social Power* (Minneapolis: U of Minnesota P, 2004). In the last see esp. David Albert Mhadi Goldberg, "The Scratch Is Hip-hop: Appropriating the Phonographic Medium," 107–44.

3. From Performing the Recorded to Dissimulating the Machine

1. In an analysis of telephone marketing, David Gautschi and Darius Jal Sabavala note that the telephone was something of an enigma for the typical U.S. household in the 1910s. Installation often seemed like an intrusion; methods of using the apparatus had to be learned; and the appropriate location of the telephone within the home required some negotiation and trial and error (76). Importantly, Gautschi and Sabavala argue that as markets evolve, "consumers learn to incorporate a product into more and more consumption activities as a result of their own initiatives, from observation of others, and at the exhortation of the supplier" (73). David A. Gautschi and Darius Jal Sabavala, "The World That Changed the Machines: A Marketing Perspective on the Early Evolution of Automobiles and Telephony," *Technology in Society* 17.1 (1995): 55–84.

Lisa Gitelman similarly argues that consumers played a crucial role in the construction of the phonograph as both a public and a domestic entertainment device. Her work covers a period roughly prior to what I consider below. While Gitelman focuses on a wide range of determinants shaping consumer encounters with the machine (their relationship to mass circulation magazines and mass culture more generally, contemporary musical culture, conventions of shopping), I am more concerned with the physical design of phonographs and the phonograph industry's public efforts to shape consumer encounters with the device. Gitelman concludes that "the phonograph provides an exemplary instance of cultural production snatched from the hands of putative producing agents," but our different perspectives strike me as compatible rather than conflicting. Lisa Gitelman, "How Users Define New Media: A History of the Amusement Phonograph," in David Thorburn and Henry Jenkins, eds., *Rethinking Media Change: The Aesthetics of Transition* (Cambridge: MIT Press, 2003), 61–79.

2. In what follows, I stress the efforts of Edison and Victor not only because of the economic dominance of these two organizations (together with Columbia), but also because they often pursued slightly different solutions to the "problem" of defining the phonograph as a consumer device.

3. Jonathan Sterne explores at some length the socially constructed nature of acoustic fidelity (*The Audible Past: Cultural Origins of Sound Reproduction* [Durham, N.C.: Duke UP, 2003], 215–86). In what follows I leave unexamined the vexed relationship between original and copy. Instead, I am interested in the ways in which public assertions *about* such a link provided tools through which users might make sense of their relationship to sound media and technological change. In his study of fidelity as a socially constructed ideal, Sterne notes how crucial are habits and practices of listening to the dissimulation of the machine. Fidelity requires ignoring the machine and "the grain of the apparatus" (258) that inevitably is part of sound reproduction. In what follows I trace how phonograph industry rhetoric sought to cultivate particular (and often ultimately conflicting) habits and practices of listening.

4. "The Test," *American Magazine* 80 (Dec. 1915): 60–61.

5. "The Test," 61.

6. "And Now Thomas A. Edison *Answers* Another Questionnaire," *National Geographic*, mid-1920s (author's collection) .

7. "Will You Join Mr. Edison in an Experiment?" *Collier's* (Feb. 19, 1921): 25.

8. "I Will Pay $10,000 for the Best Thoughts on One of My Problems," *Collier's* (June 11 1921): 21. The "Mood Change Chart" was also incorporated within Edison's Tone Test performances. See Raymond Wile, "Virginia Powell in Kansas (An Edison Artist on the Tone Test Trail)," *New Amberola Graphic* 88 (Spring 1994): 4.

9. "What Edison Likes in Music," *Country Gentleman* 86 (May 21, 1921): 21.

10. Victor was not the first to pursue this approach. Some ten years before the advent of Victor's Red Seal recordings, phonograph innovator Gianni Bettini recorded operatic titles and sold them at premium prices. Bettini's elaborate, colorful catalogs addressed an elite market segment that perhaps predated the technologi-

cal ability to supply consumer demand (Allen Koenigsberg, *The Patent History of the Phonograph, 1877–1912* (Brooklyn, N.Y.: APM Press, 1990), xxxii–xxxiii). Roland Gelatt suggests that Bettini was not seeking the mass production of his recordings, but instead catered to a small clientele consisting of the extremely wealthy (Roland Gelatt, *The Fabulous Phonograph: From Edison to Stereo*, 3d ed. [New York: Appleton-Century, 1965], 75–81). In 1903, Columbia introduced a Grand Opera Series of recordings featuring famous opera stars that were offered at a higher price indicative of the prestige associated with the product. Victor *then* responded with the inauguration of its Red Seal recordings.

11. In 1906, Edison entered the prestige recording market with grand opera cylinders. Because Victor's exclusive contracts had tied up many celebrity recording stars, Edison's advertising most often foregrounded its devices' technological precision. Marsha Siefert, "Aesthetics, Technology, and the Capitalization of Culture: How the Talking Machine Became a Musical Instrument," *Science in Context* 8.2 (1995): 440.

12. Some evidence indicates that Victor was successful in identifying its product line as the most prestigious in the phonograph field. An Edison-sponsored market research survey in May of 1910 concluded that, in northern New Jersey towns at least, "The impression is universal that the Victor is the machine for the Classes and the Edison for the Masses. Some regard the Victor as beyond the reach of any but the wealthiest, but credit for the highest quality is given the Victor" ("Digest of Conditions of Phonographic Trade in Northern New Jersey Towns," May 31, 1910). The Edison trade report is reprinted in Raymond R. Wile, "Market Research—1910 Style," *New Amberola Graphic* 90 (Autumn 1994): 10–12. Similar reports of trade conditions in New York City, Brooklyn, Newark, Jersey City, and Hoboken echoed the perception that Victor machines and recordings appealed to the better class of customers. This perception within the Edison organization that Victor had essentially cornered the market for high-quality sound reproduction provides part of the background shaping the form and rhetoric of Edison's future "Tone Test" demonstrations ("Digest of Conditions of Phonographic Trade in New Jersey Suburban Cities," [1910], reprinted in Raymond R. Wile, "Market Research—1910 Style," *New Amberola Graphic* 98 [1997]: 10–11, and "Digest of Conditions of Phonograph Business in New York City and Brooklyn" [1910], reprinted in Raymond R. Wile, "Market Research—1910 Style," *New Amberola Graphic* 92 [Apr. 1995]: 8–10).

13. Dane Yorke, "The Rise and Fall of the Phonograph," *American Mercury* 27 (Sept. 1932): 5. The full-page, conspicuously placed advertisement that announced Victor's contract with Caruso in 1904 exemplifies Victor's larger marketing strat-

egy, which relied on massive advertising expenditures. According to one account, Victor spent approximately $50 million on print advertisements from 1906 through 1929 and its merger with RCA. (Advertising expenses are reported in Koenigsberg, *Patent History*, lxv.)

In his corporate history of Victor, Benjamin Aldridge characterizes this advertising strategy as "heavy handed": "there was no effort to get maximum results at minimum cost, but rather to use every publication for which there was any justification" (Benjamin L. Aldridge, *The Victor Talking Machine Company* [New York: RCA Sales Corp., 1964]: 49). Victor's magazine advertising frequently occupied particularly conspicuous space, including back covers, color sections, double-page spreads. In addition, the company spent approximately $17 million on its own sales catalogs and brochures.

14. Aldridge, *The Victor Talking Machine Company*, 53.

15. "Both Are Caruso," *Literary Digest* (Jan. 23, 1915), reprinted in Aldridge, *The Victor Talking Machine Company*, 54; "My Victor Records," "Caruso Immortalized," in *National Geographic*, early 1920s (author's collection); "You Hear the Real Caruso," *National Geographic* (Mar. 1920).

16. See, for example, "Rachmaninoff Himself Chose the Victor," *Country Gentleman* 86 (May 21, 1921): back cover.

17. Columbia also stressed, in its 1912 advertising, a wide variety of prestigious performers. "The Columbia Phonograph Company (Exclusively) Presents," *Life* 59 (Feb. 29, 1912): back cover.

18. *Cosmopolitan* (1903), reprinted in Aldridge, *The Victor Talking Machine Company*, 50.

19. "The Greatest Music by the Greatest Artists—*Only on Victrola Records*," *The Independent* 93 (Feb. 23, 1918): inside cover. "The World's Greatest Singers Make Records for the Victor," *The Independent* 86 (Apr. 17, 1916): inside cover.

20. "Will There Be a Victrola in Your Home This Christmas?" *The Independent* 92 (Dec. 15, 1917).

21. For an example of this rhetoric, see William H. Meadowcroft, "The Story of the Phonograph," *St. Nicholas* 19 (May 1922): 692–99; see esp. 692.

22. *Collier's* 42 (Feb. 20, 1909): back cover. Also see "Jenny Lind Is Only a Memory, But the Voice of Melba Can Never Die," *The Independent* 93 (Mar. 23, 1918): inside cover (see fig. 3.2).

23. Jonathan Sterne provocatively links early discourse about the phonograph to the practice and technology of embalming, noting that "hopes that phonography would preserve the voices of the dead were only an extension of a larger, emergent

culture of preservation" (*The Audible Past*, 292). He further notes that because of the *impermanence* of recordings themselves, as objects, preservation became a "program" for sound recording rather than an accomplishment of early phonography (287–311). Announcements of electrical acoustics would also mourn the fact that Caruso had only been recorded on mechanical apparatus and died before *electrical* methods had, according to industry promotional rhetoric, fulfilled the mimetic promise of phonographic devices. Caruso's voice would be resurrected, however, in the electrical recording era as Victor released new versions of Caruso records that mixed mechanical recordings of his voice with contemporary electrical recordings of the Philadelphia Symphony Orchestra.

24. The earliest roadshow performances of phonographic devices emphasized this ability of the phonograph to eclipse spatial limits. Lyman Howe's promotional material invoked this rhetorical figure of entertainment center versus consumers-on-the-margin with Howe as a mediator, precisely the same rhetoric that would underwrite both the rise of network radio and cinema's conversion to sound.

25. *Literary Digest* (Oct. 5, 1918): 42–43.

26. *The Independent* 88 (Nov. 13, 1916): inside cover.

27. "Will There Be a Victrola in Your Home This Christmas?" *The Independent* 88 (Dec. 18, 1916): inside cover.

28. *The Independent* 92 (Nov. 10, 1917): inside cover.

29. Victor's competitors also stressed the pedagogical potential of the phonograph. Edison, for example, offered free to customers such booklets as "What Edison Likes in Music." In keeping with the promotional emphasis on "Edison" as a carefully constructed icon, the booklet featured "an intimate personal interview" with the man who had "probably listened to more music than any other man in the world's history" and included a list of Edison's twenty-five favorite compositions. "What Edison Likes in Music," 21.

30. "The World's Greatest Catalog of Music," *The Independent* 88 (Nov. 20, 1916): inside cover.

31. This mode of address also appears on the back of some Victor single-sided Red Seal recordings. The 12-inch Victor release of "O sole mio" recorded by Emilio de Gogorza (#74105) includes on the reverse side a translation of the lyrics and brief information about both Neapolitan songs and de Gogorza.

32. F. H. Martens, "Making the Music Room Attractive," *The Musician* 19 (Apr. 1914): 229. Quoted in Jessica H. Foy, "The Home Set to Music," in Jessica H. Foy and Karal Ann Marling, eds., *The Arts and the American Home, 1890–1930* (Knoxville: U of TN P, 1994), 69.

33. Jessica Foy's "The Home Set to Music" outlines the larger historical connection between American musical culture in the home and the notion of "uplift" from the 1890s into the 1930s.

34. Craig H. Roell, "The Piano in the American Home," in *The Arts and the American Home, 1890–1930*, 102. In this essay, Roell describes the "entwining of the moral value of music, the work ethic, and the ideology of domesticity." Although acknowledging that the player piano and its mechanical reproduction of music participated in such an entwining, Roell takes little account of simultaneous developments in the phonograph industry. On "musical democracy," see esp. Craig H. Roell, *The Piano in America, 1890–1940* (Chapel Hill: U of North Carolina P, 1989), 31–65, 141–60 passim.

35. Victor also cultivated prestige through the physical location of its devices within the market place. Advertising stressed that the famous Chicago musical firm of Lyon & Healy sold Victor products. Through such retailing agreements with established chains of music stores, Victor, at the level of point-of-sale, linked its phonographs with musical instruments in general and, in particular, a larger musical culture. On the importance of Lyon & Healy and other "old line" music chains like Wurlitzer and Grinnell to Victor's cultivation of its corporate image, see Aldridge, *The Victor Talking Machine Company*, 46.

36. Although Victor's, and to a lesser extent Edison's, advertising campaigns emphasized famous opera and symphonic performers, the sales of more popular music provided most of their income. In the trade paper *Talking Machine World*, Victor admitted that the cultivation of prestige was a practical business approach somewhat at odds with the company's actual sales. In a four-page advertising spread with photographs of its performers, Victor explained, "three show pictures of operatic artists, one shows pictures of popular artists. Three to one—our business is just the other way, and more, too; *but there is good advertising in Grand Opera*" (*Talking Machine World* [Oct. 15, 1905], quoted in Gelatt, *The Fabulous Phonograph*, 141–42; emphasis in original).

37. Victor's production statistics indicate that it took several years for the public to embrace concealed-horn phonographs. In 1907 Victor produced over 98,000 exposed horn models compared to approximately 3500 concealed-horn Victrolas. By 1915, the production numbers had more than reversed with a little less than 4,000 exposed-horn models produced compared to over 374,000 models with concealed horns. Aldridge, *The Victor Talking Machine Company*, 61.

38. Leon Douglass, quoted in Tim Gracyk, "Leon F. Douglass, Inventor and Victor's First Vice President," *Victrola and 78 Journal* 8 (Spring 1996), available at www.garlic.com/tgracyk/leon.htm.

39. The expense of the initial Victrolas (as much as $200) established a pricing hierarchy that quantified the "distinction" associated with the machine. By 1923, Victor's advertisements stressed the availability of a range of instruments with prices varying from $25 to $1,500. This price differentiation indicates that while the Victrola became a prestigious device, such prestige was variously packaged so as to be "universally" available.

40. Koenigsberg, *Patent History*, lxviii–lxix.

41. Yorke, "Rise and Fall," 6. Consumer desire for new cabinet styles helped new firms enter the phonograph business. Sonora, Aeolian-Vocalion, Brunswick-Balke-Collender, and others also benefited from the expiration of key basic patents governing sound technology.

42. Yorke, "Rise and Fall," 10.

43. Yorke, "Rise and Fall," 11.

44. Ibid., 10. Concerns about aesthetically integrating phonographs into the home apparently extended to market segments other than the extremely wealthy. *American Homes and Gardens* suggested in 1914 that, "The case of the talking machine is also an important consideration in its selection, which should be made with a view to its adaptability to the character and decoration of the room in which it is placed." Marie E. Camp, "The Talking Machine in the Home," *American Homes and Gardens* 11 (supplement) (Apr. 1914): vii.

45. Eleanor Hayden, "Phonographs as Art Furniture," *International Studio* 78 (Dec. 1923): 250.

46. Hayden, "Phonographs as Art Furniture," 250, 255. A similar concern with period designs informed the efforts of piano manufacturers in the 1920s.

47. Hayden, "Phonographs as Art Furniture," 256. Another example of Hayden's description should illustrate the variety in these cabinet innovations: "A cabinet designed in the Venetian manner is finished in a gray-green glaze with gold accenting the relief carving, and decorative landscapes have been painted on the doors. The top of this cabinet is a fine example of marbleizing in the Italian manner, and the interior is finished with a parchment toned glaze. This cabinet was executed after designs from the Venetian inspiration as developed in France and is therefore consistent in an interior of Venetian treatment, which now is regarded as most fitting for a town house, and also in a formal interior of Eighteenth Century French decoration" (257).

48. Jessica Foy notes that with the increasing mechanization of home music, "Artistic self-expression was transferred from music making to creating artistic, harmonious physical surroundings for the music" ("The Home Set to Music," 81). Foy's brief remarks on phonographs locate them within a larger tradition of home design in general and a concern with the aesthetics of the music room in particular.

49. "Digest of Conditions of Phonograph Business in New York City and Brooklyn," (1910): 9.

50. A similar emphasis on *both* the technology and the *uses* of that technology characterized consumers' initial encounters with the telegraph. But early promoters generated little public enthusiasm through public performances of telegraphy. Menahem Blondheim concludes, "Unlike the mechanical novelties of the time, nothing in the performance of the device helped make the scientific and technological principles it was based on intelligible, nor was the sight of the apparatus at work enthralling." Menahem Blondheim, "When Bad Things Happen to Good Technologies: Three Phases in the Diffusion and Perception of American Telegraphy," in Yaron Ezrahi, Everett Mendelsohn, and Howard O. Segal, eds., *Technology, Pessimism, and Postmodernism* (Amherst: U of Massachusetts P, 1994), 78.

51. The telephone, circa 1923, exhibited a similar process of effacing the technological apparatus and drawing consumers' attention away from technology per se and instead toward the service it provided. Popular magazines, such as *House and Garden,* counseled their readers on techniques for hiding the telephone. Like later attempts to disguise radios as lamps or globes, methods of disguising the telephone included embedding the device within other objects including globes and even dolls. Much like the innovation of the Victrola, disguising the functioning parts of the apparatus integrated telephone technology into the middle-class home in a manner consistent with prevailing design and decoration standards. See Kenneth Haltman, "Reaching Out to Touch Someone? Reflections on a 1923 Candlestick Telephone," *Technology in Society* 12.3 (1990): 344 and 348.

This impulse to disguise technology extended to such mundane domestic artifacts as radiators. See the advertisement for Tuttle & Bailey Mfg. Co. that described "radiators obtrusively setting about in all their abject utilitarian emphasis." Tuttle & Bailey Ferro-craft grilles obscured radiators, transforming them into pleasing articles of furniture, eliminating their tendency to "distract," "affront," and even "cheapen" a fireplace's effect. The Tuttle & Bailey advertisement appears on page 56 of Kate Roberts, "Fireside Tales to Fireside Chats: The Domestic Hearth," in *The Arts and the American Home, 1890–1930.*

52. Emily Thompson, "Machines, Music, and the Quest for Fidelity: Marketing the Edison Phonograph in America, 1877–1925," *Musical Quarterly* 79.1 (Spring 1995): 142.

53. *New Amberola Graphic* 73 (July 1990): 8.

54. Edison advertisement in *Cosmopolitan* (Jan. 1917), reprinted in Thompson, "Machines, Music," 147.

55. "I Will Pay $10,000 for the Best Thoughts on One of My Problems," 21.

56. Silvertone Records, the product line offered to consumers by Sears, Roebuck and Co., required a turntable speed of 80 revolutions per minute for proper reproduction. While the more expensive phonographs, such as the Victrola, included a speed regulator, allowing the consumer to vary turntable speed by as much as 15 revolutions per minute, the speed indications of such regulators weren't entirely reliable and Sears, Roebuck included on record sleeves directions for verifying turntable speed. Consumers were instructed to periodically test the speed of their machines by slipping a piece of paper under a record so that it would project far enough to brush the consumers hand while revolving and then, with watch in hand, counting the revolutions. The paper brushing against the hand facilitated this. Consumers would then make adjustments to the regulator until the machine operated at the proper 80 rpm.

57. The musicale, a largely nineteenth-century practice that continued in American homes into the twentieth century, provided an established procedure onto which the phonograph might be grafted. On the musicale in twentieth-century domestic culture, see Foy, "The Home Set to Music," 67–68.

58. *Edison Phonograph Monthly* (Nov. 1912), reprinted in Ronald Dethlefson, ed., *Edison Blue Amberol Recordings*, vol. 1: *1912–1914* (Brooklyn: APM Press, 1980), 28.

59. "Entertainment for Your Visitors," F. K. Babson Catalog (circa 1918), reprinted in Ronald Dethlefson, ed., *Edison Blue Amberol Recordings*, vol. 2: *1915–1929* (Brooklyn: APM Press, 1981), 446. The Babson catalog was directed in the first instance to rural customers.

60. "Callers Are Always Delighted," F. K. Babson Catalog (circa 1918), reprinted in Dethlefson, ed., *Edison Blue* 2:447.

61. "Your Farm Hands at Supper," F. K. Babson Catalog (circa 1918), reprinted in Dethlefson, ed., *Edison Blue* 2:135; "Keep the Boys and Girls at Home," F. K. Babson Catalog (circa 1918), reprinted in Dethlefson, ed., *Edison Blue* 2:448.

62. Babson further suggested that this form of home entertaining allowed ethnic groups to laugh at themselves. "No End of Fun and Merriment—A Regular Three-Ring Circus," F. K. Babson Catalog (circa 1918), reprinted in Dethlefson, ed., *Edison Blue*, 1:158–59.

63. "A Wonderful New Phonograph: The Aeolian-Vocalion," *American Magazine* 80 (Oct. 1915): 56–57. Jonathan Sterne notes the tension between fidelity claims (the inaudibility of the machine) and options for users to vary the tone of reproduced sound (*The Audible Past*, 269–70, 274). I see this tension as central to users' relationships to acoustic devices and consequently as a site of technical innovation and discursive struggle characteristic of subsequent sound technologies.

64. "I Went to Buy a Phonograph," *American Magazine* 80 (Dec. 1915): 56–57.

65. "In the Firelight Glow! An Evening Spent with My Aeolian-Vocalion," *American Magazine* 80 (Nov. 1915): 56–57.

66. Pierre Boucheron, "Advertising Radio to the American Public," *The Radio Industry: The Story of Its Development* (New York: A. W. Shaw, 1928; rpt., New York: Arno P, 1974), 265. I specifically use the term *home engineer* to distinguish the associated sets of practices, discourses, technological devices, and listening behavior from those of the radio amateur as described by Susan Douglas in "Amateur Operators and American Broadcasting: Shaping the Future of Radio," in Joseph J. Corn, ed., *Imagining Tomorrow: History, Technology, and the American Future* (Cambridge: MIT Press, 1986), 35–57. Douglas's *Inventing American Broadcasting, 1899–1922* (Baltimore: Johns Hopkins UP, 1987) provides greater detail on the roles played by amateurs in the innovation of wireless. The "amateur" figured prominently in debates about radio in the 1910s and 1920s as a specific identity with respect to electrical acoustics that was variously hailed as democratically heroic and denigrated as a public menace to sea-bound safety. Radio amateurs were included in Hoover's initial radio conferences and in certain provisions of early radio regulation. They formed their own advocacy group (the Amateur Radio Relay League) and were identified as a market segment in the radio patent pool of 1919–20. Radio amateurs' activity in reciprocal wireless communication caused the U.S. Navy to publicly foster the perception that they disrupted commerce and maritime safety. I use the term *home engineer* to refer instead to a set of practices, a mode of address, and an identity subsequent to the redefinition of wireless as broadcasting. The term *home engineer* more specifically references a relationship to the apparatus, a kind of engineering of radio receivers, the goal of which was not reciprocal communication as it was for radio amateurs, but instead the reception of geographically dispersed broadcast signals. The home engineer designed, developed, and perpetually redesigned (most often) his own radio receiver, listening to far-flung signals rather than seeking ethereal communication with like-minded colleagues. Radio amateurs explored the technology through reciprocal communication. Home engineers technologically wandered through the ether, seeking to conquer geographical distance by listening to and logging distant signals. All of which is not to say that the two identities did not in some instances apply to the same individual. Douglas examines what I call "home engineers" in *Listening In: Radio and the American Imagination* (New York: Times Books, 1999), esp. 55–82. Importantly, she links such "exploratory listening" to constructions and enactments of masculinity within the domestic sphere (see esp. 65–68).

67. Boucheron, "Advertising Radio," 266, reprints one of these advertisements.

68. Radio journals during the period include numerous contests involving the reception of distant signals. See, for example, "How Far Have You Heard?" *Radio Broadcast* 2 (Dec. 1922): 114–17. On the desirability of frequent announcements of station identification, see "A Suggestion for Announcers," *Radio Broadcast* 2 (Apr. 1923): 448–49. The article notes that, "One correspondent mentions the fact that many, if not most, radio enthusiasts are more interested in picking up a distant station, even if faint, than in listening to even a good programme from a near-by station." For a period account in which DX and its pleasures are compared to fishing for trout, see Rev. H. F. Huse, "Radio Angling and Fisherman's Luck," *Radio Broadcast* 3 (Aug. 1923): 316–18.

69. Although the political economy of U.S. radio solidified around broadcasting's identity as a domestic entertainment service, the activities of home engineers were never entirely eclipsed. Some consumers continued to build and rebuild their own receiving sets, experiencing radio as primarily the enactment of applied science and only secondarily as the reception of broadcast signals. A journal like *Radio Craft* in 1938 and 1939 continued to include editorial content addressing home radio engineers. Radio manufacturers, retailers, specialized publications, and even spectrum allocations continued (and continue today) to facilitate a marginalized place for this experience of radio.

70. Radio Corporation of America (RCA), *Radio Enters the Home* (New York: RCA, 1922; rpt., Vestal, N.Y.: Vestal Press, 1980), 7.

71. RCA, *Radio Enters*, 5.

72. RCA, *Radio Enters*, 5.

73. Lynn Spigel locates the emergence of first the radio and then the television set within shifting definitions of bourgeois domesticity and its related values. Lynn Spigel, *Make Room for TV: Television and the Family Ideal in Postwar America* (Chicago: U of Chicago P, 1992); see esp. 11–35.

74. Many of these improvements upon phonograph performance were not yet available with RCA's introduction of the Aeriola Grand in 1922. That receiver, for example, provided the ability to control volume only through the removal of one or more of the device's amplifying vacuum tubes. RCA, *Radio Enters*, 29.

75. "5 Days from New York Debut to Chicago Dance Floor!" *American Magazine* (Mar. 1922), reprinted in *New Amberola Graphic* 55 (Winter 1986): 13.

76. In late 1922, one commentator arguing for the skill required to operate a radio set noted that "a vacuum tube is more like a violin than a victrola." W. H. Worrell, "Do Brains or Dollars Operate Your Set?" *Radio Broadcast* 2 (Nov. 1922): 13.

77. Radio innovators frequently announced the commercial results of research and development in both technical journals (such as *Proceedings of the Institute of Radio Engineers*) and more popular radio magazines (*Radio Broadcast* and *Wireless Age*). Some retrospective accounts usefully condense this material, although often assuming a linear evolution: Leslie J. Page, Jr., "The Nature of the Broadcast Receiver and Its Market in the United States from 1922 to 1927," *Journal of Broadcasting* 4.2 (Spring 1960): 174–82; William O. Swinyard, "The Development of the Art of Radio Receiving from the Early 1920's to the Present," *PIRE* 50.5 (May 1962): 793–98; Charles Buff, "Radio Receivers—Past and Present" *PIRE* 50.5 (May 1962): 884–91. Of particular interest is Arthur P. Harrison, Jr., "Single-Control Tuning: An Analysis of an Innovation," *Technology and Culture* 20.2 (Apr. 1979): 296–321. Harrison's essay is a model for future, detailed work on radio receivers as artifacts.

78. Orrin E. Dunlap, Jr., "What Is New in Radio?" *Scientific American* 137 (Nov. 1927): 424–26.

79. Dunlap, "What Is New," 425.

80. Sydney de Brie, "The Call of the Cabinet," *Country Life* 46 (Aug. 1924): 84.

81. Lower prices facilitated greater market penetration and encouraged consumers to purchase multiple sets for a single home or apartment, much like the telephone in the 1920s.

82. *American Magazine* (Oct. 1925): 157 (Irwin Cobb); *American Magazine* (Nov. 1925): 102 (George Ade); *American Magazine* (Dec. 1925): 132 (Booth Tarkington).

83. *American Magazine* (Oct. 1929): 131.

84. Norma B. Kastl, "New Cabinets for the Radio," *House Beautiful* 72 (Oct. 1932): 229.

85. RCA, "Chosen by Victor and Brunswick," *American Magazine* (Oct. 1925): 91. Also see "The Radio Receiver of the Victor Company," *Radio Broadcast* 7 (Aug. 1925): 487–88.

86. "Victor Co. Announces Plans for Radiola Installation in Victor Talking Machines," *Talking Machine World* (June 15, 1925), available at www.sfmuseum.org/hist2/radiola.html. Radio might have entered the home in a number of different ways. Throughout the mid-1920s, radio journals speculated on the possibility of wired rather than over-the-air transmissions of broadcast signals. See for example "Is Wired Wireless the Future of Broadcasting?" *Radio Broadcast* 3 (Oct. 1923): 457. Some engineers within AT&T actively developed wired broadcasting transmission and even demonstrated its possibilities during 1925 in New York City using the power transmission lines of the New York Edison Company as an acoustic transmission medium. Such an arrangement would have allowed AT&T to offer broad-

casting as a service paid for by listeners rather than as an advertising supported medium. "Lloyd Espenschied Interview [June 2, 1973]," IEEE History Center, www. ieee.org/organizations/history_center (accessed Feb. 21, 2000).

Some municipalities and individual apartment building owners set up centralized antennas to receive radio signals that were then distributed to subscribers or individual apartments. See Grayson L. Kirk, "Supplying Broadcasts Like Gas or Electricity," *Radio Broadcast* 3 (May 1923): 35–37, and the photo of a Newark (New Jersey) apartment building in *Radio Broadcast* 3 (June 1923): 99. American publications also reported on the use of coin-operated broadcast receivers in Paris. The so-called "Radio-Automatic," located in cafes, movie theaters, and hotels allowed customers to listen for a fee to broadcasts through attached headphones. See *Radio Broadcast* 3 (Sept. 1923): 364.

87. As innovators redesigned radio receivers to ease their entrance into American homes, they simultaneously pursued a design impulse toward ever greater portability, thereby contributing to the increasing ubiquity of electrically augmented sounds throughout the mid- to late-1920s and into the 1930s. For a detailed account of the development of portable radio receivers, see Michael Brian Schiffer, *The Portable Radio in American Life* (Tucson: U of Arizona P, 1991).

88. Louis Carlat, "'A Cleanser for the Mind': Marketing Radio Receivers for the American Home, 1922–1932," in Roger Horowitz and Arwen Mohun, eds., *His and Hers: Gender, Consumption and Technology* (Charlottesville: U of Virginia P, 1998), 113–37. While Carlat describes the "re-gendering" of the radio receiver and its transformation into a domestic device, he strikingly fails to account for the tremendous influence on this transformation exerted by the prior redefinition of the phonograph into a domestic device. For a period account of "radio-set building" as a gendered activity, see Robert Oliver, "Radio Is Expensive for the Married Man," *Radio Broadcast* 3 (July 1923): 202–04.

89. This section appeared in a different form as Steve J. Wurtzler, "Standardizing Professionalism and Showmanship: The Performance of Motion Picture Projectionists During the Early Sync-Sound Era," in Matthew Tinkcom and Amy Villarejo, eds., *Keyframes: Popular Cinema and Cultural Studies* (London: Routledge, 2001), 359–76. From most perspectives, the "consumers" of early sound films and sound-film technology were, of course, the audiences in newly converted theaters. My concern in this section is with a different social group, a different set of *users* of new technology, namely those required to adapt their professional working lives to technological change. I am suggesting a perspective on the "interface" between technology and user that, while it includes "listening," is not limited merely to lis-

tening. Domestic consumers both listened to mechanical phonographs and radio receivers and, by necessity, operated them. While motion picture projectionists "listened" to reproduced sound, their primary vocational concern was the *interface* with new technology, its operation. That labor, I suggest, had profound implications for the other, numerically larger, group of listeners, namely the film audiences.

90. F. H. Richardson, "Lessons for Operators: Chapter IX. The Picture," *Moving Picture World* (May 9,1908): 413. The Silent Film Bookshelf's Web site provides a useful introduction to and overview of the debates surrounding projection speeds for silent film. Richardson's article is reprinted there as are a number of other period documents; *see* http://cinemaweb.com/silentfilm/bookshelf (accessed May 26, 2000). Interestingly, the Society of Motion Picture Engineers eventually established standards of 60 feet per minute for camera operation and 80 feet per minute for projection operation immediately prior to the industry's conversion to sound, thereby institutionalizing in word if not in deed the disparity between filming and projection speeds.

91. F. H. Richardson, "Projection Department," *Moving Picture World* (Dec. 2, 1911): 721.

92. Some examples of this historical work include Rick Altman, "The Silence of the Silents," *Musical Quarterly* 80.4 (1996): 648–718; Mary Carbine, " 'The Finest Outside the Loop': Motion Picture Exhibition in Chicago's Black Metropolis, 1905–1928," *camera obscura* 23 (May 1990): 9–41; Gregory A. Waller, *Main Street Amusements: Movies and Commercial Entertainment in a Southern City, 1896–1930* (Washington, D.C.: Smithsonian Institution Press, 1995); and Gregory A. Waller, "Hillbilly Music and Will Rogers: Small-town Picture Shows in the 1930s," in Melvyn Stokes and Richard Maltby, eds., *American Movie Audiences: From the Turn of the Century to the Early Sound Era* (London: BFI, 1999), 164–79. Of particular interest in this context is Rick Altman's *Silent Film Sound* (New York: Columbia UP, 2004).

93. Tim Anderson, "Reforming 'Jackass Music': The Problematic Aesthetics of Early American Film Music Accompaniment," *Cinema Journal* 37.1 (1997): 12.

94. William F. Canavan, "Motion Picture Projection," *Projection Engineering* 1.4 (Dec. 1929): 31.

95. Canavan, "Motion Picture Projection," 31–32.

96. Canavan, "Motion Picture Projection," 31.

97. The effect of Hollywood's conversion to sound on the hierarchical division of labor within the film production component of the industry has been documented elsewhere. Historians like James Lastra and Donald Crafton have recently revised the received wisdom that technological change affected Hollywood labor by requir-

ing "quiet on the set" and abruptly terminating the careers of several silent-era stars whose voices were found to be unsuitable to recording. James Lastra, "Standards and Practices: Aesthetic Norm and Technological Innovation in the American Cinema," in Janet Staiger, ed., *The Studio System* (New Brunswick, N.J.: Rutgers UP, 1995), 200–25; James Lastra, *Sound Technology and the American Cinema: Perception, Representation, Modernity* (New York: Columbia UP, 2000), 154–79; Donald Crafton *The Talkies: America's Cinema's Transition to Sound, 1926–1931* (New York: Scribner's, 1999).

98. See, for example, F. H. Richardson, "Why Expert Knowledge and High Grade Intelligence Is Essential in the Theater Projection Room," *Transactions of the SMPE* 11.31 (1927): 500–511. Richardson's concern with the labor of projectionists dated to at least 1908 and the inauguration of his regular column in *Moving Picture World*. As noted above, he was a frequent advocate of controlling silent film projection speeds.

99. J. McGuire, discussion of F. H. Richardson's "Why Expert Knowledge and High Grade Intelligence Is Essential," 508. McGuire strikingly echoes both Richardson's silent-era rhetoric about projection speeds and the arguments of those concerned with reforming nickelodeon-era musical accompaniment. Although it is an idea that goes largely undeveloped in his essay, Tim Anderson notes that one site of intervention in the problem of live accompaniment was the professionalization of the musician (Anderson, "Reforming 'Jackass Music,'" 9, 17–18). Anderson acknowledges that for some exhibitors the desire for professionalism in musical performance clashed with the need to maximize profits, and that more costly, unionized professional musicians were therefore excluded from some early film exhibition. Recorded sound eliminated musicians from theater orchestra pits and thereby offered exhibitors a financial incentive to embrace technological change.

100. F. H. Richardson, "The Importance of Good Projection to the Producer," *Transactions of the SMPE* 12.34 (1928): 360–61.

101. Richardson, "The Importance," 359–60. Anderson also cites at least one letter to *Moving Picture World* that addressed the lack of praise and appreciation received by appropriately disciplined nickelodeon accompanists (Anderson, "Reforming 'Jackass Music,'" 15). Again, many of the issues surrounding concerns about film projection in the early sync-sound era echo similar issues surrounding live accompaniment in the nickelodeon era.

102. Crafton notes that the conversion to sound, in the short run at least, allowed projectionists to demand increases in wages and additional staff in projection booths. Crafton, *The Talkies*, 217–18.

103. H. Rubin, "Some Problems in the Projection of Sound Movies," *Transactions of the SMPE* 12.35 (1928): 867–71.

104. Much like advocates of professional standards for projectionists some twenty years later, nickelodeon music reformers also cited the necessity of rehearsal. See Anderson, "Reforming 'Jackass Music,'" 14. Reel changes required projectionists to shift from one projector to another in order to guarantee the continuous presentation of a film. A successful reel change dissimulated the labor of the projectionist, rendering invisible and inaudible the very fact that a shift from one projector to another had taken place. Idiosyncratic marking mechanisms such as the punching of holes in a film (the consequent flash of white light on the screen could cue the projectionist when to start the next projector and prepare for the changeover) not only undermined the industry goal of standardization but also damaged the film print that would be shipped from one venue to the next. Without standardized methods of indicating changeovers, film prints would bear a physical trace of one projectionist's labor, carrying that trace to subsequent projection rooms and film venues.

105. Warren Nolan, "Talking Pictures and the Public," *Transactions of the SMPE* 13.37 (1929): 132–33.

106. Harold B. Franklin, *Sound Motion Pictures: From the Laboratory to Their Presentation* (Garden City, N.Y.: Doubleday, Doran, 1929), 361.

107. James P. Kraft provides a useful account of the conversion to sound from the perspective of musicians and film industry employment practices. See James P. Kraft, *Stage to Studio: Musicians and the Sound Revolution, 1890–1950* (Baltimore: Johns Hopkins UP, 1996). See esp. ch. 2, "Boom and Bust in Early Movie Theaters," 33–58.

108. E.C. Wente, "Contributions of Telephone Research to Sound Pictures," *Journal of the SMPE* 27 (Aug. 1936): 188–94.

109. I.W. Green and J.P. Maxfield, "Public Address Systems," *Bell System Technical Journal* 2 (1923): 138–39.

110. Coke Flannagan, "Servicing Sound Picture Projection Equipment in Theatres," *Transactions of the SMPE* 13.38 (1929): 294.

111. S.K. Wolf, "The Western Electric Reproducing System," in Lester Cowan, ed., *Recording Sound for Motion Pictures* (New York: McGraw-Hill, 1931), 301.

112. R.H. McCullough, "Practices and Problems of Sound Projection," in Cowan, ed., *Recording Sound*, 330.

113. McCullough, "Practices and Problems," 331.

114. I am grateful to Martin Barnier for providing me with copies of the Projectionist Cue Sheet for *Warming Up*. Barnier apparently obtained the docu-

ment at the Film Study Center, Museum of Modern Art, New York ("Sound Files 1927–1935"). The Cue Sheet is appended to a memo from M. H. Lewis entitled "Additional Instructions Regarding Sound Pictures and Discs," July 26, 1928.

115. F. H. Richardson, "The Effect of Sound Synchronization Upon Projection," *Transactions of the SMPE* 12.35 (1928): 874.

116. Flannagan, "Servicing Sound," 294.

117. "Volume Too Loud," *Projection Engineering* 2.10 (Oct. 1930): 28.

118. Flannagan, "Servicing Sound," 294; J. Garrick Eisenberg, "Technique of Sound-Picture Projection," *Projection Engineering* 1.4 (Dec. 1929): 25.

119. Haviland Wessells, "Sound—As the Customers Hear It," *Projection Engineering* 1.3 (Nov. 1929): 13. Here, volume variations echo the rationale for variations in projection speed during the silent era, a practice that drew similar scorn from the production end of the industry. In 1920 even Richardson admitted that "there are occasional exceptions where a scene may actually be improved by moderate overspeeding of projection. Such scenes are, however, rare. As a rule they are those where speeding automobiles are involved, with no animate figures other than those in the machines. Such scenes merely form the exception which proves the rule." F. H. Richardson "The Various Effects of Over-Speeding Projection," *Transactions of the SMPE* 10 (1920), reprinted on the Silent Film Bookshop's Web site (see note 90).

120. Flannagan, "Servicing Sound," 295.

121. Samson Qualpensator, advertisement, *Projection Engineering* 2.4 (Apr. 1930): 1.

122. John O. Aalberg, "Theater Reproduction by RCA Photophone System," in Cowan, ed., *Recording Sound*, 311.

123. Hardwick, Hindle Volume Faders, advertisement, *Projection Engineering* 2.1 (Jan. 1930): 35.

124. Eisenberg, "Technique of Sound-Picture Projection."

125. Eisenberg's presumption that a cut to a close-up should be accompanied by a concomitant increase in volume was not universally shared throughout the film industry. In fact, during this period representational issues like the relationship between shot scale and sound volume were the subject of intense debate within the film industry. For analyses of these debates, see Rick Altman, "Sound Space," in Altman, ed., *Sound Theory/Sound Practice* (New York: Routledge, 1992), 46–64, and James Lastra, "Reading, Writing, and Representing Sound," in Altman, ed., *Sound Theory/Sound Practice*, 75–78. Also see Lastra, *Sound Technology and the American Cinema*, esp. 180–215. I briefly discuss this debate among Hollywood sound practitioners in "'She Sang Live, But the Microphone Was Turned Off': The Live, the

Recorded, and the *Subject* of Representation," in Altman, ed., *Sound Theory/Sound Practice*, 96–99, and revisit this issue in chapter 5.

126. Eisenberg, "Technique of Sound-Picture Projection," 24.

127. Ibid., 22. Similar associations between projection and artistry also characterized film industry rhetoric during the projection speed debates: "The operator 'renders' a film, if he is a real operator, exactly as does the musician render a piece of music, in that, within limits, the action of the scene being portrayed depends entirely on his judgement" (Richardson, "Projection Department," 721–22).

128. See also J. Garrick Eisenberg, "Mechanics of the Talking Movie," *Projection Engineering* 1.3 (Nov. 1929): 22–24.

129. Eisenberg, "Mechanics of the Talking Movie," 24.

130. Donald McNicol, "Editorial: Picture Rehearsals in the Theater," *Projection Engineering* 2.12 (Dec. 1930): 4.

4. Making Sound Media Meaningful: Commerce, Culture, Politics

1. Kate Lacey, "Radio in the Great Depression: Promotional Culture, Public Service, and Propaganda," in Michele Hilmes and Jason Loviglio, eds., *The Radio Reader: Essays in the Cultural History of Radio* (New York: Routledge, 2002), 28–29.

2. Lacey, "Radio in the Great Depression," 29.

3. U.S., Committee on Public Information, *Complete Report of the Chairman of the Committee on Public Information* (Washington, D.C.: GPO, 1920; rpt., New York: Da Capo Press, 1972), 60.

4. Details on these film-related activities can be found in the Creel Committee's final report: domestic income, p. 7; international distribution, p. 103; short subjects, pp. 57–59; newsreel footage, p. 53; theater curtains, p. 44. For a brief account of the film industry's relationship with the Creel Committee, see Steven J. Ross, *Working-Class Hollywood: Silent Film and the Shaping of Class in America* (Princeton: Princeton UP, 1998), 123–27. Also see Steven Smith, "Personalities in the Crowd: The Idea of the 'Masses' in American Popular Culture," *Prospects* 19 (1994), esp. 229–46.

5. U.S., Committee on Public Information, *Complete Report*, 29.

6. The "service" provided by the film industry in improving musical standards, of course, elided the financial savings afforded to theater chains and the consequent sudden unemployment of thousands of local musicians.

7. Hays's remarks in the Vitaphone short are transcribed in Will H. Hays, *See and Hear: A Brief History of Motion Pictures and the Development of Sound* (New York: MPPDA, 1929), 48–49.

8. "Help 'Em Laugh!" *Exhibitors Herald-World* (Dec. 18, 1926): 37.

9. Elliot N. Sivowitch, "A Technological Survey of Broadcasting's 'Pre-History,' 1876–1920," *Journal of Broadcasting* 15.1 (Winter 1970–71): 1–20.

10. For a description of DX listening in 1925 and a discussion of the practice of "silent nights," see Kingsley Welles, "Do We Need 'Silent Nights' for Radio Stations?" *Radio Broadcast* 7 (Oct. 1925): 751–54. Welles notes that "by a kind of gentleman's agreement during the past year, Chicago broadcasters have kept off the air on Monday nights to allow the natives to indulge their urge for distance" (753). Welles also describes an effort by the Broadcast Listener's Association to have a bill introduced into Congress that would divide the nation into six broadcast zones and assign to each an evening in which broadcasting was suspended for the sake of local DX listeners (753). Finally, Welles articulates the tension between two definitions of radio in describing a clash between those who sought programming and those who sought primarily the conquest of distance: "If radio really is entertainment, then the program is the important thing. If that is true, the source matters less than, in a manner of speaking, the bone and sinew of the program itself. But, on the other hand, if radio reception is a kind of elaborate animated geography lesson, then every effort ought to be bent to give those devotees the chance to hear distant names whisked through the microphone" (754).

11. Leland L. Smith, "An SOS in the Jungle of Indo-China," *Radio Broadcast* 3 (May 1923): 42–51; Burnham McLeary, "In Touch with the World from the Arctic," *Radio Broadcast* 3 (Aug. 1923): 282–88 (on Donald B. MacMillan's expedition to the Arctic and the role radio was expected to play in keeping the expedition entertained by relieving their isolation and allowing them weekly coded communication back to civilization). MacMillan's various expeditions were frequently described from the so-called "radio angle": E. F. McDonald, Jr., "With MacMillan and Radio, North of Civilization," *Radio Broadcast* 3 (Oct. 1923): 500–507; and Austin C. Cooley, "With MacMillan to the Arctic," *Radio Broadcast* 10 (Apr. 1927): 551–54.

Both DX listening and the definition of radio as the conquest of distance still had their advocates in late 1927. In an article celebrating the ongoing advocates of DX, John Wallace wrote, "In short, wired radio and, to a lesser degree, chain broadcasting, summarily renounce the fundamental principle upon which radio was founded—space annihilation" (John Wallace, "The DX Listener Finds a Champion," *Radio Broadcast* 12 [Dec. 1927]: 140). Wallace went on to claim that, "Radio's unique contribution to what is drolly referred to as the 'progress' of civilization is the conquering of space" (140).

12. Accounts of a broadcast by President Harding on June 21, 1923, stressed radio's ability to provide a national mode of address. "President Harding Over Wire and Radio," *Radio Broadcast* 3 (Sept. 1923): 359–60. For a representative statement advocating the broadcast of congressional proceedings, see "Why Does Congress Refuse to Broadcast Its Proceedings?" *Radio Broadcast* 7 (June 1925): 198–99.

13. The increasing hegemony of this identity for broadcasting influenced research initiatives as corporations sought through various means to enable simultaneous national broadcasts. AT&T, not surprisingly, advocated the use of augmented long-distance telephone lines to link geographically dispersed stations, while Westinghouse experimented with short-wave transmissions to form a network of broadcasters. RCA, on the other hand, advocated the use of increasingly high-powered transmitters to enlarge the audience of any given station. In her history of radio, Susan Smulyan describes these different technological efforts to provide national radio coverage. Susan Smulyan, *Selling Radio: The Commercialization of American Broadcasting, 1920–1934* (Washington, D.C.: Smithsonian Institution Press, 1994), 42–57.

For accounts of Westinghouse's experiments with short-wave transmissions, see W. W. Rodgers, "Is Short-wave Relaying a Step Toward National Broadcasting Stations?" *Radio Broadcast* 3 (June 1923): 119–22; and "The Possibility of Re-Broadcasting," *Radio Broadcast* 3 (July 1923): 187–88. Also see Steven P. Phipps, "The Commercial Development of Short Wave Radio in the United States, 1920–1926," *Historical Journal of Film, Radio, and Television* 11 (1991): 215–27. In 1925, radio engineer Carl Dreher, a frequent contributor to radio journals like *Radio Broadcast*, advocated the use of higher-powered transmitters to improve the distance of transmission without static; see Carl Dreher, "Radio Power and Noise Level," *Radio Broadcast* 7 (Sept. 1925): 615–17.

14. For a period account of network radio and such a national experience, see John Wallace's 1927 description of NBC's broadcasts of Memorial Day Ceremonies at Arlington National Cemetery and the celebration of Lindbergh's return (John Wallace, "Where Broadcasting Reigns Supreme," *Radio Broadcast* 11 [Sept. 1927]: 287). One account of the Dempsey-Tunney boxing match of September 1926 speculated that the network of over thirty broadcasting stations addressed the largest audience of any broadcast to that time, and thus "the largest audience in the history of the world" (John Wallace, "The Dempsey-Tunney Fight," *Radio Broadcast* 10 [Dec. 1926]: 161–62). Another account suggested that radio's national mode of address might not always be put to good use ("The Dempsey-Tunney Fight—A Dangerous Precedent," *Radio Broadcast* 10 [Dec. 1926]: 149). The "danger" was that a commer-

cial concern, the Royal Typewriter Company, paid a reported $35,000 for the rights to sponsor the broadcast of the fight. Wallace, on the other hand, characterized the commercialism of the broadcast as "well-handled," noting that although the name of the sponsor was mentioned multiple times the fact that the identity of the sponsor was never mentioned during the 45-minute fight demonstrated a "surprising and commendable exhibition of restraint" (Wallace, "The Dempsey-Tunney Fight," 161–62).

15. During 1925, some commentators suggested that radio advertising improved the quality of programming; however, such advertising took the form of little more than sponsorship, or rather the identification of an organization as a program sponsor. James C. Young, "New Fashions in Radio Programs," *Radio Broadcast* 7.1 (May 1925): 83–89. "Indirect advertising" or, in a phrase of the day, "selling by publicity," stopped short of overt, direct-address sales pitches. While Young credited the practice with improving programming quality, he also raised fears that the ether might be tainted by unchecked commercialism: "We may conceive of a new expedition to the pole so that the explorer shall describe to us how comfortable he was in some particular brand of knit underwear, while he drank a special blend of tea and munched upon a soda cracker of national reputation" (89).

16. Arguably, the political democracy facilitated by radio was not all that different from the consumer democracy proffered by national advertisers. Both offered merely the illusion of choice and both addressed listeners as consumers rather than active participants in politics or culture. This absence of actual and existing participation was successfully elided through various orchestrated rituals of participatory behavior. Even knowledgeable and informed political participation that only led up to and stopped at voting was, as a type of participation, every bit as theatrical as the deliberation that led to the selection of a brand of detergent or household coal.

17. Jeff Land, *Active Radio: Pacifica's Brash Experiment* (Minneapolis: U of Minnesota P, 1999), 22.

18. "Cinemadvertising," *Time* 17 (May 25, 1931): 58; and Marsh K. Powers, "How Far Can Commercial Sponsorship Be Extended?" *Printers' Ink* (Apr. 30, 1931): 49–52.

19. Darwin Teilhet, "Propaganda Stealing the Movies," *Outlook and Independent* 158 (May 1931): 126. Jeanne Allen briefly mentioned sponsored films in her 1980 essay on interwar cinema's relationship to consumer culture, linking the origin of these films to "financially hard-pressed studios" and their termination to the public's objections (Jeanne Allen, "The Film Viewer as Consumer," *Quarterly Review of Film Studies* 5 [1980]: 489–90). Charles Eckert dealt with sponsored films more thoroughly and located these films within a tradition of cinematic product placement in

the 1910s. While I do not disagree with this interpretation, I see the sponsored film experiment as an intermedia phenomenon in which the film and radio industries similarly sought to navigate the relationship between entertainment and a broader consumer culture. Interestingly, Eckert characterized the commercial relations linking Hollywood, radio, and advertisers by the late 1930s as "a kind of symbiosis which blurred the outlines of both media" (19). Charles Eckert, "The Carole Lombard in Macy's Window," *Quarterly Review of Film Studies* 3 (1978): 1–21.

20. "Talkies Adopt Radio Methods in New Sponsored Programs," *Business Week* (July 30, 1930): 8.

21. Warner Bros. Industrial Pictures, Inc., "If the World Is Your Market We'll Tell the World," *Fortune* (Nov. 1930): 142.

22. "$5 Per Thousand Admissions Is Plan for Warner Industrial Shorts," *Exhibitors Herald-World* (Dec. 6, 1930): 18.

23. Peter B. Andrews, "The 'Sponsored' Movie," *Advertising & Selling* (Dec. 10, 1930): 38. Two period accounts of sponsored films even suggested that additional theater acquisitions by the major studios in 1929 and 1930 should be understood in light of the commercial possibilities offered by sponsored films (Andrews, "The 'Sponsored' Movie," 38; and Teilhet, "Propaganda," 126). Some advertising and business commentators viewed these theater acquisitions as enhancements to the movie studios' "networks" that would facilitate an ever wider, more pervasive national mode of commercial address.

24. "Cinemadvertising," 58; and "Business Anxiously Watches Reaction to 'Sponsored' Shorts," *Business Week* (Apr. 8, 1931): 14.

25. "Business Anxiously," 14.

26. RKO would eventually release a series of twelve short subjects produced by General Electric called "The House of Magic"—although announcements of the series were careful to indicate that the films should be classified with the nonadvertising shorts of other producers. "General Electric in Movie Production," *Projection Engineering* 3.2 (Feb. 1931): 22.

27. In a 1928 article Paul Bliss outlined procedures for successfully incorporating films into community fund-raising efforts. He cited examples from Minneapolis and St. Louis where such fund-raising films received local theatrical distribution. Paul S. Bliss, "Making and Showing a Campaign Movie," *The Survey* 59 (Mar. 15, 1928): 787–88; see also "Selling with Sound Effects," *Business Week* (Oct. 26, 1929): 37.

Studebaker's two films submerged promotion of their product line in staged sporting events: one depicted a race up Pike's Peak and the other a 30,000-mile race between four Studebaker stock models. D. G. Baird, "The Talking Picture—An Important New Sales Tool," *Sales Management* 21 (Feb. 1930): 320.

28. "Talkies Adopt Radio Methods," 8.

29. Teilhet, "Propaganda," 113.

30. While sponsored films were particularly suited to vertically integrated film companies, organizations such as Theatre Service, Metropolitan, Visugraphic, and Pathé began in 1931 to solicit the participation of independent theaters in the sponsored film experiment. They contracted with independent theaters and small regional chains to create larger releasing associations, in the words of one period commentator, "very much like the Columbia Broadcasting Company, and, to a lesser extent, the N.B.C., have done with strategically located independent broadcasting stations" (Teilhet, "Propaganda," 126).

31. "Study Conducted on Use of Films for Advertising," *Projection Engineering* 3.2 (Feb. 1931): 14.

32. Andrews, "The 'Sponsored' Movie," 36.

33. Teilhet, "Propaganda," 126.

34. Importantly, Universal did not own a theater chain and thus was precluded from competing directly with those who did. W. D. Canaday, "How Lehn & Fink Are Using Talking Picture 'Shorts,'" *Printers' Ink* (Mar. 19,1931): 17 and 19.

35. "Screen Advertising Banned; Mr. Schenk Points Out Evils," *The Loew-Down* 2.2 (May 25, 1931): 1.

36. Canaday, "How Lehn & Fink," 17.

37. Ibid., 18.

38. *Motion Picture Herald* (Mar. 14, 1931): 71.

39. *Motion Picture Herald* (Feb. 21, 1931): 47.

40. Andrews, "The 'Sponsored' Movie," 36.

41. Teilhet, "Propaganda," 112.

42. Powers, "How Far?" 50.

43. Today, in a world in which corporations bid to apply their names to sports stadiums—until, of course, national scandal (Enron) causes that name to be removed and quickly replaced by the brand of a different corporation (Minute-Maid)—in which the halftime, pregame, and postgame segments of broadcast sporting events and even individual statistics each bear a different corporate logo, Powers' suggestion seems particularly resonant. For a discussion of these issues today, see Naomi Klein, *No Logo: Taking Aim at Brand Bullies* (New York: Picador, 1999).

44. "Feist Arranges Chesterfield Plug for Latest Song," *Motion Picture Herald* (Jan. 31, 1931): 69.

45. "Newspapers Put an End to Advertising Movies," *Business Week* 96 (July 1931): 22–23.

46. With the end of the sponsored film experiment, commercially oriented films that promoted national products did not entirely disappear from American theaters. Although largely displaced to public venues like World's Fairs and the semipublic realm of club meetings and schools, commercial films intermittently appeared in local theaters. Manufacturers continued to produce promotional films that included a commercial message within entertainment (or often educational) short subjects. For example, in 1936 the Norton Company, a manufacturer of abrasives, produced the 18-minute "The Alchemist's Hourglass." Designed to be shown as part of regular programs in movie theaters, the film traced the history of abrasives from medieval times to the present. "Norton's Experience with Motion Picture Advertising," *Industrial Marketing* 21 (Mar. 1936): 17–18, 40.

But without the involvement of the vertically integrated major studios, distribution of these films became irregular. Some organizations did offer to advertisers and to manufacturers dispersed "chains" of locally owned theaters, and for a time in the mid-1930s ERPI offered a nontheatrical "chain" of various community groups, clubs, and lodges that classified audiences into seventeen categories for either local or regional exhibition. But the possibility of centralized distribution on a national scale that linked a "network" of movie houses disappeared with the sponsored film experiment. "Movie Merchandising," *Business Week* (June 15, 1935): 18–20.

47. In 1934, Richard H. Grant, vice president of General Motors, concisely summarized the philosophy of such product placement: "If you blatantly put up your product with a great big trade-mark on it, the picture man doesn't want the business and the public will be bored, but you can put your car into such beautiful surroundings that you get the interest of the public in looking at the surroundings and incidentally get in the advertising of your car. You cannot do a thing like that unless you submerge your product. The main thing in a public place must be entertainment." Richard H. Grant, "How G-M Uses Motion Pictures," *Printers' Ink* (Feb. 1934): 19.

48. "Films Now Seeking Manufacturers' Products for Use as 'Props,'" *Sales Management* 33 (Oct. 1, 1933): 316.

49. "National Advertisers Tie Up with Warner Feature Films; Quaker Oats Latest to Try It," *Sales Management* 35 (Sept. 1, 1934): 195. In the article, Einfeld noted that the studio only pursued cooperative arrangements with "important nationally advertised products," so that Warner Bros. would avoid "the mass of small advertisers

who for years have capitalized on the endorsement of movie stars with ultimate injury to both stars and the motion picture business."

50. "National Advertisers," 195.

51. Will H. Hays, "The Movies Are Helping America," *Good Housekeeping* (Jan. 1933): 44–45. Page numbers for subsequent references to this article are given in the text.

52. Bruce Lenthall, "Critical Reception: Public Intellectuals Decry Depression-era Radio, Mass Culture, and Modern America," in Hilmes and Loviglio, eds., *The Radio Reader*, 41–61.

53. William S. Paley, *Radio as a Cultural Force* (New York: Columbia Broadcasting System, 1934), 1.

54. Paley, *Radio*, 14.

55. Ibid., 15.

56. Susan Hegeman, *Patterns for America: Modernism and the Concept of Culture* (Princeton: Princeton UP, 1999), 22.

57. Hegeman, *Patterns*, 23.

58. Henry Jenkins, "'Shall We Make It for New York or for Distribution?'": Eddie Cantor, *Whoopie*, and Regional Resistance to the Talkies," *Cinema Journal* 29.3 (Spring 1990): 32–52; Henry Jenkins, *What Made Pistachio Nuts? Early Sound Comedy and the Vaudeville Aesthetic* (New York: Columbia UP, 1992). Unfortunately, Jenkins fails to address the role played by radio in this process.

59. The farmer also figured prominently in assertions regarding radio's capacity to disseminate news instantaneously. Here, radio broadcasting could provide crucial information (both weather and, importantly, market reports) so as to facilitate the farmer's maximization of profits. Radio's transformation of rural isolation not only promised uplift and edification but also the more efficient interpolation of rural producers into the market economy. See, for example, Christine Frederick, "Exit the Jonas Hayseed of 1880," *Wireless Age* 12.2 (Nov. 1924): 34–35.

60. Frank Biocca tracks this shift in the publicly expressed attitudes of the United States' musical elites with regards to radio. Frank Biocca, "Media and Perceptual Shifts: Early Radio and the Clash of Musical Cultures," *Journal of Popular Culture* 24.2 (Fall 1990): 1–15.

61. Biocca, "Media and Perceptual Shifts," 9.

62. Ibid., 11.

63. Levine provides an overview of this history. Lawrence W. Levine, *Highbrow/Lowbrow: The Emergence of Cultural Hierarchy in America* (Cambridge: Harvard UP, 1988).

64. Lawrence W. Levine, "Jazz and American Culture," *The Unpredictable Past: Explorations in American Cultural History* (New York: Oxford UP, 1993), 174–75.

65. Levine notes that critiques of jazz appropriations of preexisting musical forms extended beyond concerns regarding the Western classical canon and included African-American leaders who decried jazz versions of spirituals. Levine, "Jazz and American Culture," 180.

66. Jennie Irene Mix, "How Classical Music Should Be Played," *Radio Broadcast* 7.1 (May 1925): 67.

67. Jennie Irene Mix, "What Happened at WTAM," *Radio Broadcast* 7.1 (May 1925): 66–67.

68. The inaugural program also included two Vitaphone performances by the New York Philharmonic Orchestra and short subjects with concert violinist Mischa Elman and a duet featuring violinist Efrem Zimbalist and pianist Harold Bauer performing Beethoven. The only "popular" performer represented in the initial Vitaphone shorts program was Roy Smeck playing "Hawaiian" guitar and ukelele.

69. With release 802 (recorded April 1929), the Warner's catalog began to include these categories. Much of what follows is based on two specific sources: a copy of the "Vitaphone Release Index" that I obtained from Rick Altman, the original for which apparently came from the Academy of Motion Picture Arts and Sciences Library—the catalog was a six-ring binder into which exhibitors could add new pages for each subsequent installment of Vitaphone shorts—and Roy Liebman, *Vitaphone Films: A Catalogue of the Features and Shorts* (Jefferson, N.C.: McFarland, 2003). Recording dates for the Vitaphone films are estimates from Liebman's book.

70. "Vitaphone Release Index," 234.

71. This was in fact Martinelli's final Vitaphone short. "Vitaphone Release Index," 240, 264, 282, and 309.

72. Charles Wolfe, "On the Track of the Vitaphone Short," in Mary Lea Bandy, ed., *The Dawn of Sound* (New York: Museum of Modern Art, 1989), 37.

73. See note 36, chapter 3 (this volume).

74. Wolfe, "On the Track," 38.

75. Short-subject series, of course, had the added economic benefit of more or less guaranteeing distribution of subsequent installments once the initial installments were booked—a process that could be enforced through contractual arrangements with exhibitors.

76. In her book *The Making of Middlebrow Culture*, Joan Shelley Rubin describes two notable spokesmen for Culture in the early 1930s who blurred

the distinction between culture and entertainment. In radio shows ostensibly about books, former Yale professor William Lyon "Billie" Phelps and bon vivant Alexander Woolcott (called by one critic an "intellectual crooner") combined variety show formats with often-anecdotal discussions of ideas. Adopting a mode of address that appropriated radio's apparent intimacy, their shows advocated less the perpetuation of culture than the cultivation of personality. Joan Shelley Rubin, *The Making of Middlebrow Culture* (Chapel Hill: U of North Carolina P, 1992), 266–329.

77. Jennie Irene Mix, "Good National Radio Programs Prove 'What the Public Wants,'" *Radio Broadcast* 7.1 (May 1925): 62.

78. Mix, "Good National Radio Programs," 64.

79. Young, "New Fashions," 83–89.

80. John Wallace, "Radio Is Doing a Good Job in Music," *Radio Broadcast* 11.4 (Aug. 1927): 218.

81. "The Advance in Radio," *Etude* 44.11 (Nov. 1926): 1.

82. Whiteman's own account of the Aeolian Hall performance appears in his autobiography. Paul Whiteman and Mary Margaret McBride, *Jazz* (New York: J. H. Sears, 1926; rpt., New York: Arno, 1974), 87–111.

83. Gerald Early contextualizes the concert and locates Whiteman's career both in terms of his personal class anxieties and the racial politics of the United States during the early twentieth century. Gerald Early, "Pulp and Circumstances: The Story of Jazz in High Places," *The Culture of Bruising: Essays on Prizefighting, Literature, and Modern American Culture* (Hopewell, N.J.: Ecco Press, 1994), 163–205.

84. Whiteman, *Jazz*, 153.

85. Ibid., 151.

86. Early, "Pulp and Circumstances," 178–79.

87. Susan Hegeman describes a process one decade later that is strikingly similar to the discursive construction of Paul Whiteman. Hegeman's analysis of painter Thomas Hart Benton describes the crafting of a regional identity, linked to middlebrow culture, in which the painter's aesthetic ("antielitist, antimodernist, decentralized, and politically and socially conservative") provided an alternative opposed to modernist aesthetics (Hegeman, *Patterns*, 138). Rejecting modernist art and the political left, Benton's brand of regionalism posited an American art practice (conceptually and spatially aligned with the Midwest) in opposition to "foreigness, elitism and charlatanism" (ibid., 143). In Hegeman's words, "the middlebrow position operated centrally against 'foreigness'—a quality that could be variously inhabited by, for example, blacks, Jews, and intellectuals" (145).

Paul Whiteman's attempted synthesis of symphonic music and jazz might be viewed similarly in that it too attempted to create a middlebrow alternative to both a perceived elitist European (hence, foreign) high-culture tradition and a so-called "debased" (that is, African-American) popular tradition. That amalgam was then hailed as uniquely "American" in character. Although less spatially aligned to a specific geographic region (as Benton's work would be a decade later), Whiteman's assertion of a uniquely "American" music served and hence reinforced a growing perceived need for a shared national acoustic culture that accompanied the new media of sound.

88. "Music Land" (1935), *Silly Symphonies* (Walt Disney Home Entertainment DVD 23087).

89. Debates surrounding Culture often involved the issue of national identity and sometimes revealed a sense of American inferiority that might be overcome through emerging cultural forms. Levine notes that Culture was often perceived to be Eurocentric with the United States lagging behind and needing to emulate the enobled products of the Old World (Levine, "Jazz and American Culture," 175–77). Within this context, jazz could be celebrated as a native-born mode of American expression (ibid., 180).

90. Hugo Riesenfeld, "Music as a Vocation for Americans," *The Outlook* 138 (Oct. 29, 1924): 331.

91. Whiteman, *Jazz*: "American essence" (132–33), "soul of America" (132), "American character" (118), "American noise" (267), "America's first cultural contribution" (275).

92. U.S., Committee on Public Information, *Complete Report*, 7–8.

93. J.M. McKibbin, Jr., "The New Way to Make Americans," *Radio Broadcast* 2 (Jan. 1923): 238–39. McKibbin's views would be echoed nearly a decade later by Arthur Garbett, director of Education for the Pacific Division of NBC. Arthur S. Garbett, "Radio and Musical Education," *Etude* 50.3 (Mar. 1932): 165–66. Also see Frederick, "Exit the Jonas Hayseed," 60.

94. McKibbin, "The New Way," 239.

95. Robert Olyphant, "Radio in the Work of Americanization," *Wireless Age* 9.10 (July 1922): 27.

96. A posited link between listening and assimilation was not an entirely new development. Before the innovation of electrical methods, the phonograph had been enlisted to "solve the immigrant problem." In 1920, Victor published a booklet extolling the ability of sound recording to foster assimilation. *The Victrola in*

Americanization (Camden, N.J.: Educational Dept., Victor Talking Machine Co., 1920).

97. While local Chicago radio stations did present black jazz and blues through live broadcasts from nightclubs, African-Americans in the city did not participate in their own aural representation, particularly when compared to other local groups. Derek W. Vaillant, "Sounds of Whiteness: Local Radio, Racial Formation, and Public Culture in Chicago, 1921–1935," *American Quarterly* 54.1 (Mar. 2002): 25–66.

98. Vaillant, "Sounds of Whiteness," 26.

99. George Sanchez describes the important role played by music during this period in maintaining an ethnic identity among Mexican immigrants in Los Angeles. Sanchez suggests that large phonograph companies profited from this process, exploiting a niche market with the help of several local entrepreneurs who functioned as cultural brokers. In radio, some national products (he cites Folgers Coffee) used radio to address Spanish-speaking consumers, although local businesses were the predominant sponsors of Spanish-language radio programming. After an increase in Spanish-language broadcasts in the late 1920s, Sanchez describes how many stations curtailed this programming in the early 1930s. George J. Sanchez, "Familiar Sounds of Change: Music and the Growth of Mass Culture," *Becoming Mexican American: Ethnicity, Culture, and Identity in Chicano Los Angeles, 1900–1945* (New York: Oxford UP, 1993), 171–87.

100. Vaillant, "Sounds of Whiteness," 29. Vaillant productively expands upon Liz Cohen's study of Chicago's working classes and their use of consumption, particularly listening to the radio, to both maintain existing ethnic, religious, and community identities while also identifying common interests across such lines (Lizabeth Cohen, *Making a New Deal: Industrial Workers in Chicago, 1919–1939* [New York: Cambridge UP, 1990]). Michele Hilmes stresses the primacy of radio as an instrument of cultural assimilation in the United States, suggesting that "early radio took on the problem of cultural assimilation and diversity head on, in the form of programs dealing directly with both *difference* . . . and with the delineation of the *norm*" (313). Michele Hilmes, "Invisible Men: *Amos 'n' Andy* and the Roots of Broadcast Discourse," *Critical Studies in Mass Communication* 10 (1993): 301–21.

101. Program Schedule, *WCFL Radio Magazine* 2.3 (Summer 1929): 72–73.

102. WCFL's Log, *WCFL Radio Magazine* 3.2 (Spring 1930): 64–66. At the time, WCFL's studios were located in the Brunswick-Balke-Collender building, and I suspect that much of the ethnic musical programming originated from Brunswick's phonograph recordings.

103. Nathan Godfried, *WCFL: Chicago's Voice of Labor, 1926–1978* (Urbana: U of Illinois P, 1997), 113–16, esp. 114. In addition to its ethnic and labor-related programming, WCFL also addressed rural listeners within the range of its signal from a progressive, leftist perspective that was otherwise absent from the commercial and cultural assimilationist mode of address adopted by corporate broadcasters.

104. WCFL was not the only radio voice of the left. The Socialist Party operated a radio station (WEVD in New York) from late 1926 into the early 1930s and faced some of the same obstacles as WCFL. The history of WEVD is related in Nathan Godfried, "Legitimizing the Mass Media Structure: The Socialists and American Broadcasting, 1926–1932," in Roland C. Kent, Sara Markham, David R. Roedigger, and Herbert Shapiro, eds., *Culture, Gender, Race, and US Labor History* (Westport, Conn.: Greenwood Press, 1993), 123–49. Also see Paul F. Gullifor and Brady Carlson, "Defining the Public Interest: Socialist Radio and the Case of WEVD," *Journal of Radio Studies* 4 (1997): 203–17.

Chicago is in some ways itself a special case, in that the city was characterized by both greater ethnic diversity and greater ethnic segregation than many urban centers, as well as by a comparatively greater number of broadcasters. But such "special cases" often exemplify in magnified form larger conflicts characteristic of the nation as a whole. The number of radio stations in or near Chicago and the resulting perception that its broadcasting situation was particularly cacophonous made the city a particular focus of FRC concern during the struggle over frequency allocations in the 1927–28 period. See Vaillant, "Sounds of Whiteness," 51–55.

105. Godfried, *WCFL*, 113; and WCFL Program Schedule (Summer 1929).

106. "Theatre Receipts," *Motion Picture Herald* (Jan. 10, 1931): 26, and (Feb. 2, 1931): 20.

107. The Little Theater was indeed "little," seating a few less than three hundred in "competition" with, for example, the 2,500-seat Keith's or the 3,500-seat Loew's Stanley, and showing in previous weeks the French film *Under the Roofs of Paris* and the Soviet film *Turksib* ("Theatre Receipts," *Motion Picture Herald* [Feb. 7, 1931]: 19, and [Feb. 21, 1931]: 19).

108. Because my concern here is the relationship between technology and the rhetorical, economic, and regulatory constructions of emergent media, I am necessarily leaving unaddressed the transaction between actually existing listeners and any particular recording, broadcast, or soundtrack.

109. As part of the 1908 presidential campaign, both William Jennings Bryan (Democrat) and William Howard Taft (Republican) recorded short speeches for all three of the major phonograph companies (Edison, Victor, and Columbia).

Recordings from each candidate often addressed similar issues. Bryan recorded "The Trust Question" for Victor, and Taft "responded" with the Victor recording "What Constitutes an Unlawful Trust"; or Taft's "Labor and Its Rights" spoke to similar issues as Bryan's "The Labor Question." While largely intended for home use, some evidence indicates that commercial phonograph parlors staged recorded "debates" between the candidates. (See, for example, the photograph of a New York phonograph parlor in which the foyer is decorated with life-sized models of the two candidates that face each other from adjacent podia, reprinted in *New Amberola Graphic* 46 [Autumn 1983]: 13.) During the 1912 presidential campaign, all three of the major party candidates again presented speeches for commercial phonograph companies (Woodrow Wilson, William Howard Taft, and Theodore Roosevelt).

Recordings from both the 1908 and the 1912 elections are available on the CD "In Their Own Voices: The U.S. Presidential Elections of 1908 and 1912," Annenberg School for Communication and Marston Records (2000). The recordings are described, some of them transcribed, and publicity material is reproduced in John S. Dales, "'Foreign Missions' Columbia 2-Minute Cylinder, no. 40554," *New Amberola Graphic* 46 (Autumn 1983): 12–13; "New Columbia Records by W. J. Bryan," *New Amberola Graphic* 66 (Oct. 1988): 8; and "Bryan Speaks to Millions through the Edison Phonograph," *New Amberola Graphic* 82 (Oct. 1992): 18–19. Between 1918 and 1920 two series of political phonograph recordings were released as "The Nation's Forum: Orations by Notable Americans." For a discography of the Nation's Forum recordings, see David Goldenberg, "Nation's Forum: A History and Discography," *The Discographer* 2 (n.d.): 195–200. Excerpts from many of the recordings are available at the Library of Congress's American Memory Web site.

110. For a partial list of political phonograph recordings from the early electric era, see *Phonograph Monthly Review* 3.7 (Apr. 1929): 234. The list includes a "speech to the American people" by Benito Mussolini, a speech by King George, and— less overtly political—Calvin Coolidge's speech welcoming Charles Lindbergh. Victor released in October 1928 election-related performances by Amos 'n' Andy (*Phonograph Monthly Review* 3.1 [Oct. 1928]: 33). Trucks equipped to project syncsound films were used extensively in a New York City mayoral campaign in 1929 (Baird, "The Talking Picture," 320).

111. "Sound Trucks Deliver Campaign Addresses for NY Politicians," *Motion Picture Herald* (Oct. 18, 1930): 24.

112. See, for example, "The Misuse of a Municipal Broadcast Station," *Radio Broadcast* 7 (Oct. 1925): 739–41.

113. "What Stations Shall Be Eliminated?" *Radio Broadcast* 11.3 (July 1927): 139–40.

114. On radio demagogues, see Alan Brinkley, *Voices of Protest: Huey Long, Father Coughlin, and the Great Depression* (New York: Vintage, 1983).

115. Gullifor and Carlson, "Defining the Public Interest," 204.

116. William G. Shepherd, "The Stump in Your Living-Room," *Collier's* 80 (Sept. 24, 1927): 8–9, 32.

117. J. L. Simpson, " 'Selling' the Public on Better City Government," *Radio Broadcast* 3 (Aug. 1923): 300.

118. Douglas B. Craig, *Fireside Politics: Radio and Political Culture in the United States, 1920–1940* (Baltimore: Johns Hopkins UP, 2000), 167–85.

119. Craig, *Fireside Politics*, 173.

120. Ibid., 170.

121. Shepherd, "The Stump," 9.

122. "Even Americans Will Rebel," *The Nation* (Oct. 19, 1932): 341, quoted in Peter Baxter, *Just Watch! Sternberg, Paramount, and America* (London: BFI, 1993), 32.

123. "Political Talks a Pain to Radio," *Billboard* 44 (Oct. 15, 1932): 3.

124. "Obligatos," *Billboard* 44 (Oct. 22, 1932): 17.

125. "Broadcast Miscellany," *Radio Broadcast* 10.5 (Mar. 1927): 476.

126. On the use of media and entertainers by the left, see Michael Denning, *The Cultural Front* (New York: Verso, 1998).

127. Terry Ramsaye, "The Short Picture: The Cocktail of the Program," *Motion Picture Herald* (Mar. 14, 1931): 59.

128. "Fox Warns Mgrs. To Keep an Eagle Eye on Newsreels," *Motion Picture Herald* (Mar. 7, 1931): 54.

129. *Billboard* (Oct. 22, 1932): 18, and (Oct. 29, 1932): 13.

130. Radio researchers Hadley Cantril and Gordon Allport concluded in 1935 that radio was presenting public policy issues in clearly delineated, binary terms: "prohibition or repeal, Republican or Democrat, pro-strike or anti-strike, Americanism or Communism, this or that. One would think that the universe were dichotomous." Hadley Cantril and Gordon Allport, *The Psychology of Radio* (New York: Harper, 1935), quoted in Susan J. Douglas, *Listening In: Radio and the American Imagination* (New York: Times Books, 1999) 156.

131. The relationship between entertainment media and politics was, of course, not finally and ultimately determined during this period. Instead, for example, toward the end of the decade Hollywood would be called to task for allegedly presenting anti-isolationist propaganda, and the subsequent history of radio would be marked by intermittent political controversies. The barrier between entertainment

and politics would prove to be semipermeable, but the notion that there should properly be such a barrier was dramatically reinforced during this period. Further, both a rhetoric of public service that would be invoked in succeeding debates and the process by which political performance adapted to and came to resemble entertainment media had crucial precedents in this era. Susan Douglas has begun to address subsequent struggles over radio's relationship to U.S. politics, particularly in terms of broadcast journalism during World War II. Douglas, *Listening In*, 161–98.

5. Transcription Versus Signification: Competing Paradigms for Representing with Sound

1. The debate is particularly vivid in the journal *Radio Broadcast*, especially in the monthly column "The Listener's Point of View."

2. John Wallace, "The Listener's Point of View," *Radio Broadcast* 10.6 (Apr. 1927): 569.

3. Wallace, "The Listener's Point of View," 568.

4. Carl Dreher, "Broadcasting and Shows," *Radio Broadcast* 10.4 (Feb. 1927): 390–91.

5. My use of the terms "document" and "construct" elaborate on Paul Théberge's argument in "The 'Sound' of Music: Technological Rationalization and the Production of Popular Music," *New Formations* 8 (Summer 1989): 99–111. Théberge examines shifts in recording techniques from the 1950s to the late 1980s and concludes, in part, that popular music was influenced by "a shift in recording aesthetics away from the 'realistic' documentation of a musical event to the *creation* of one" (104). This chapter tracks similar shifts long before the period examined by Théberge.

6. These conflicting paradigms resonate with Jonathan Sterne's assertion that, "Both the discourse of true fidelity and an alternative discourse of artifice in sound reproduction develop out of the social configurations embodied in new sound media." Jonathan Sterne, *The Audible Past: Cultural Origins of Sound Reproduction* (Durham, N.C.: Duke UP, 2003) 223.

7. See Rick Altman, "The Material Heterogeneity of Recorded Sound," in Altman, ed., *Sound Theory / Sound Practice* (New York: Routledge, 1992), 15–31. Also see James Lastra's chapter "Sound Theory" in his *Sound Technology and the American Cinema: Perception, Representation, Modernity* (New York: Columbia UP, 2000) 123–53. Jonathan Sterne argues that it is only through the possibility of reproduction that a notion of an original sound takes on a meaning.

8. Sterne suggests something similar when he asserts that, "Sound fidelity is much more about faith in the social function and organization of machines than it is about the relation of a sound to its 'source.'" Sterne, *The Audible Past*, 219.

9. See Emily Thompson, "Machines, Music, and the Quest for Fidelity: Marketing the Edison Phonograph in America, 1877–1925," *Musical Quarterly* 79.1 (Spring 1995): 131–71; Marsha Siefert, "The Audience at Home: The Early Recording Industry and the Marketing of Musical Taste," in James S. Ettema and D. Charles Whitney, eds., *Audiencemaking: How the Media Create the Audience* (Thousand Oaks, Calif.: Sage, 1994), 186–214; and Marsha Siefert, "Aesthetics, Technology, and the Capitalization of Culture: How the Talking Machine Became a Musical Instrument," *Science in Context* 8.2 (1995): 417–49.

10. Columbia Phonograph Company, *Columbia Disc Records Catalogue* (Bridgeport, Conn.: American Graphophone, 1905), hereafter, Columbia. Other phonograph catalogs that indicate similar recording practices include Kenneth M. Lorenz, *Two-Minute Brown Wax and XP Cylinder Records of the Columbia Phonograph Company* (Wilmington, Del.: Kastlemusick, 1981); Ronald Dethlefson, ed., *Edison Blue Amberol Recordings, 1912–1914* (Brooklyn: APM Press, 1980); Ronald Dethlefson, ed., *Edison Blue Amberol Recordings*, vol. 2, *1915–1929* (Brooklyn: APM Press, 1981).

11. Columbia, 22.

12. Columbia, 26.

13. Ibid., 52.

14. Columbia's "Grand Opera" Records were so perfect, or so they claimed, that "at times it is difficult to realize they are only mechanical reproductions of the human voice" (Columbia, 76).

15. "Address by the Late President McKinley at the Pan-American Exposition" (Columbia, 67); "Dutch Dialect Series" (i.e., "Schultz on Christian Science" or "Schultz on Malaria") (Columbia, 67); recordings by famous vaudevillians Weber and Fields (Columbia, 30); thirty-one different recordings in the "Uncle Josh Weathersby's Laughing Stories" series (Columbia, 67–68); "The Transformation Scene from 'Dr. Jekyll and Mr. Hyde'" (Columbia, 66); "Ghost Scene from Hamlet" (Columbia, 65); "Flogging Scene from Uncle Tom's Cabin" (Columbia, 64). The "Dr. Jekyll and Mr. Hyde" recording was described as follows: "This tragic scene from the last act of the play depicts the final transformation of Dr. Jekyll into the demon Hyde, and his subsequent death by his own hand. The ringing of the chimes and pealing of the organ lend realism to the intensely thrilling climax" (Columbia, 66).

16. Columbia, 7.

17. Ibid., 16.

18. The description for recording number 1563, "Departure of a Hamburg-American Liner," refers consumers to their own experiences watching the departure of an ocean liner and promises an acoustic experience of "striking realism" (Columbia, 15). "Those who have witnessed the departure from our shores of one of the great ocean greyhounds, will be impressed with the striking realism of this record. The big whistle sounds the time of the departure; the windlasses hoist the last of the late arriving luggage. 'All ashore going ashore.' The band plays popular airs from the Fatherland; friends on the pier shout farewells and adieus; and the giant liner backs into midstream, while the excitement increases. The music of the band is drowned in the cheering" (Columbia, 15).

19. Susan Douglas refers to such listening as "dimensional listening," in which audiences create "in [their] mind's eye three-dimensional locales," and she suggests that multiple types of radio content would evoke this type of listening. Susan J. Douglas, *Listening In: Radio and the American Imagination* (New York: Times Books, 1999), 33.

20. Existing between transcriptions of performances and the signification of fictional acoustic scenes, several Columbia disks combined these two paradigms. Many of its musical recordings included the use of sound effects. " 'Taint No Disgrace to Run When You're Skeered," a "coon" song, was categorized as a baritone solo with orchestral accompaniment but included "terrifying 'chill' effects by orchestra, introducing Thunder Crash effects" (Columbia, 36). Similarly, the baritone solo with piano accompaniment for "Carrie Nation in Kansas" was augmented, not surprisingly, with an axe effect (Columbia, 38). These recordings demonstrate that the precursors to electrical acoustics were also characterized by "hybrid" texts that combined transcription and signification paradigms for the work of sound technology.

21. Lastra, *Sound Technology and the American Cinema*, 85 and 88. Jonathan Sterne argues something similar when he asserts that "recording has always been a studio art." Sterne, *The Audible Past*, 235–37.

22. Milton Metfessel, "The Collecting of Folk Songs by Phonophotography," *Science* 67 (Jan. 13, 1928): 28.

23. Carl E. Seashore, "Three New Approaches to the Study of Negro Music," *Annals of the American Academy of Political and Social Science* 140 (Nov. 1928): 191–92.

24. Carl E. Seashore, "Phonophotography in the Measurement of the Expression of Emotion in Music and Speech," *Scientific Monthly* 24 (May 1927): 467.

25. Seashore, "Phonophotography," 468.

26. Metfessel, "Collecting," 28.

27. Ibid., 28.

28. Seashore, "Phonophotography," 471.

29. Ibid., 466.

30. I have selected Potamkin not for his direct influence on the daily practices of the American film industry; with some important exceptions (such as articles in *American Cinematographer, Billboard, Musical Quarterly, Phonograph Monthly Review*), his work was predominantly published in so-called "little magazines" with comparatively small circulations. But Potamkin provides a particularly vivid, consistent, and articulate presentation of a conceptualization of electrical acoustics as an instrument of signification. The fact that the aesthetic model he described in his criticism was most dramatically enacted in some Soviet, European, and avant-garde films attests to the wider influence of this conceptual paradigm.

31. Harry Alan Potamkin, "New York Notes: II," in Lewis Jacobs, ed., *The Compound Cinema: The Film Writings of Harry Alan Potamkin* (New York: Teachers College Press, 1977), 372. Potamkin was explicitly refuting an assertion made by fellow critic Gilbert Seldes that assumed the microphone possessed some privileged, direct access to the real. Such assumptions, especially when counterposed to similar assertions about the inherent mediation of image technologies, have been quite common in writing on the sound film. See for example the long list of quotations that begins James Lastra's essay "Reading, Writing, and Representing Sound," in Altman, ed., *Sound Theory / Sound Practice*, 65.

32. Potamkin, "New York Notes: II," 374.

33. Potamkin, "New York Notes: I," in Jacobs, ed., *The Compound Cinema*, 359.

34. Potamkin, "New York Notes: II," 371. Potamkin's polemical position against sound and image redundancy was also the cornerstone of his critique of many of the conventional practices of musical accompaniment for silent films. See his essay "Music and the Movies," written in 1927 although not published until 1929, and reprinted in Jacobs, ed., *The Compound Cinema*, 95–109.

35. Harry Alan Potamkin, "The Phonograph and the Sonal Film," *Phonograph Monthly Review* 4 (Aug. 1930): 374. The Soviet sound manifesto is S. M. Eisenstein, V. I. Pudovkin, and G. V. Alexandrov, "A Statement," in Elisabeth Weis and John Belton, eds., *Film Sound: Theory and Practice* (New York: Columbia UP, 1985): 83–85.

36. Harry Alan Potamkin, "Phases of Cinema Unity: II," in Lewis Jacobs, ed., *The Compound Cinema*, 22.

37. Potamkin, "Phases of Cinema Unity: II," 24.

38. Potamkin, "The Phonograph and the Sonal Film," 374, and Potamkin, "Playing with Sound," in Jacobs, ed., *The Compound Cinema*, 86–87.

39. Potamkin, "Playing with Sound," 87.

40. On fabrication of sounds, see "Playing with Sound," 88; on ignoring the duration of acoustic events, see Potamkin, "New York Notes: III," in Jacobs, ed., *The Compound Cinema*, 380; on the use of tone test records, see Potamkin, "The Phonograph and the Sonal Film," 374.

41. See, for example, Edward T. Lee, "The Use of Sound Films in Trial Courts," *Notre Dame Lawyer* 7 (1932): 209–16. In 1936, the New Jersey State Police, in cooperation with RCA, developed a method of recording sound films of criminals to aid in identification and apprehension of scofflaws. See J. Frank, Jr., "The Schwarzkopf Method of Identifying Criminals," *Journal of the Society of Motion Picture Engineers,* hereafter *JSMPE* (Feb. 1937): 212–17.

42. See, for example, Richard S. James, "Avant-Garde Sound-on-film Techniques and Their Relationship to Electro-Acoustic Music," *Musical Quarterly* 72.1 (1986): 74–89; Arthur Edwin Krows, "Sound and Speech in Silent Pictures," *JSMPE* 16.4 (1931): 427–36; Eisenstein, Pudovkin, and Alexandrov, "A Statement," 83–85; and Rene Clair, "The Art of Sound," in Weis and Belton, eds., *Film Sound,* 92–95. Frank Biocca describes the work of some composers who used the phonograph as a musical instrument (altering its speed, playing recordings backwards, altering the grooves of the disk, etc.), thereby employing the device as a producer of sounds rather than as a reproducer of a prior performance. Frank A. Biocca, "The Pursuit of Sound: Radio, Perception, and Utopia in the Early Twentieth Century," *Media, Culture, and Society* 10 (1988) 61–79. Also see Peter Manning, "The Influence of Recording Technologies on the Early Development of Electroacoustic Music," *Leonardo Music Journal* 13 (2003): 5–10.

43. Lee De Forest, "The Phonofilm," *Transactions of the SMPE* 16 (May, 1923): 73.

44. De Forest, "The Phonofilm," 75.

45. Ibid., 74.

46. Ibid., 74. De Forest interestingly did not foresee the synchronous-sound accompaniment of the actual sounds of newsreel events, precisely one of the techniques the Fox Film Corporation would use to innovate its competing sound-on-film system in 1927. Whether the result of technical limitations of De Forest's 1923 system (his later films would be shot under the acoustically controlled conditions of a studio) or other factors, De Forest clearly envisioned the transcription paradigm for synchronous-sound films as exclusively a matter of recording human voices and performances in synchronization with their bodies. Events existing in phenomenal

reality worthy of being transcribed were limited to those involving human performance.

47. De Forest, "The Phonofilm," 74.

48. Ibid., 73. This method of incorporating the voice as a kind of special effect within recorded scores would characterize some of Warner Bros.' initial uses of synchronous sound (discussed later in the chapter).

49. De Forest, "The Phonofilm," 72.

50. André Bazin, "The Myth of the Total Cinema," in *What Is Cinema?* trans. and ed. Hugh Gray (Berkeley: U of California P, 1971), 17–22. David Bordwell, Janet Staiger, and Kristin Thompson, *The Classical Hollywood Cinema: Film Style and Mode of Production to 1960* (New York: Columbia UP, 1985). See esp. Bordwell's chapter, "The Introduction of Sound," 298–308.

51. "Bringing Sound to the Screen," *Photoplay* (Oct. 1926), reprinted in Miles Kreuger, ed., *The Movie Musical from Vitaphone to "42nd Street"* (New York: Dover, 1975), 2–3.

52. Charles Wolfe, "Vitaphone Shorts and *The Jazz Singer*," *Wide Angle* 12 (July 1990): 67.

53. Wolfe characterizes the Vitaphone shorts as akin to "phonograph records with visual accompaniment" ("Vitaphone Shorts," 62). This characterization is particularly appropriate since the preexisting transcription paradigm on which the initial Vitaphone shorts were based was drawn by the cinema from the phonograph industry. In effect, the Vitaphone shorts put into wide practice De Forest's 1923 prediction that the synchronous-sound film could successfully overcome many of the phonograph's "limitations." By the summer of 1928, Warner Bros. offered an ever-growing back catalog of short subjects to exhibitors. See Charles Wolfe, "On the Track of the Vitaphone Short," in Mary Lea Bandy, ed., *The Dawn of Sound* (New York: Museum of Modern Art, 1989), 38.

54. For Lastra, the presence of such pseudo-synchronous events in the early Vitaphone features suggests a trace of silent film sound effects conventions and serve both to "foreground the spectacle of technology" and to display some of the potentials offered by Vitaphone technology (accurate synchronization). Lastra, *Sound Technology and the American Cinema*, 119–20.

A variety of practices displayed the representational possibilities offered by new sound media. In the late 1920s, Victor selected and released special recordings particularly suited to demonstrate the expanded frequency response of electrical phonographs. In Fox's earliest Movietone newsreels, subjects seem selected to demonstrate the range of sounds the apparatus could reproduce in synchronization with images. The first nationally released issue of Fox's sound newsreel contained footage of the Vatican's choir singing at the Tomb of the Unknown Soldier in Arlington, a recording of an obsolete Maryland dam being dynamited, and the Army versus Navy football game. Combining visual spectacle with a variety of readily identifiable sounds (choir, explosion, crowd noise) demonstrated the range of sound qualities that could be recorded using the Movietone system. In the second regularly released reel of December 1927, footage of federal agents destroying the cargo of captured bootleggers was accompanied by a printed intertitle instructing audience members to "listen to it gurgle." Described in Raymond Fielding, *The American Newsreel, 1911–1967* (Norman: U of Oklahoma P, 1972), 163–64.

55. Wolfe, "Vitaphone Shorts," 66.

56. The film is briefly discussed in Henry Jenkins, *What Made Pistachio Nuts? Early Sound Comedy and the Vaudeville Aesthetic* (New York: Columbia UP, 1992), 137–41. Jenkins describes a series of early sync-sound comedies and their short-lived use of a "vaudeville aesthetic." Jenkins makes no attempt to locate these performers and this performance style in a broader context that includes their simultaneous presence on radio and phonograph recordings. James Lastra mentions *The Big Broadcast* in passing, taking particular note of an opening sequence that exhibits some of the "isomorphic" techniques of cartoon sound. While Lastra sees the film as evidence that Hollywood's sound practices included a variety of disparate, marginalized strategies some of which harkened back to diverse pre-sync sound techniques, I hear instead in the film evidence of an ongoing struggle within Hollywood during the early 1930s between different ways of conceptualizing the work of the soundtrack. Lastra, *Sound Technology and the American Cinema*, 120–21.

57. While *The Big Broadcast* is unusual for the dramatic manner in which it alternates between representational practices premised on transcription and signification paradigms, the clash enacted within the film's final thirty minutes is exemplary of larger ongoing struggles surrounding the practical deployment of electrical acoustics.

58. Setting a film in a radio station so as to present filmed performances by radio stars was common practice in early sound short subjects. Warner Bros. released at least two series of such shorts, both called "Ramblin' Round Radio Row," in 1932 and 1934. Tiffany's 1929 series of shorts entitled "The Voice of Hollywood" used the fictional radio station "STAR" to present brief performances by second-tier Hollywood performers like Marjorie Kane, Lillian Rich, and a singing performance

by Ken Maynard and his horse. The practice continued with shorts like "Listening In" (1934) and "Rah, Rah, Radio" (1935).

59. *Billboard* (Oct. 22, 1932): 18.

60. *Variety*'s list of film grosses is reprinted in Tino Balio, *Grand Design: Hollywood as a Modern Business Enterprise, 1930–1939* (New York: Scribner's, 1993), 405.

61. *Variety* (Oct. 18, 1932): 14.

62. The issue of incorporating performances within Hollywood films was particularly salient prior to and concurrent with the release of *The Big Broadcast*. In his work on early Warner Bros. sound films, Charles Wolfe traces *The Jazz Singer*'s status as a "hybrid text," focusing less on the film's combination of "talking" and "silent" sections than on the division "between the (filmic) presentation of a musical performance and the harnessing of that performance for a narrative fiction" (Wolfe, "Vitaphone Shorts," 67). As *The Big Broadcast* illustrates, the "competition among various strategies for representing vocal acts on film" (Wolfe, "Vitaphone Shorts," 71) continued to inform conflicting Hollywood practices long after the electrical acoustic apparatus was diffused throughout the industry.

63. Arguably, this strategy of representing musical performance looks to the same traditions as, for example, radio's critically valorized "Eveready Hour" (c. 1925) in which musical cues functioned to impose a continuity of time, place, and action instead of, in the words of one radio critic, "the present jumble of odds and ends put together on the general pattern of Joseph's coat." James C. Young, "New Fashions in Radio Programs," *Radio Broadcast* 7.1 (May 1925): 89.

64. This strategy also characterized significant portions of the 1933 United Artists' release *Hallelujah, I'm a Bum*. Other films of the early 1930s, like *Strange Interlude*, *Hallelujah, I'm a Bum*, and *Blessed Event*, considered either individually or collectively, can only from the broadest aesthetic perspective be considered exemplary of a consistent Hollywood practice. Each film dramatically illustrates that as late as 1932 *The Big Broadcast* was not unique in its embodiment of conflicting conceptions of recorded sound's role in film representation. Besides the sung dialogue in *Hallelujah, I'm a Bum*, *Strange Interlude* includes voice-over narration of characters' thoughts, and *Blessed Event* awkwardly attempts to interpolate a crooning performance by Dick Powell.

65. Burns and Allen's radio appearances essentially involved a gradual shift from the transcription of overt performance events to their careful containment within narrative frameworks in which diegetic integrity resulted from the careful orchestration of sounds according to the signification paradigm. These shorts

include *Fit to Be Tied* (1930), *The Antique Shop* (1930), *Pulling a Bone* (1930), *Once Over, Light* (1931), *100% Service* (1931), and *Oh, My Operation* (1931). *The Big Broadcast*'s method of incorporating Burns and Allen was repeated in some of their subsequent feature film appearances (*International House* and *The Big Broadcast of 1936*). However, once their characters were well established nationally through radio appearances and short subjects, Burns and Allen interpolated this style and their characters' relationships into a recurring situation-comedy radio show. This movement from more or less direct audience address, vaudeville-like situations to performances interpolated within narratives based on recurring relationships characterized one trend in radio comedy during the later 1930s. It also character-ized later Burns and Allen films such as Paramount's *Six of a Kind* (1934) and *Here Comes Cookie* (1935).

66. Alan Williams, "The Musical Film and Recorded Popular Music," in Rick Altman, ed., *Genre: The Musical* (London: Routledge & Kegan Paul, 1981), 156. In his book-length study of Busby Berkeley, Martin Rubin traces a tradition of nonin-tegrated approaches to representing performance in a variety of the film musical's precursors. See especially chapter 1 of *Showstoppers: Busby Berkeley and the Tradition of Spectacle*, (New York: Columbia UP, 1993), 11–32.

67. All musical reviews appearing in *Photoplay* are reprinted in Miles Kreuger, ed., *The Movie Musical from Vitaphone to "42nd Street."*

68. See, for example, *Show of Shows* (Warner Bros., 1929), *Happy Days* (Fox, 1930), *Roman Scandals* (Goldwyn, 1933), *Wonder Bar* (Warner Bros., 1934).

69. These intermittent moments of diegetic sound predominantly function to bridge transitions from Leslie's urban quest to the WADX studio and the transition as well between two radically different approaches to representing sound.

70. Only a brief comic moment in the film's initial five minutes exhibits an approach to soundtrack construction so fully conceptualized according to a signi-fication paradigm. An exaggerated ticking clock, a cat whose paws squeak, and a sliding whistle as the cat cartoonishly squeezes under a door further underscore the conflict I'm identifying within this film. This is the sequence that Lastra cites as an example of "isomorphic" sound.

71. Lastra, "Reading, Writing"; and Lastra, *Sound Technology and the American Cinema*, 138 passim; and Rick Altman, "Sound Space," in Altman, ed., *Sound Theory/ Sound Practice* (New York: Routledge, 1992), 46–64. Also see Mary Ann Doane, "Ideology and the Practice of Sound Editing and Mixing," in Weis and Belton, eds., *Film Sound*, 54–62. Emily Thompson has also examined these sound practice de-bates. Emily Thompson, *The Soundscape of Modernity: Architectural Acoustics and the*

Culture of Listening in America, 1900–1933 (Cambridge: MIT Press, 2002), 273–85. I address her conclusions below.

72. Lastra, *Sound Technology and the American Cinema*, 138–39. Lastra contextualizes these issues in terms of a clash between "inscription" and "signification" tropes in nineteenth-century discourses about photography and phonography. He later summarizes what I am defining as transcription versus signification on page 195.

73. Lastra, "Reading, Writing," 77.

74. Altman, "Sound Space" ("recording/construction," 53; "realism/intelligibility," 64).

75. Altman asserts that early sound film referred to other systems of representation, including "radio for sound perspective." Altman, "Sound Space," 55.

76. Altman points to the influence on early cinema sound of preexisting radio conventions (sound must carry "all the information necessary to reconstruct the 'real' space of the scene" [Altman, "Sound Space," 55]). Other scholars attribute to radio's influence precisely the countervailing "intelligibility" impulse. While apparently mutually exclusive, neither historical position is wrong. Instead, radio conventions for representation simultaneously involved both impulses—that is, radio's representational protocols were informed by both transcription and signification paradigms.

77. Altman, "Sound Space," 64.

78. This minor point of contention with Altman's claims in no way challenges his important insight that new technologies look to, and borrow their initial conventions of realistic representation from, preexisting and well-established reality codes of other technologies (see Altman, "Sound Space," 62–63). As this chapter illustrates, cinema originally looked to the phonograph and its (rhetorical, at least) embodiment of a transcription paradigm to develop initial protocols of representing sound. At times, Lastra too seems to suggest that such paradigms were linked to particular media. "Like the phonograph and the radio industries from which it emerged, the Vitaphone short was institutionally and economically bound to an aesthetic that privileged an absolutely faithful rendering of a sonic performance conceived of as wholly prior to the intervention of the representational devices" (Lastra, *Sound Technology and the American Cinema*, 196). I would suggest instead that while one or the other of these paradigms might have become institutionally *dominant* in the practices of a given medium, they both simultaneously informed different practices in the radio and phonograph industries. While claims about acoustic fidelity to an original event functioned powerfully in rhetorical assertions

about media identity, in actual practice—as radio certainly reveals and "Descriptive Discs" suggest—representational practices in the mid- to late 1920s (and earlier) were actually quite diverse. Lastra seems to acknowledge this in his discussion of Edison's manipulation of the pro-phonographic event as part of the process of recording (86–89).

79. In their SCOT model for understanding technological change, Trevor Pinch and Wiebe Bijker observe that different social groups define new and emerging technologies in different ways and therefore also define differently the *problems* posed for the social uses of technology (Trevor J. Pinch and Wiebe E. Bijker, "The Social Construction of Facts and Artifacts: Or, How the Sociology of Science and the Sociology of Technology Might Benefit Each Other," in Wiebe E. Bijker, Thomas P. Hughes, and Trevor J. Pinch, eds., *The Social Construction of Technological Systems: New Directions in the Sociology and History of Technology* [Cambridge: MIT Press, 1987], 17–50). In what follows I view Hollywood's sound practice debates as occupying a *single* social group within which there was substantial disagreement as to how those "problems" are formulated and framed. Lastra has described Hollywood's conversion in terms of a clash between competing professional identities and related protocols of representation. See Lastra, *Sound Technology and the American Cinema*, 154–79.

80. J. B. Engl, "A Study in Acoustics," *American Cinematographer* (July 1928): 13–14, 28, advocated the use of two recording channels relying on two directional microphones to reproduce the experience of binaural hearing. In a 1935 article, Leopold Stokowski indicated that some still experienced monaural recording and reproduction as a problem to be solved. Leopold Stokowski, "New Vistas in Radio," *Atlantic Monthly* 155 (Jan. 1935): 6.

81. For example, the "solution" is raised and then refuted in each of the following: Stokowski, "New Vistas"; H. F. Olson and F. Massa, "On the Realistic Reproduction of Sound with Particular Reference to Sound Motion Pictures," *JSMPE* 23 (Aug. 1934): 63–81; Wesley C. Miller, "The Illusion of Reality in Sound Pictures," in Lester Cowan, ed., *Recording Sound for Motion Pictures* (New York: McGraw-Hill, 1931), 214. The notion of a cinema sound-reproduction system relying on individual reproducers for each audience member is not as improbable as it may at first appear. In this context, it is useful to remember that consumers' encounters with the electrical-acoustic apparatus initially involved the use of headphones. Accessories to home radio receivers in the 1920s allowed multiple family members to listen to the same broadcast through separate headphones. Further, once electrical acoustics was diffused through American film exhibi-

tion, a variety of electrical manufacturers marketed headphone-based sound-augmentation devices for hearing-impaired theater patrons (for example, RCA's Audiophone device).

Through the filter of historical hindsight, it may appear obvious, natural, or even inevitable that the reproduction of sound in movie theaters would use loudspeaker technology. But the decision to innovate cinema sound using loudspeakers was precisely a decision, a choice (albeit an overdetermined one) among a range of potential technological manifestations. Had binaural reproduction been considered a *necessary* prerequisite for the application of electrical acoustics (had the model of human binaural perception been the *basis* of cinema sound apparatus) rather than one goal among several, application-driven research and development and the conceptual conservatism surrounding technological innovation could have taken the practices of American sound cinema and the resultant sound of American films in entirely different directions.

82. Harvey Fletcher, "Transmission and Reproduction of Speech and Music in Auditory Perspective," *JSMPE* 22 (May 1934): 314–29.

83. John L. Cass, "The Illusion of Sound and Picture," *JSMPE* 14 (Mar. 1930): 325.

84. Cass, "Illusion," 325.

85. Porter H. Evans, "A Comparison of the Engineering Problems in Broadcasting and Audible Pictures," *PIRE* 18 (Aug. 1930): 1324; Joe Coffman, "Art and Science in Sound Film Production," *JSMPE* 14 (Feb. 1930): 174.

86. Michael Chanan, Mary Ann Doane, and Lucy Fischer all take Coffman's claims about the influence on Hollywood by radio engineers at face value: Michael Chanan, *Repeated Takes: A Short History of Recording and Its Effects on Music* (New York: Verso, 1995), 75; Doane, "Ideology," 60; Lucy Fischer, "*Applause*: The Visual and Acoustic Landscape," in Weis and Belton, eds., *Film Sound*, 244. It is particularly striking that this single article by Coffman, and in fact a specific passage from that article, would have such a profound and pervasive impact on the way in which contemporary scholars understand the complex relationship between Hollywood sound practice and radio.

87. Carl Dreher, "This Matter of Volume Control," *Motion Picture Projectionist* (Feb. 1929): 11.

88. Dreher, "This Matter," 11.

89. Ibid.

90. *Motion Picture Projectionist* (Nov. 1929). For more on this issue of standardizing film projection, see chapter 3. In October of 1930, the Projection Committee of the Society of Motion Picture Engineers advocated adjustments to volume through

such a remote control as a routine procedure to overcome deficiencies in recording (i.e., ground noise) and to compensate for audience vocal reactions. They also advocated its use for the creation of certain audible effects. "Such effects include the momentary increase of volume level to bring about more natural effects during the firing of a cannon, the slamming of a door, or other temporary noises which cannot be recorded and reproduced at the desired levels by the ordinary process." They pointed to an effect in the recorded score for *Sunrise*, a horse's whinny that punctuates a tense, otherwise silent moment of the film. "The effect upon the audience of the whinny is somewhat startling even as recorded and reproduced by normal methods; but by the use of a remote control device such as described, the whinny of the horse can cause the audience to freeze in their seats and cause some members to cry out in alarm" (558).

91. Carl Dreher, "Stage Technique in the Talkies," *American Cinematographer* 10 (Dec. 1929): 16. In another article published first in *Radio News* and then in *Projection Engineering*, Dreher outlined a system that involved mixing the outputs from several microphones as a last resort. Here, close miking is a function of technological limitations of microphones rather than an application of radio technique to the cinema. That limitation, a 10-foot maximum distance between dialogue source and microphone, is repeated elsewhere during the period. Carl Dreher, "When Camera Meets 'Mike': Radio Technique as Applied to Talking Movies," *Projection Engineering* 1 (Dec. 1929): 11–14. The maximum 10-foot distance is repeated in Gordon S. Mitchell, "The Beam Microphone for Sound Recording," *Projection Engineering* 2 (Nov. 1930): 16.

92. For example: J.P. Maxfield, "Technique of Recording Control for Sound Pictures," in Cowan, ed., *Recording Sound*, 252–67; J.P. Maxfield, "Acoustic Control of Recording for Talking Motion Pictures," *JSMPE* 14 (Jan. 1930): 85–95; J.P. Maxfield, "Some Physical Factors Affecting the Illusion in Sound Motion Pictures," *Journal of the Acoustic Society of America* 3 (July, 1931): 69–80.

93. In some of his essays Maxfield included charts demonstrating appropriate microphone placement in relation to lens focal length and camera distance (i.e., "Some Physical Factors"). The charts are reprinted and discussed in Altman, "Sound Space," 52–55. The sheen of scientific precision implicit in these charts demonstrates the sometimes uncomfortable tension surrounding technological change in aesthetic forms. The typical commonsense boundaries separating applied science from aesthetic practice become temporarily fluid. During such moments the respective worlds of Carl Seashore and Harry Alan Potamkin clash, but they also may converse. While Carl Seashore, through the phonophotographic study of vibrato, anticipated isolating and quantifying component factors of the

beautiful, Maxfield sought to quantify and specify through scientific measurement the aesthetic production of realism. This tension between art and science appeared elsewhere in the development of electrical acoustics as voices from some quarters expressed concern that the radio engineer operating controls that mixed multiple microphones had usurped control of musical production from the orchestra conductor and his baton (for example, the Radio Editor of the *Cleveland News* in *Radio Broadcast* 10 (1927): 360).

94. Dreher, "Stage Technique," 3.

95. Albert W. DeSart, "Sound Recording Practice" in Cowan, ed., *Recording Sound*, 273.

96. Coffman, "Art and Science," 176. Some evidence indicates that Maxfield's model was applied by Hollywood studio sound engineers, such as Universal's Charles Felstead. Charles Felstead, "Monitoring Sound Motion Pictures, Part II," *Projection Engineering* 2 (Dec. 1930): 9–11, 20; see esp. 10–11.

97. Evans, "A Comparison," 1324–25.

98. Carl Dreher, "Recording, Re-recording, and Editing of Sound," *JSMPE* 16 (June 1931): 756.

99. Maxfield, "Acoustic Control," 87.

100. Maxfield, "Technique of Recording," 257.

101. Olson and Massa, "On the Realistic Reproduction of Sound," 63–81. Interestingly Harvey Fletcher's demonstration of a binaural electrical-acoustic system preceded Olson and Massa's presentation. Although they stress the benefits of such a binaural system, their remarks address the monaural system already in place.

102. For a discussion of such sound collectors, see Mitchell, "The Beam Microphone for Sound Recording," 16. Mitchell, an electrical engineer in the sound department at Universal Pictures Corporation, describes the new "beam microphone" being developed at RKO Studios in Hollywood. The device involved locating a condenser microphone at the focal point of an ellipsoidal mirror. Mitchell claimed that the apparatus overcame the problem of collecting adequate sound from distant microphone placement necessitated by extreme long shots and/or shots containing many extras. Whereas the previous maximum microphone distance for adequate dialogue recording was ten feet, Mitchell claimed the apparatus was capable of adequately recording dialogue at fifty feet, a claim that may well have been exaggerated.

103. Dreher, "Recording, Re-recording," 756.

104. L. E. Clark, "Accessory and Special Equipment," in Cowan, ed., *Recording Sound*, 128. Clark also describes different microphone baffles (125–26).

105. Clark, "Accessory and Special Equipment," 126–27.

106. Ibid., 127. Clark refers his readers to two publications providing "the necessary formulas and derivations for the calculation of filters to produce any desired effect in a simple and straightforward manner" (127).

107. On looping, see page 151 of Kenneth F. Morgan, "Dubbing," in Cowan, ed., *Recording Sound*, 145–54.

108. DeSart, "Sound Recording Practice," 280–83.

109. Morgan, "Dubbing,," 150.

110. See Morgan, "Dubbing,"148; and Evans, "A Comparison," 1324. It is indicative of the confusion of the time that Evans, a strong advocate for what Lastra has labeled the "telephonic" model of recording (arguably a signifying practice), would resist the further potential aesthetic choices provided by dubbing. Under such circumstances, Lastra's "telephonic" and "phonographic" models might best be viewed as two competing rhetorical positions invoked to rationalize and make sense of Hollywood's use of sound as signification.

111. "Progress Report," *JSMPE* 14 (Feb. 1930): 231. For a period revelation of this practice, see Mark Larkin, "The Truth About Voice Doubling," in Kreuger, ed., *The Movie Musical*, 34–37. Larkin's article originally appeared in *Photoplay* in July of 1929. It would seem at first that the public, like Maxfield and other advocates for realist protocols, clung to their epistemological faith in electrical acoustics as an apparatus characterized by mimetic objectivity. I've drawn the phrase "mimetic objectivity" from Chanan, *Repeated Takes*, 138–39.

112. *Film Daily Year Book* (New York: Film Daily, 1931), 563.

113. Coincident with this formal codification of voice doubling, the practice became the basis for self-reflexive jokes. MGM's 1930 short subject "The Dogway Melody" featured an all-dog cast with, among other scenes, the voices of the players in *The Hollywood Revue* dubbed for the dogs during a canine re-creation of the "Singing in the Rain" sequence. The film is reviewed in *Billboard* (Nov. 1, 1930): 38. In MGM's 1931 feature *The Prodigal* (also released as *The Southerner*) the aging British "Doc" (played by Roland Young) sings in a powerful resonant baritone. When the baritone voice stops and Doc continues singing in a squeaky murmur, camera reframing and character blocking reveal the actual source of the baritone to be opera star Lawrence Tibbett. The moment's humor hinges on a misrecognition of sound source—aided by the illusion of synchronization with Doc's lips—and on the apparent disjunction between sound quality (baritone) and apparent sound source (Doc). The joke is, of course, relatively common in sound films, particularly animation. I point to it here to suggest that by the early 1930s film audiences may well have begun to recognize the signifying function of the soundtrack.

114. "Progress Report," 231–32.

115. Richard Barrios, *A Song in the Dark: The Birth of the Musical Film* (New York: Oxford UP, 1995), 65. Electrical-acoustic prerecording was used in the theater some three years earlier. In 1926 a device called the "Super-panatrope," installed in New York's Globe Theatre, provided recorded vocal performances for Florenz Ziegfeld's chorus girls. "With this electrical voice, the burden of the vocal rendition by the chorus may be entirely eliminated leaving the chorus girls free to concentrate their efforts on dancing or acting. So realistic is the electrical voice, that the audience is virtually convinced that the chorus girls are doing all the singing." "Giving the Chorus Girl a Lift," *Scientific American* (Oct. 1926): 288.

116. Kreuger, ed., *The Movie Musical*, 186–87.

117. Lastra comes close to neatly defining what I have in mind by the phrase "signifying fidelity": "the primary ideological effect of sound recording might rather be in creating the *effect* that there is a single fully present 'original' independent of its representation at all." Later, he notes, "sound engineers ultimately jettisoned the all-defining 'original sound' and the aesthetic of duplication it entailed and, for better or worse, set about the business of manufacturing sonic worlds whose parameters were judged mainly in terms of their internal coherence and representational functions" (Lastra, *Sound Technology and the American Cinema*, 134 and 178–79). Similarly, Jonathan Sterne points to an RCA prizefight broadcast and notes, "the point of the artifice is to connote denotation, to construct a realism that holds the place of reality without being it." Or, stated differently, "to fashion an aesthetic realism worthy of listeners' faith" (Sterne, *The Audible Past*, 245–46).

118. Edward W. Kellogg, "Some Aspects of Reverberation," *JSMPE* 14 (Jan. 1930): 106.

119. Lastra's account of Hollywood's development of a foreground/background system of soundtrack construction is one component of what I refer to as "signifying fidelity." Briefly, while Hollywood sound practices privileged the voice (or narratively important sound effects), placing them in the forefront of sonic representation, engineers simultaneously supplemented these sounds with a sonic background running continuously across cuts. While the foreground would carry the "enunciative" function of sounds (their narrative importance), this background signified a spatially coherent world. Lastra, *Sound Technology and the American Cinema*, 180–215, esp. 205–206.

Emily Thompson notes that the technologies and techniques associated with electrical acoustics allowed greater flexibility in sound recording, sound manipulation, and soundtrack construction. But this ability was actually used sparingly, she

suggests, because of what she identifies as the pervasive desire for direct, nonreverberant sounds. Hollywood sound practices joined building materials, auditorium
design, acoustic theory, and other media practices in promulgating this new standard for sound. Thompson argues that Hollywood soundtracks ultimately mimicked the larger soundscape of the time with sounds that lacked spatial signature
and that seemed directly addressed to auditors. "The sound track," Thompson concludes, "epitomized the sound of modern America" (Thompson, *The Soundscape of
Modernity*, 284). Thompson is of course correct that emerging Hollywood sound
practices emphasized the intelligibility of dialogue and thereby acoustically resembled auditors' encounters with other modern soundscapes. But as Lastra's
description of Hollywood's foreground/background system suggests, such an emphasis—such an imagining of sound as a signal—was not to the exclusion of other
sounds. Thompson's conclusion, I would suggest, threatens to neglect the other
acoustic manipulations that produced a sonic background and provided both continuity and, in Sterne's words, "an aesthetic realism worthy of listeners' faith" to
Hollywood's soundtracks.

120. Both the field-recording efforts of the Library of Congress and sound-
collection efforts by ethno-musicologists continued to rely upon an epistemological
faith in electrical recording's ability to transcribe aural phenomena.

121. Western Electric, "How the Sounds Get Into Your Record by the Electrical
Process," *Phonograph Monthly Review* 1 (Oct. 1926): 5–6.

122. Western Electric, "How the Sounds Get Into Your Record," 6.

123. This process continues with newer acoustic innovations such as multitrack
stereo, automatic dialogue replacement, Foley effects, etc. Note that in the case of
Dolby Noise Reduction systems, like the 1931 introduction of Western Electric's
Noiseless Recording, the rhetoric announcing Dolby technology stressed fidelity
and invoked a transcription paradigm for conceptualizing sound technology's function. After the initial innovation of electrical acoustics, successive sound technologies frequently were framed as part of the inexorable asymptotic move of technologically augmented representation toward the thing itself.

Conclusions/Reverberations

1. William Uricchio, "Historicizing Media in Transition," in David Thorburn
and Henry Jenkins, eds., *Rethinking Media Change: The Aesthetics of Transition*
(Cambridge: MIT P, 2003), 32.

2. Uricchio, "Historicizing Media," 35.

3. Paul Théberge, *Any Sound You Can Imagine: Making Music/Consuming Technology* (Hanover, N.H.: 1997); Timothy D. Taylor, *Strange Sounds: Music, Technology, and Culture* (New York: Routledge, 2001).

4. James Rorty, *Order on the Air!* (New York: John Day, 1934), 24.

5. Robert W. McChesney and John Nichols, *Our Media Not Theirs* (New York: Seven Stories, 2002).

6. Hamilton, quoted in Rorty, *Order on the Air!*, 8.

7. Roland Marchand, *Advertising the American Dream: Making Way for Modernity, 1920–1940* (Berkeley: U of California P, 1985), 217–22.

8. Reebee Garofalo, "From Music Publishing to MP3: Music and Industry in the Twentieth Century," *American Music* (Fall, 1999): 318–54.

9. Note, for example, the disproportionate attention paid to Martha Stewart's insider trading, a crime with far less impact on citizens than the collapse of Enron, but a story far more conducive to a melodrama of the celebrity laid low. Stewart's tale of an all-American, "post-feminist," Horatio Alger success story-gone-wrong functions as more than corporate media prosecuting its own (although it is *also* that). The tale is far easier to narrativize, to render media-genic, than a story that would implicate the structural logic of corporate hegemony.

Index

electrical-acoustic innovations of, 52–53, 305*n*73; license for wired telegraphy/telephony applications, 23; licensing of radio transmitters, 37, 296*n*12, 299*nn*36–38; ownership of WEAF, 36–37, 82, 299*n*40; patent control and corporate power of, 29; patent pool formation and, 22; public performance of technology, 81–86; public relations strategies of, 106–107; radio network connections and, 57; sound technology change and, 4; synchronous-sound research of, 305*n*73; on wired broadcasting, 39, 328*n*86, 336*n*13; withdrawal from radio broadcasting, 39–40, 57

Atwater Kent radios, 149–50

The Audible Past (Sterne), 9–10

Auditorium Orthophonic Victrola, 74–75

Aylesworth, Merlin, 171, 217

Babson (F. K.) catalog, 138–39, 146, 325*n*62

baffles, microphone, 271

Ball, Don, 258

Barnouw, Erik, 8

Barrymore, John, 50

Barthelmess, Richard, 97, 98f, 314*n*61

Bauer, Harold, 50, 342*n*68

Bazin, André, 243

Bell, Alexander Graham, 76

Bellamy, Madge, 97, 314*n*61

Bell Laboratories: advertising technological innovation of, 106; binaural sound system of, 263–64; electrical recording system of, 41; fidelity-based innovations of, 41; sound technology change and, 4; theory of matched impedance, 40–41

Bell System: approach to competition of, 26–27; coordinated industrial research of, 40; exhibit at 1933 World's Fair, 117–18; patent position of, 28–29; phonograph recording innovations of, 40–41; public relations use of motion pictures by, 312*n*39. *See also* AT&T

Bell Telephone Co., 22–23

Bensman, Marvin R., 307*n*86

Benton, Thomas Hart, 343*n*87

Bernhardt, Sarah, 127

Bettini, Gianni, 318*n*10

B. F. Keith-Albee Vaudeville Exchange, 56

B. F. Keith Corporation, 56

Big Bands, 65

The Big Broadcast (1932), 355*nn*56–58; in film history, 251–52; integration of performance and narrative in, 253–59, 356*n*62; representation of sound phenomena in, 258–60, 357*nn*69–70; as succession of talking shorts, 253, 257; transcription vs. signification in, 251–60, 278

The Big Broadcast of 1936 (film), 356*n*65

Big Five, 54–55

Bijker, Wiebe, 11–13, 68, 295*n*5, 298*n*24, 359*n*79

Billboard, 220, 224

binaural sound: simulation of perception, 263–64, 274, 359*nn*80–81; transmission of, 116–18

in, 29; phonograph as model for, 122; politics and, 110–12, 216, 219, 288, 316n87; professionalization of, 57, 62; as public sphere, 31–32, 298n29; quality and standardization in, 175; regulation of, 7, 57, 58, 60–61, 62–63, 307n84; scheduling conventions of, 182; *Scientific American* coverage of, 90; service rhetoric in, 225; "Silent Nights," 179, 335n10; spatial metaphors for, 178–79, 182, 225, 335n11; special interest, 216–17; spectrum allocation for (*see* spectrum allocation); transcription vs. signification in, 277; uncontrolled nature in 1920s, 25; utopian visions of, 107–112, 146–47, 218; wired vs. wireless transmission in, 328n86; women's image in, 35–36
Radio Broadcast (magazine), 142
Radio Conferences, 58, 59
Radio Craft (journal), 327n69
radio engineers, 265, 360n86. *See also* home engineers
Radio Enters the Home, 143–44, 145
radio enthusiasts. *See* amateur broadcasting
Radiola radio, 144, 151
radio networks: affiliate relations with, 39, 174, 182; AT&T and development of, 34, 39, 57; competing visions for, 336n13; cultural enhancement via, 205; political conventions and formation of, 108–109; as public performance of technology, 82–83, 311n29; rise of, 2; serving isolated populations, 198, 341n59

radio-phonograph combinations, 45, 149, 150–51, 302nn55–56
radio receivers: as artifacts, 121, 151–52; cabinetry design for, 144–45, 145f, 149–50; decreasing size of, 149–50, 329n87; design linkage to phonograph, 44–46, 144–46; pricing of, 328n81
radio stations: in African American communities, 36; expansion of, 24; first station, identity of, 178; licensing of, 57, 177–78; as setting for films, 355n58; of Socialist Party, 346n104. *See also specific station call letters*, e.g., WCFL
Ramsaye, Terry, 222–23
"Rathergate," 280
RCA (Radio Corporation of America): control of sound technology innovation, 2; establishment of NBC radio network, 57; formation as patent-holding company, 22; license for commercial applications of wireless, 23; as multimedia conglomerate, 64, 307n91; promotional rhetoric for radio, 143–44; purchasing controlling interest in Victor, 151; as vertically integrated company, 56; vision of networked radio, 336n13
Read, Oliver, 8, 299n41
recording engineers, 276–77, 364n119
Red Seal recordings, 126, 135, 318n10
regenerative circuits, 23, 148
regional identity: crafting of, 343n87
Reich, Leonard, 8
Reichman, Frank, 221
Reisenfeld, Hugo, 209

Pre-Code Hollywood: Sex, Immorality, and Insurrection in American Cinema, 1930–1934
Thomas Doherty

Sound Technology and the American Cinema: Perception, Representation, Modernity
James Lastra

Melodrama and Modernity: Early Sensational Cinema and Its Contexts
Ben Singer

Wondrous Difference: Cinema, Anthropology, and Turn-of-the-Century Visual Culture
Alison Griffiths

Hearst Over Hollywood: Power, Passion, and Propaganda in the Movies
Louis Pizzitola

Masculine Interests: Homoerotics in Hollywood Film
Robert Lang

Special Effects: Still in Search of Wonder
Michele Pierson

Designing Women: Cinema, Art Deco, and the Female Form
Lucy Fischer

Cold War, Cool Medium: Television, McCarthyism, and American Culture
Thomas Doherty

Katharine Hepburn: Star as Feminist
Andrew Britton

Silent Film Sound
Rick Altman

Home in Hollywood: The Imaginary Geography of Hollywood
Elisabeth Bronfen

Hollywood and the Culture Elite: How the Movies Became American
Peter Decherney

Taiwan Film Directors: A Treasure Island
Emilie Yueh-yu Yeh and Darrell William Davis

Shocking Representation: Historical Trauma, National Cinema, and the Modern Horror Film
Adam Lowenstein